Principles of Animal Behavior

Principles of Animal Behavior: Mechanisms, Ecology, and Applications in Veterinary Science is designed for advanced undergraduate and postgraduate students in veterinary science, animal science, and zoology, and offers an integrated view of behavioral biology, bridging classical ethology, cognitive neuroscience, and ecological adaptation with practical veterinary applications. The initial chapters provide foundational aspects such as orientation, motivation, and communication before moving on to core themes of learned behavior, cognition, social dynamics, reproduction, and courtship. The book further addresses applied behavioral contexts in veterinary practice, including stress and welfare indicators, the impact of hormones and drugs, and human–animal interactions. The final chapters explore ecological and evolutionary behavior, including movement, habitat selection, and behavioral responses to altered ecosystems. Emphasizing both theoretical depth and practical application, the book provides students and professionals with a modern, interdisciplinary perspective on the role of behavior in animal health, welfare, and survival.

Key features

- Provides an overview of animal behavior, covering foundational principles, learning and cognition, social dynamics, reproductive strategies, and ecological interactions
- Addresses complex behavioral phenomena such as social dominance, reproductive strategies, maternal investment, altruism, and host–parasite interactions within both naturalistic and managed settings
- Explains the evolutionary and physiological basis of behavior, drawing upon current understanding in genetics, endocrinology, neuroethology, and environmental biology
- Examines behavioral responses to environmental pressures, including migration, habitat selection, anthropogenic disturbance, and ecosystem alterations
- Incorporates the impact of exogenous influences such as hormone therapy, drug administration, and human–animal interactions on behavioral outcomes

Principles of Animal Behavior

Mechanisms, Ecology, and Applications in Veterinary Science

Edited by Tanmoy Rana

CRC Press
Taylor & Francis Group
Boca Raton London New York

CRC Press is an imprint of the
Taylor & Francis Group, an **informa** business

Designed cover image: Shutterstock

First edition published 2026
by CRC Press
2385 NW Executive Center Drive, Suite 320, Boca Raton FL 33431

and by CRC Press
4 Park Square, Milton Park, Abingdon, Oxon, OX14 4RN

CRC Press is an imprint of Taylor & Francis Group, LLC

ISBN: 978-1-041-19145-2 (hbk)
ISBN: 978-1-041-19146-9 (pbk)
ISBN: 978-1-003-71027-1 (ebk)

DOI: 10.1201/9781003710271

Typeset in Times
by Deanta Global Publishing Services, Chennai, India

Contents

Preface

This book depicts principles of animal physiology with morphological information in relation to the cells, tissues, and organs involved in the physiological system. It describes the physiological functions of all systems with the goal of providing students with a better understanding. The book is a valuable learning resource with an interrelationship to veterinary medicine. The text emphasizes the relation of cellular physiology to body system function, homeostasis, and immune function. The book also contains an introductory chapter presenting an overview of animal physiology in this regard. It elaborately describes cellular interaction with various functions of tissues and organs. The physiological processes with clinical correlations are nicely represented. The book will be a very useful tool for veterinary students, researchers, academics, and veterinary nurses to better understand physiology in the mechanisms of veterinary disease, clinical diagnosis, and the treatment of animals. The contributors read many textbooks, reference books, and websites to gather information and incorporate it into this book. Suggestions of ways to improve this text in subsequent editions are always welcome.

Editor

Tanmoy Rana earned his bachelor's of veterinary sciences and animal husbandry (BVSc and AH), and master's of veterinary medicine, ethics, and jurisprudence (VMEJ) from West Bengal University of Animal and Fishery Sciences, Kolkata, India. He earned his PhD in veterinary science from the University of Calcutta, Kolkata, India. Dr. Rana is currently an assistant professor in the Veterinary Clinical Complex at West Bengal University of Animal and Fishery Sciences, Kolkata, India. Previously, he worked as a veterinary officer in the Animal Resources Development Department, Government of West Bengal, India. He is actively engaged in teaching and clinical practices in veterinary medicine and research related to animal health, production, and disease monitoring regimes.

Dr. Rana's research interests involve arsenic toxicity, molecular diagnosis, molecular toxicology and medicine, oxidative stress, immunopathology, nanoparticles, echinococcosis, and microbes. He has published several articles in reputed international and national journals, along with review articles in international journals. He is an editorial board member of *BMC Veterinary Research*, associate editor of *Frontiers in Veterinary Science*, and reviewer of international and national journals. Dr. Rana is a member of many international scientific societies and organizations, including the West Bengal Veterinary Council (WBVC), the Indian Society for Veterinary Medicine (ISVM), the Association of Public Health Veterinarians, and the Indian Science Congress Association (ISCA). He is also an associate of the West Bengal Academy of Science & Technology, West Bengal, India.

1 Orientation, Navigation, and Communication in Animals

Chetna Mahajan, Abhishek Gupta, Neeti Lakhani and Tanmoy Rana

1.1 INTRODUCTION

Orientation, navigation, and communication are closely interconnected aspects of animal behaviour that play crucial roles in their survival and daily activities. Understanding how animals orient themselves, navigate through their environments, and communicate with conspecifics is fundamental to appreciating the complexity and diversity of animal behaviour. It showcases the remarkable ways in which different species have evolved to interact with their surroundings and each other.

Orientation involves the ability of an animal to determine its position and direction in relation to its environment. It can be characterised as the animal's orientation with respect to gravity or a resource in order to reach it (Chandra, 2023). Another way to think of it is as an organism's way of responding to a stimulus, taking up a specific stance. Movement in a specific direction is hence referred to as orientation. The orientation bears a strong correlation to the spatial characteristics of the surroundings, and an animal's behaviour can be classified into six types based on the orientation.

Orientation of position: Positional orientation with respect to gravity and substrate is how the majority of mobile animals keep their spatial position. In contrast to lighter species, such as flies walking on roofs, larger creatures stay upright by depending more on the substrate's position than on gravity. Together with the primary sense organs in the head, less specialised gravity sensing organs are found in the intestine of pigeons in order to maintain their position during flight. A lot of insects, such as dragonflies and locusts, employ dorsal light orientation to keep their location in flight. The halteres, a unique structure found in flies, are stick-like organs that have microscopic sensory organs at their bases that detect aberrations. Along with gravity, many aquatic organisms also rely on light direction to stay in one place. Daphnia, nauplius larvae, and other small aquatic animals have a longitudinal body axis that is vertical, but larger aquatic animals prefer to have it horizontal. Because swimming and flying animals might be disoriented by air or water currents, positional orientation also aids in preventing displacement. Making sure the locomotory organ is positioned on the substrate is another benefit of the positional orientation. This type of orientation also provides a certain amount of camouflage. For example, aquatic animals with light bellies and dark backs are difficult to find against a background when viewed from below or above.

Object orientation: Object orientation is the capacity to locate, approach, and steer clear of stressors. Generally, there are two stages to object orientation: approach and search. The first phase of the search, known as range, starts in a bigger area and consists mostly of straight lines with sporadic turns. In the event that a source is found, the search is referred to as local search and is conducted in a narrower region, primarily involving turns and minimal straight travel. When something leads to a resource, it is pursued; when it leads to a source of stress, it is steered clear of.

Strato orientation is the process of moving through a habitat's vertical levels. Because smaller animals can't move as quickly and find it difficult to travel long distances horizontally to encounter

DOI: 10.1201/9781003710271-1

variations in physical factors, less vertical movement results in the same changes. This makes vertical movement more important for smaller forms. Because all animals can detect the earth's gravity, which serves as a reference, the sensory input for strato orientation is rather simple. Light can serve as a reference as well. The vertical migration of aquatic zooplankton is the greatest illustration of strato orientation. Numerous creatures of this kind migrate up for the night and down for the day on a daily basis. Gravity and the direction and intensity of the light drive this movement. Animals' own biological clocks are also actively involved in this. The avoidance of predators during the day and the utilisation of an abundance of food in the upper layer of water during the night are the obvious advantages of this type of conduct.

Zonal orientation is caused by elements such as changes in the ground level and level of forestation. The clam Donax lives on marine beaches, where wave action produces an abundance of organic material for food. The tides shift this habitat back and forth. In order to remain in this zone, the clams dig themselves out of the sand and move with the outgoing tide and dig in again at the proper place to get optimum supply of food. A crustacean known as Talitrus saltator seeks out beaches with a particular humidity level; it travels downhill on dry sand and uphill on wet sand in search of these conditions. Compass orientation, which uses some directional or celestial signals, frequently replaces this basic form of zonal orientation response. Zonal orientation is also important to avoid sources of stress. A frog would jump into water if a predator approaches. Similarly, a rodent would jump into its burrow in such a situation.

Topographic orientation: Topographical orientation is the ability of an animal to follow its path back to its starting position. It's a complicated type of orientation that needs a lot of different elements, including taught knowledge and orientation by land markings, directed orientation, and distant orientation. In many creatures, such as certain spiders and crabs, it is just the final element that contributes to topographical orientation. They turn around and travel the same distance after turning 180 degrees, as if they somehow knew how far it was going outward. But still all the components are used by wasps, ants, and bees in their topographical orientation. The distance is measured by the bee and replicated based on the energy expended during movement. To stay on the right path to their nest or feeding locations, they employ compass orientation. They also make use of landmarks, which is useful, particularly in cloudy skies.

Geographical orientation: Numerous animals that travel long distances need to be oriented geographically. These migratory movements, as they are known, are connected to seasonal variations. This behavioural style is seen in a variety of animal species, including birds, bats, fish, insects, turtles, and whales. Huge swarms of the locusts Schistocerca gregaria and Locusta migratoria allow themselves to be swept over a great distance by wind currents. In autumn, many butterflies such as Pieris brassicae and Pieris rapae fly towards the sun, while in spring, they fly away from it. The migrating birds use a kind of vector geographical orientation to move over great distances for nesting or wintering. Invertebrates have statocysts, while vertebrates have membrane labyrinths, which help them maintain an upright posture against gravity (Studocu, 2023).

On the other hand, object orientation occurs when an animal approaches an object, such as food or water. Strato-orientation is the term for the vertical movement of aquatic animals in ponds and lakes. It is referred to as zonal orientation when animals attempt to migrate from grasslands to forests, deserts, or mountains. Long-distance migrants typically have a topographical or geographic orientation. It is known that kinesis, or undirected movement, and taxis, or moving in response to a stimulus, are components of orientation.

1.2 KINESIS

Animal kinesis is the ability to move in reaction to external stimuli. Depending on the stimulus source, it might travel in an orientated or undirected manner. The stimulus intensity may influence the reaction in a proportionate manner.

1.2.1 Klinokinesis

Klinokinesis is the term for the direction change that occurs during movement and can vary depending on the intensity of the stimulus. Receptor cells are found throughout the surface of various organisms, particularly in the dorsal anterior region. Thus, the animal moves right and left alternately to compare the direction of the stimulus to achieve the correct orientation. Animals having a single receptor also show alternate movements. When migrating, caterpillars and maggots search for pupation sites.

Orthokinesis: Orthokinesis shows the animal's speed of locomotion, which is correlated with the stimulus's intensity and the animal's build-up of action-specific energy. The animal's entire body is involved. For instance, burrowing organisms like lamprey larvae's Ammocoete dig in sand far from sources of light. Cockroaches migrate from areas of light to those of darkness. Different kinds of kineses are named according to the stimulus; for example, chemokinesis is named according to chemical stimuli, photokinesis is named according to light gradients, and hygrokinesis is named according to humidity, as in isopods.

Taxis: The animal's orientation in relation to the direction of a stimulus in space is known as taxis. Movement can be towards or away from the stimulus, and depending upon the stimulus, it can be named as follows: hygrotaxis (humidity); geotaxis (gravity); chemotaxis (taste or odour); thermotaxis (temperature); anemotaxis (air current); rheotaxis (water current); phototaxis (light intensity); phonotaxis (sound waves); astrotaxis (sun, moon, and stars); menotaxis (angle to the stimulus); and mnemotaxis (based on memory).

Klinotaxis: Klinotaxis is a behaviour seen in animals with a single receptor, such as the euglena, which uses alternating lateral movements to compare the stimulus's intensity. Comparably, in Dipteran maggots, the light-sensitive organ consists of a group of cells situated above and below the mouth. The body flexes in a negative manner in reaction to light.

Tropotaxis: Animals with paired receptors, similar to eyes in planaria, exhibit tropotaxis. The animal can move straight towards or away from light since it receives equal input on both receptors. An insect will circle towards the side of its painted eye if one of its eyes is painted black.

Telotaxis: When an animal is given the option between positive and negative stimuli, or when its input on the two receptors is not balanced, telotaxis occurs. Fixing the image on one side requires turning the head and making a decision in order to change orientation. When a honey bee sees two light sources, it chooses which one to fly to.

Menotaxis: Keeping an angle with respect to the stimulus source is a requirement of menotaxis. To fly parallel to the ground, nocturnal moths have developed the habit of maintaining a straight angle between their body and the light source, which is typically the moon or stars. They have to fly in circles, though, when they do the same thing with artificial light that is too close. The scout bees' wiggling dance indicates the way honey bees fly from their colonies to the flowers: they keep a consistent angle towards the light. The angle to the sun is remembered by foraging bees while watching the dance on the vertical surface of the comb. Foraging bees then fly towards the food source maintaining the same angle to the sun.

Mnemotaxis: Kühn (1919) provided the first description of mnemotaxis. Niko Tinbergen (1951) investigated this memory-based orientation using a digger wasp experiment. The wasp can precisely orient itself and find its way back to the nest because it carries a memory map of the nest and its surroundings and loops around it. This is also known as geographical orientation or zonal orientation, and it involves the use of landmarks, distance, and direction to help an animal find its way to its nest.

1.3 NAVIGATION AND HOMING

Navigation: Since almost all animals move in a directed manner, navigation is more than just that—it refers to the capacity of many creatures to locate their location with accuracy even in the absence of maps or other tools (Gould, 2014). Navigation involves the neural processing of sensory inputs to determine a direction and perhaps distance. Migratory animals are known to cover long distances accurately over oceans, deserts, forests, and mountains either to reproduce or to escape from the harsh climate. Similarly, fishes, birds, and many invertebrates possess extraordinary capabilities to cross oceans, deserts, and mountains in order to reach their destination. Furthermore, a variety of invertebrates, including crustaceans, amphipods, ants, bees, and wasps, have excellent homing and navigational instincts. They use the local topography, the sun, moon, and stars to help them navigate safely in the accurate route. Moreover, honey bees use the sun's position to guide them accurately. For example, if they needed to find food south of their hive, they would leave their home with the sun to their left in the morning and to their right in the afternoon. Many other species can navigate efficiently over shorter distances, but certain species, including the Arctic tern, fish, and insects, frequently migrate thousands of miles to and from their breeding areas. To put it most simply, the aim of navigation is to be able to get to a place that is distinct from where you are right now. The ability to navigate is the most advanced kind of spatial awareness, requiring an animal to have both a sense of its physical environment and a means of self-orientation. Thus, only more sophisticated animals are capable of navigation. Three different animal navigational strategies are known to exist, according to our understanding of their behaviour. Pilotage, for example, refers to following a specific path by identifying well-known landmarks along the way. Compass orientation is the ability of certain animals to move in one direction (north or south) without the need of a landmark. Furthermore, certain animals have the ability to sense minute geomagnetic effects, such as a magnetic compass. It's interesting to note that homing pigeons, regardless of the orientation of their compass, can fly in any direction towards their home. Animal navigation is influenced by a number of elements that have been recognised, including air pressure, infrasound (which is inaudible to humans), electromagnetic radiation, odour, position of the stars, sun, polarised light, and geographical markers. To comprehend the mechanics underlying the use of natural assistance in long-distance navigation, it is necessary to investigate the migration patterns of the animals. The term "migration" refers to the recurring large-scale movement of birds or other creatures between their seasonal habitat and other areas. It could be done in part to adapt to shifting climatic conditions and the availability of food, or it could be done in part to support reproduction and the needs of the young. Birds are the best example of migration; many of them show mass migration from a breeding and nesting area to a feeding area and back. The majority of bird species in the northern hemisphere migrate, while the ranges of individual species vary significantly. Birds have colonised most of the world, even islands that are deserted by other animals. This colonisation is only possible because birds can fly. Few bird species may be found in the most extreme locations; only those that have adapted to the hard conditions typically live there (i.e. Arctic zones, deserts, etc.) These birds can dwell there indefinitely because they can tolerate the variations in temperature and find enough food. However, seasonal variations can make an unforgiving habitat perfect for avian life.

Because it can fly over huge distances at high speeds, the bird is able to take advantage of these seasonal variations by flying from one region to another, avoiding certain areas in bad weather and using them in favourable weather. Birds that move backwards and forwards in a regular seasonal sequence are called migratory birds.

Migratory birds include:

1. Individuals that shift their location north or south in relation to variations in day and night duration as well as temperature.
2. Those who migrate up and down mountains are referred to as altitude migrants.
3. Those who visit the land periodically to reproduce (e.g. penguins).
4. Nomads are birds that travel without following set routines; an example of this would be desert birds that move in response to rainfall as it falls.

There are still some amazing avian navigational feats that defy explanation. To migrate, most birds rely on both internal and environmental stimuli. Natural selection also prunes back errors in migration, leaving only the best navigators to propagate the species.

1.4 THE DISTANCE OF MIGRATIONS

Migrations can involve travelling short distances or long (across half the world). The Arctic tern flies 8,000 km from the Arctic to the Antarctic and back every year. These long-distance travellers rely on their fat reserves as fuel, stopping along the way to eat. There are still a lot of unsolved questions about avian migration. The exact moment when a migratory bird begins its journey and the amount of time it will last before it needs to rest and feed are known to be influenced by weather patterns. It is also known that birds use navigational aids such as star patterns, the sun, topographical features they fly over, and the magnetic grid of the earth. The need and capacity for migration appear to be innate, or instinctual, to all individuals who migrate, much like so many other behavioural patterns in fish, mammals, and birds (ACS, 2023).

Clock-shift experiments: A study to investigate whether animals can determine their orientation using a time-compensated sun compass. By holding animals under altered day–night regimes for a few days, their inner clock can be reset either clockwise or anticlockwise (i.e. "sunrise" no longer is at the time of natural sunrise, but some hours earlier (anticlockwise) or later (clockwise shift). Animals' internal clocks can be reset clockwise or anticlockwise by keeping them under modified day–night schedules for a few days. Due to their perception that it is a different time and the sun is therefore in a different direction, animals employing time-compensated sun compasses experience a predictable inaccuracy in orientation. In the case of a slow shift (clock delayed), the projected shift in orientation at high latitudes can be roughly represented by a 15° clockwise shift per hour clock shift, and by the same amount of shift in the anticlockwise direction in the case of a quick shift (clock-enhanced).

Experiments on cue conflict and cue calibration: In these two types of experimental manipulations, two orientation cues are tested against each other to assess which cue is dominant over the other. In cue conflict experiments, the animals are exposed to the cue conflict while being tested for orientation, whereas in cue calibration experiments, the animals are exposed to cue conflict not during the actual orientation experiment, but beforehand. The former experimental tool is better suited to test for a preference between cues when the animals determine their orientation direction. The latter experimental tool is used to test which cue calibrates other cues, and has the advantage that the animals can be tested for changes in either cue separately.

Magnetic coil experiments: The most crucial experimental instruments to determine whether an animal utilises a magnetic compass and what kind it has (inclination vs. polarity compass) are magnetic coils (Helmholtz coils, large enough to accommodate experimental cages in the middle). It is possible to intentionally alter or eliminate the magnetic field that surrounds the testing equipment by using magnetic coils. By purposefully inverting the vertical component of the magnetic field, one can test whether an animal utilises a polarity or inclination compass. This reverses the inclination but maintains the polarity, flipping the magnetic field vector. The inclination angle of the magnetic field is used by animals that react to such an inversion by 180° shifting their orientation to identify the correct side of the axis; animals that do not respond to an inversion but only to a shift in the horizontal component of the magnetic field use the polarity of the magnetic field lines to determine their orientation.

The capacity for spatial orientation, navigation, and memory is a well-recognised indicator of evolved intelligence. Significant progress has been achieved in the last few decades in characterising vertebrate spatial memory and navigation (Cheng and Spetch 1998; Cheng et al. 2013; Kelly and Spetch 2012; Tommasi et al. 2012). One basic and recurring problem that mobile animals encounter is the capacity to orient themselves and maintain a desired heading in order to reach destination places. However, invertebrate species also demonstrate remarkable spatial cognitive capacities and are capable of performing navigational feats that are quite similar to those of vertebrate navigators by employing comparable navigational tactics (Chapman et al. 2011; Cheng 2022; Freas and Cheng 2022). In insects, navigation and orientation are dependent on the unique sensory ecology of each species, which includes the interplay and balancing of multiple guidance systems operating in parallel (Büehlmann et al. 2020; Wehner 2020). The navigational systems of insects, as well as vertebrates, are characterised by goal-directed servomechanisms, which are self-regulatory control mechanisms driven by negative feedback. Although these systems are customised for each species, common mechanistic principles are evident in a diverse range of insect species (Cheng 2022; Freas and Cheng 2022). Various representations of goal locations can be used, such as learned cues from past travels, innately preferred compass directions, or visually appealing cue gradients to attract migrating species. According to these representations, the goal is an end state where animals are moved by navigational systems in a way that minimises errors or differences between the cues at their current location and the goal site (Cheng 2022). The function of navigational servomechanisms is to decrease the amount of error between these places by modifying the animal's path (Cheng 2022). These navigational servomechanisms are thought to function on oscillators, or periodically cycling behaviours, in more recent times (Cheng 2022). Both the desired location and the individual's current position are indicated by these guidance systems using a combination of internal and external cue sets across several sensory modalities, with visual cues being a common cue across many insect species (Büehlmann et al. 2020; Freas and Cheng 2022; Heinze et al. 2018; Schultheiss et al. 2020; Wehner 2020). The behavioural research on visually mediated spatial cognition in insects has been reviewed, which aimed to highlight some of the challenges with navigation that insects encounter, along with the behavioural and cognitive strategies that have been found in a few model species to address these challenges. There are three general types of navigational problems. The first challenge is keeping a straight path over short distances, with or without a predetermined destination. Secondly, heading maintenance over extended distances occurs when the objective site is outside the navigator's sensory range and their navigational expertise, such as during migratory voyages or during dispersal without a specified goal direction. Lastly, there is the homing problem, where a navigator returns to a beginning point, usually their nest, after visiting a designated goal site, which is frequently a known resource area. The organism needs some sort of navigation to find its way back, regardless of how far it is travelling. These techniques for navigating consist of visual cues/landmarks, magnetic fields, solar navigation, star navigation, and chemical navigation and are discussed below.

Visual cues/landmarks: This is where visual signals and landmarks are remembered by organisms. To find their nesting places, birds frequently follow the topography, or shape, of a coastline. Some wasp species are able to identify the landmarks surrounding their nest, so even if they shift it, they will eventually return to it. The wasps will search for their nests here if the landmarks are moved and rearranged nearby, or if a comparable arrangement of landmarks is placed nearby (this is an example of homing as the wasps return each day to the same nest).

1.5 MAGNETIC FIELDS

The ability to sense the earth's magnetic field is necessary for this. Animals that use the earth's magnetic field as a compass for navigation frequently have iron-rich magnetite crystals in their brain cells that can orient themselves in relation to the magnetic field. We humans also possess them. Furthermore, certain studies have indicated that humans might also possess a restricted capacity to align themselves with the earth's magnetic field. It is believed that whale stranding results from disorientation due to iron-bearing land features interfering with their compass.

Solar navigation (birds and bees): Timing is also necessary when using the sun as a compass because the sun moves during the day, and the creature must modify its course accordingly. Particularly when it comes to calibrating the compasses of creatures, polarised light at dawn and sunset may be crucial. Bar-tailed godwits are said to use a sun compass as their primary navigational aid during their week-long, non-stop migration from Alaska to New Zealand. Godwits travel to the Alaskan tundra to breed, most likely attracted by the region's plentiful clam supply and dearth of predators. But the godwits fly all the way back down to New Zealand to spend the winter there, avoiding the severe Alaskan winter. The godwits nearly double in size in preparation for flight, accumulating fat for the journey across during which they don't even stop to eat. The genes of those birds whose strength prevents them from travelling will remain out of the gene pool since they will not have an opportunity to procreate.

Every day, bees journey outside of the hive in pursuit of food, a process known as homing. Not only can these critters find their way back to the colony, but they can also communicate directions to other bees, helping them find a food source. In fact, bees use dance as a form of communication. A bee will perform "the round dance" to indicate that food is nearby (less than 50 metres away), but it does not indicate which way the food is. A bee may, however, perform the "waggle dance" if a food source is a bit farther away. This is a figure-of-eight dance where the distance to the food source is indicated by the amount of waggles and the speed of the straight run between the two loops. The angle of this straight run (where it does its "waggles") and the vertical corresponds to the angle between the sun and the food source. As if that were not hard enough, the whole dance is done in the dark! This is truly a remarkable feat for such small creatures.

Star navigation (sometimes called a Star Compass) is a navigation method that makes use of time and the star positions as a compass. The creature must have a sense of timing since the stars will migrate during the course of the night, and it must be able to modify its course accordingly. Stars are used by birds, like the indigo bunting, to guide them. This was demonstrated in an experiment where the birds were housed in sizable cages with an ink pad bottom and enormous blotting paper cones. The sides of the cone showed where the birds would fly if they attempted to flee. The birds were kept in cages inside a planetarium, and when the stars overhead were displayed, the researchers saw that the birds constantly positioned themselves towards what they believed to be "south."

1.6 CHEMICAL NAVIGATION

This is the chemical sensitivity, or gradient sensitivity, that enables an organism to locate its way back to its breeding grounds. It frequently calls for an enhanced sense of smell in relation to a chemical pheromone, which is a chemical that an animal releases to affect the behaviour of other members of its own species. In addition to providing a chemical trail (called a pheromone) for other ants to follow, ants also use it to guide themselves back to the nest. Ants follow one another to a feeding place, which explains why they are frequently observed in trails. An ant's pheromone is only released for around two minutes. This guarantees that other ants will only find new routes leading to adjacent food sources. An increased number of ants marching along a track indicates that more pheromone is secreted and the trail will persist much longer after a suitable food source has been identified. In order to demarcate their domains and create more durable pathways to dependable food sources, ants might also employ a different pheromone. The acidic pheromone that ants release is what gives them their peculiar taste and smell. Animals employ a range of reference cues to ensure correct orientation and navigation. Different senses may be needed to recognise these navigational markers, which can be cues of various kinds, including visual items, odours, or even the geomagnetic field. A given navigation job may dictate the significance or selection of a specific cue or cues, ranging from short-range navigation that may be applicable for a few-meter radius to long-distance navigation during, say, bird migration. Animals navigate in a variety of ways, so it makes sense that a mammal navigating underground would use different senses than a bird searching for a hidden food cache, and birds migrating at night most likely use a slightly different navigational strategy than those migrating during the day.

1.7 COMMUNICATION BEHAVIOUR

Animal communication is first and foremost about signal transmission and aims to understand how communication occurs. It is a field that has helped other fields—from neuroscience to information technology—and been inspired by them in its quest to develop ever-better techniques for listening in on the "message" that lies at the heart of communication (Kaplan 2014). Animals use communication to express information about their wants, intents, and surroundings. It is an essential part of animal behaviour. When an animal communicates with another, the receiving animal changes in some way as a result of receiving the information. Communication is usually between animals of a single species, but it can also happen between two animals of different species. Certain communication skills come naturally and don't need to be learned. For instance, when a parent bird is nearby, freshly hatched birds may open their lips and make beseeching sounds. Animals use signals to communicate. These messages might be chemical, involving pheromones; visual, auditory, or sound-based; tactile, or touch-based. There have been good records of vision (Martin 2007; Kelber et al. 2003), auditory ability (Tian et al. 2001; Romanski 2002), and olfaction (Kemp 2012; Menini 2010), as well as research results for electrical (Stoddard and Markham, 2008) and chemical (Hay 2009) perception and communication in animals. These methods are aligned with the five basic senses: **vision**, **hearing**, **taste**, **smell**, and **touch**, where taste and smell are combined into the olfactory category. Communication behaviours can help animals find mates, establish dominance, defend territory, coordinate group behaviour, and care for young.

1.7.1 Visual

Visible signals are used in visual communication. Gestures, facial emotions, body postures, and colours are a few examples of these messages. Posture and gestures are common visual cues. A cat arching its back or a dog wagging its tail are examples of submissive behaviour. Chimpanzees can display a threat by, for example, striking the ground, lifting their arms, or glaring at one another. Certain creatures also use their facial expressions to communicate information. For example, the baby chimpanzee below displays what's known as a "fear grin," which denotes submission. When a young chimpanzee approaches a dominant male in their troop, they employ this expression to show that they respect the male's dominance. Certain behavioural reactions to visual stimuli are inherited; e.g. chickens will cower in terror if they spot a picture of a hawk in flight. Certain visual reactions are taught. For example, when someone approaches an animal during their expected meal period, it makes them rush towards the food bowl, and female birds are attracted to the colourful plumage of their male counterparts. Certain animals employ visual cues, including imitation or camouflage, to help them survive. Certain insects imitate the appearance of dangerous species to ward off predators, whereas chameleons alter their colour to communicate and blend in with their surroundings.

1.7.2 Auditory

Auditory communication—communication based on sound—is widely used in the animal kingdom. Birds, which employ sounds to communicate warnings, entice mates, defend territories, and coordinate group behaviours, are especially noteworthy examples of animals that communicate through sound. A species' vocalisations, known as birdsong, are characteristically long and melodic and can be specific to certain birds as well.

1.8 MANY NON-BIRD SPECIES ALSO COMMUNICATE USING SOUND

When a predator is approaching, monkeys will shout out, giving the other troop members an opportunity to flee. Even their sounds differ to denote various predators, based on observations of vervet monkeys. In order to get female frogs to mate, bullfrogs croak. Some frog species have sounds that are audible up to a mile away. Gibbons mark their territory with calls in order to deter possible rivals. Males and females in pairs, as well as their young, may make calls together. Female farm animals react to their young's distress sounds. Bats fly in the dark by using ultrasonic echolocation.

Sound waves can travel through water just like they can in the air, and marine life uses sound to communicate. For example, dolphins make a variety of sounds, such as clicks, chirps, and whistles, and they arrange these sounds into intricate patterns.

1.8.1 Pheromones

A pheromone is a chemical signal that an individual of the same species secretes that causes another to react. Pheromones are particularly prevalent in social insects like bees and ants. Aside from attracting the other sex, pheromones can also set off an alert, identify a food trail, and cause other, more complicated, behaviours. Pheromones secreted by glands or urine are responsible for defining territorial boundaries in many mammals, including dogs. The n-butyl mercaptan of the skunk is a well-known animal pheromone. In general, pheromones regulate sexual conduct. For example, they make male canines more attracted to bitches during oestrus. They control maternal behaviour, or the rejection of another animal's baby by its mother, and hasten the puberty of young rodents. A key component in pheromone detection is the vomeronasal organs and the olfactory mucosa. Cats and other ungulates are known to have the species-specific curling of the upper lip called flehmen, which aids the VNO (vomeronasal organ) in detecting pheromones.

Illustrations: Ever wonder how ants follow seemingly undetectable trails that lead to food sources? When you take your male dog for a stroll, why does it mark its territory by urinating on shrubs and lampposts?

Pheromone trails left by ants point other members of the colony in the direction of food sources. Ants will leave and return pheromone trails behind them when they find rich food sources, which will attract additional ants to the area. The ants will cease producing pheromone on their return trip and allow the trail to disappear when the food supply is likely to run out. Pheromones are also used by ants to convey their social standing or position within the colony; hence, ants belonging to distinct "castes" may react differently to identical pheromone signals. Additionally, a crushed ant will produce a burst of pheromones that alert other ants in the area to the danger and may even cause them to swarm and sting. Dogs use pheromones to communicate as well. In addition to releasing many of the compounds into their urine, they smell each other to gather this chemical information. A dog can stake its claim to adjacent territory by marking its identity with a pee mark on a shrub or post, which other dogs can read.

1.8.2 Tactile Signals—Touch

Because tactile signals require two creatures to be in close proximity to one another in order to be detected, their range is more constrained than that of other communication types. Nevertheless, many species use these signals as a significant component of their communication repertoire. In insects, tactile messages are rather common. For example, a honey bee forager will communicate the location of food by engaging in a complex series of gestures known as a waggle dance. Because this dance takes place inside the nest in complete darkness, the other bees mostly use touch to understand it. In social relationships, tactile cues are equally crucial. For example, in many species of primates, group members groom each other, getting rid of parasites and doing other personal hygiene chores. Additionally, extremely immature organisms depend on tactile inputs for survival. Newborn puppies, for example, will knead at their mother's mammary glands instinctively, which releases the hormone oxytocin and promotes the production of milk.

Electrocommunication: Electric signals are used by certain aquatic creatures, such as electric fish, for navigation and communication. Information concerning identity, sex, and reproductive status can be sent by these signals.

Mechanical communication: Certain creatures, such as woodpeckers, communicate by drumming on surfaces. Others, like spiders, alert their prey to danger or announce courtship through vibrations in their webs.

1.8.3 Benefits of Communication

Animals use a variety of signals for communication, as seen by the aforementioned instances, and they do so in a variety of settings. The following are a few of the most typical uses for communication:

Obtaining mates: When it comes to mating, many animals engage in complex communication behaviours. These behaviours can include luring a partner or battling with other possible suitors for the attention of mates.

Establishing dominance or defending territory: In many species, staking out a territory or asserting authority in a social hierarchy depends on communication behaviours.

Coordinating group behaviours: When it comes to social animals, communication is essential for both preserving group cohesiveness and organising group tasks like food gathering and defence.

Caring for young: In animals where offspring are cared for by their parents, communication serves to coordinate the behaviours of both parents and young, increasing the likelihood of the offspring surviving.

As a result, communication facilitates interactions between organisms that are necessary for them to survive, find partners, and raise their progeny.

1.9 INTEGRATION OF ORIENTATION, NAVIGATION, AND COMMUNICATION

In the animal kingdom, orientation, navigation, and communication are critical behaviours that enable individuals to live and prosper in their surroundings. Orientation and navigation can be supported by successful communication, and vice versa; hence, these behaviours are frequently closely related.

Migratory behaviour: Numerous species travel great distances during long-distance migrations, utilising communication, navigation, and orientation.

Social communication during migration: Some animals, like certain birds, employ social cues like V formation flying to coordinate their migration efforts.

Parent offspring communication in navigation: Young animals may pick up navigational skills and foraging locations by using their parents' communication signals.

In migratory species, the integration of these behaviours is more noticeable. For instance, during migration, birds may use the sun or stars for orientation, the earth's magnetic field for navigation, and vocalisations or visual displays for flock communication. These behaviours collectively contribute to the adaptability and survival of animals in diverse ecosystems and play a crucial role in maintaining the balance of ecosystems.

REFERENCES

ACS 2023, ACS distance education, UK. Navigation (https://www.acsedu.co.uk/info/pets/animal-behaviour/navigation.aspx).

Buehlmann, Cornelia, Michael Mangan, and Paul Graham. "Multimodal interactions in insect navigation." *Animal Cognition* 23 (2020): 1129–1141. https://doi.org/10.1007/s10071-020-01383-2

Chandra, G. Orientation navigation and homing in animals. Zoology and entomology articles- The Indian administrative service zoology, 2023. https://www.iaszoology.com/orientation/

Chapman, Jason W., Raymond HG Klaassen, V. Alistair Drake, Sabrina Fossette, Graeme C. Hays, Julian D. Metcalfe, Andrew M. Reynolds, Don R. Reynolds, and Thomas Alerstam. "Animal orientation strategies for movement in flows." *Current Biology* 21, no. 20 (2011): R861–R870.

Cheng, Ken. "Oscillators and servomechanisms in orientation and navigation, and sometimes in cognition." *Proceedings of the Royal Society B* 289, no. 1974 (2022): 20220237.

Cheng, Ken, Janellen Huttenlocher, and Nora S. Newcombe. "25 years of research on the use of geometry in spatial reorientation: a current theoretical perspective." *Psychonomic Bulletin & Review* 20 (2013): 1033–1054.

Cheng, Ken, and Marcia L. Spetch. "Mechanisms of landmark use in mammals and birds." In S. Healy (Ed.), *Spatial representation in animals*. Oxford: Oxford University Press, pp. 1–17, 1998.

Freas, Cody A., and Ken Cheng. "The basis of navigation across species." *Annual Review of Psychology* 73 (2022): 217–241.

Gisela Kaplan. WIREsCognSci2014. https://doi.org/10.1002/wcs.1321

Gould, James L. "Animal navigation: a map for all seasons." *Current Biology* 24, no. 4 (2014): R153–R155.

Hay, M. E. "Marinechemical ecology: chemical signals and cues structure marine populations, communities, and ecosystems." *Annual Review of Marine Science* 1 (2009): 193–212.

Heinze, Stanley, Ajay Narendra, and Allen Cheung. "Principles of insect path integration." *Current Biology* 28, no. 17 (2018): R1043–R1058.

Kelber, Almut, Misha Vorobyev, and Daniel Osorio. "Animal colour vision–behavioural tests and physiological concepts." *Biological Reviews* 78, no. 1 (2003): 81–118.

Kelly, Debbie M., and Marcia L. Spetch. "Comparative spatial cognition: encoding of geometric information from surfaces and landmark arrays." In *Oxford handbooks online*. Oxford University Press, 2012. https://doi.org/10.1093/oxfordhb/9780195392661.013.0020

Kemp, C., and G. Kaplan. "Olfactory cues modify and enhance responses to visual cues in the common marmoset (*Callithrix jacchus*)." *Journal of Primatology* 1 (2012): 102. https://doi.org/10.4172/jpmt.1000102.

Kühn, A. *The orientation of animals in space*. Gustav Fischer, Jena, 1919.

Martin, Graham R. "Visual fields and their functions in birds." *Journal of Ornithology* 148, no. Suppl 2 (2007): 547–562.

Menini A. "The neurobiology of olfaction." *Bioscience Net Base* 1 (2010): 450.

Romanski, L. M., and P. S. Goldman-Rakic. "An auditory domain in primate prefrontal cortex." *Nature Neuroscience* 5 (2002): 15–16.

Schultheiss, P., A. Wystrach, M. Lihoreau, and S. Schwarz. Insect navigation. In J. Vonk and T. Shackelford (Eds.), *Encyclopedia of animal cognition and behaviour*, p. 1, 2020.

Stoddard, Philip K., and Markham, M.R. "Signal cloaking by electric fish." *Bioscience* 58, no. 5 (2008): 415–425.

Studocu, 2023 (https://www.studocu.com/in/document/mahatma-jyotiba-phule-rohilkhand-university/development-biology-and-animal-behavior/orientation-in-different-animals/41892663#).

Tian, Biao, David Reser, Amy Durham, Alexander Kustov, and Josef P. Rauschecker. "Functional specialization in rhesus monkey auditory cortex." *Science* 292, no. 5515 (2001): 290–293.

Tinbergen, N. *The Study of Instinct*. Clarendon Press. 1951

Tommasi, Luca, Cinzia Chiandetti, Tommaso Pecchia, Valeria Anna Sovrano, and Giorgio Vallortigara. "From natural geometry to spatial cognition." *Neuroscience & Biobehavioral Reviews* 36, no. 2 (2012): 799–824. https://doi.org/10.1016/j.neubiorev.2011.12.007

Wehner, Rüdiger. *Desert navigator: the journey of an ant*. Harvard University Press, 2020.

2 Motivational Behaviour

Deepika Diana Jesse A., Jyotsana Shakkarpude, Pramod Kumar and Rajesh Kumar

2.1 INTRODUCTION

Behaviour is a co-ordination between the nervous system and locomotion, i.e. the response of an animal to external stimuli that can be seen and analysed by other individual animals. It can also be defined as an expressed sum of individual muscle contractions and hormonal secretions. The limbic system is that part of brain which is involved in behavioural and emotional responses. The structures of limbic system are buried deep within the brain, underneath the cerebral cortex and above the brainstem. The thalamus, hypothalamus and basal ganglia are involved in the actions of the limbic system, but two of the major structures are the hippocampus and the amygdala. Behaviour alters an animals' relationship to its external world, and one of the principal consequences of behaviour is adjustment of the sensory input (Grandin, 2002).

Motivation is the will to do and the will to do well. Motivation may be self-directed, or one may be motivated by others and by the environment. One can motivate others as well. The motivation of an individual depends on the strength of their motives. Motives are needs, wants, drives or impulses within the individual. Motives are directed towards goals, which may be conscious or subconscious. Inference about motivation can explain many behaviours, such as changes in responsiveness, goal-directed behaviour and spontaneous behaviour. Motivation can be called a reflex response to cues or signals that correspond, for example, to a prey, an enemy, or a social or sexual partner. Motivation usually leads to certain goal-directed behaviour. For example, if there is a drive for quenching thirst, an animal explores for water; if it is found, the goal will be achieved and water will be consumed. The drive for water is thus reduced.

2.2 FACTORS AFFECTING MOTIVATION

Motivation is affected by memories of past experiences and the context created by other stimuli. Other factors include level of arousal, level of autonomic activity, hormone levels and emotion-producing functions within the limbic system.

Autonomic activity:
The heart is controlled by the sympathetic and parasympathetic nervous systems, while the autonomic nervous system's activity is one of the physiological indices. In general, the sympathetic nervous system promotes cardiac activity while the parasympathetic nervous system suppresses it, and resting heart rate is preponderantly controlled through parasympathetic nervous system activity (Hori *et al.*, 2017a). An increase in sympathetic nervous system activity and the suppression of parasympathetic nervous system activity is generally observed when negative emotions are aroused (Hori *et al.*, 2017a).

Decreased parasympathetic activity has been reported in individuals with psychiatric disorders such as schizophrenia and major depression (Agelink *et al.*, 2002; Montaquila *et al.*, 2015). The decreased ability to recover from a stress response due to vagal tone deficits has also been reported in patients with schizophrenia (Montaquila *et al.*, 2015). Hence, the index of the autonomic nervous system's activity should be used for advancing tailored psychiatric rehabilitation (Hori *et al.*, 2017b).

 DOI: 10.1201/9781003710271-2

A prior study showed that there was a positive correlation between depression and anxiety and the parasympathetic nervous system during stressful tasks, while the impact of passive coping was considered a possible explanation for the said correlation (Brugnera *et al.*, 2017). In a task presented as a "passive coping condition" requiring no effort, a "conservation-withdrawal"-type response was observed, which is characterised by increased parasympathetic tone (Bosch *et al.*, 2001).

2.3 HORMONE LEVELS

There are four main hormones that are primarily responsible for creating those feel-good emotions and sensations. They moderate our feelings of well-being and are influenced by stress levels and self-care. They are endorphins, which primarily help one deal with stress and reduce feelings of pain; serotonin, which is a mood stabiliser like well-being and happiness; dopamine, which plays motivational role in the brain's reward system; and oxytocin, which creates a bonding of love and trust.

Hypocretins, or orexins, are other neurochemicals that are released from the hypothalamus and have a function in reward and motivation. Orexin neurons directly stimulate dopamine neurons due to their direct synapses but also indirectly via glutamate. These orexinergic activities enhance glutamate release onto dopamine neurons and at the same time enhance the probability of the firing of dopamine cells, facilitating glutamatergic signalling. This suggests that orexins play a central role in motivated behaviours by having their action on the dopamine neurons.

2.4 EMOTION-PRODUCING FUNCTIONS

An emotion is something that motivates. For example, if someone is happy, they will be energised to do something that they believe will help to maintain the happy feeling. If angry, they may be motivated to act aggressively against the animal or individual who made them angry. Some behaviours motivated by emotion are grand displays involving words and a complex series of actions. Tone of voice, body movements and facial expressions can all help to convey the emotions that we are experiencing.

2.5 THEORIES OF MOTIVATION

Motivation theories seek to explain what drives individuals to initiate, direct and sustain behaviour towards achieving certain goals. They explore the underlying needs, desires and incentives that influence these behaviours and how external factors interact to motivate individuals in different contexts. These theories are Maslow's hierarchy of needs theory, Herzberg's two-factor theory, expectancy theory, goal-setting theory and equity theory.

2.5.1 Maslow's Hierarchy of Needs Theory

This theory came into existence in 1943, ranking human needs hierarchically in a five-level pyramid, where people seek to satisfy their needs from bottom to top, starting from physiological, safety, belongingness, and esteem, and moving finally to self-actualisation needs.

2.5.2 Herzberg's Two-Factor Theory

Theory was postulated by Herzberg in 1959; this distinguishes between factors preventing dissatisfaction and affecting motivation: hygiene and the motivation factors. Hygiene factors are needed to ensure an employee is not dissatisfied. Motivation factors are needed to motivate one to higher performance.

2.5.3 Expectancy Theory

Theory was postulated by Vroom (1964), and it advances that motivation is the product of an individual's "expectancy" that a certain effort will lead to the intended performance, the "instrumentality" of this performance to achieve a certain result and "valence" which is the desirability to achieve that result.

2.5.4 Goal-Setting Theory

According to this theory of Locke (1984), individuals' goals must be aligned with the overall goals of the organisation. Individual goals must be realistic and mutually agreed. The more difficult a goal is, the more motivation it produces. Motivation through goals is influenced by such factors as clarity, challenge, commitment, feedback and task complexity.

2.5.5 Equity Theory

Equity theory of motivation, introduced by Adams in 1963, is influenced by the sense of "fairness" that workers possess after comparing their efforts and rewards ratio with others who are in the same situation. Adams used the terms "inputs" and "outputs" instead of efforts and rewards to describe as "inputs" the qualifications, effort and competence, and as "outputs" the pay, prospects, benefits and recognition.

2.6 NEURAL SUBSTRATES FOR APPETITE AND THIRST

Peripheral stimuli associated with taste, smell, stomach contractions, metabolic rate and body temperature are important in the regulation of appetite and thirst. In the brain, a group of neurons in the lateral hypothalamus seems to promote the drive for eating, and the neurons in the ventromedial hypothalamus promote satiety. Appetite and thirst are regulated by noradrenergic and cholinergic transmitter systems. Utilisation of glucose by hypothalamic cells is a primary influence on appetite regulation. Thirst drive is affected by the octapeptide hormone, angiotensin II, which acts on the anterior hypothalamus (Figure 2.1).

2.7 NEURAL SUBSTRATES FOR AGGRESSION

The amygdala and hypothalamus are parts of the limbic system that can influence the display of aggressive behaviour. Discrete lesions in various areas of the limbic system reveal that a variety of emotional reactions can ensue, including rage, fear, escape, unemotional attack and defense.

2.8 NEURAL SUBSTRATES FOR REWARD AND PUNISHMENT

The medial forebrain bundle regulates reward, and the periventricular system regulates punishment.

2.8.1 Positive Reinforcement Systems

Animals repeat behaviours that are positively reinforcing; that is, they repeat behaviours that satisfy some drive or need. The neuronal basis for such reinforcing behaviours resides in several areas of the limbic system. The effects of positive reinforcement result from electrical stimulation of such limbic areas as the lateral hypothalamus, the septum and the medial forebrain bundle. The neurotransmitter systems most involved in positive reinforcement seem to be catecholaminergic (Kakkos and Panagiotis, 2011).

There may be multiple reinforcement systems. The endogenous opiate peptide system is reinforcing, and it does not seem to depend on dopamine or norepinephrine. In addition, systems that

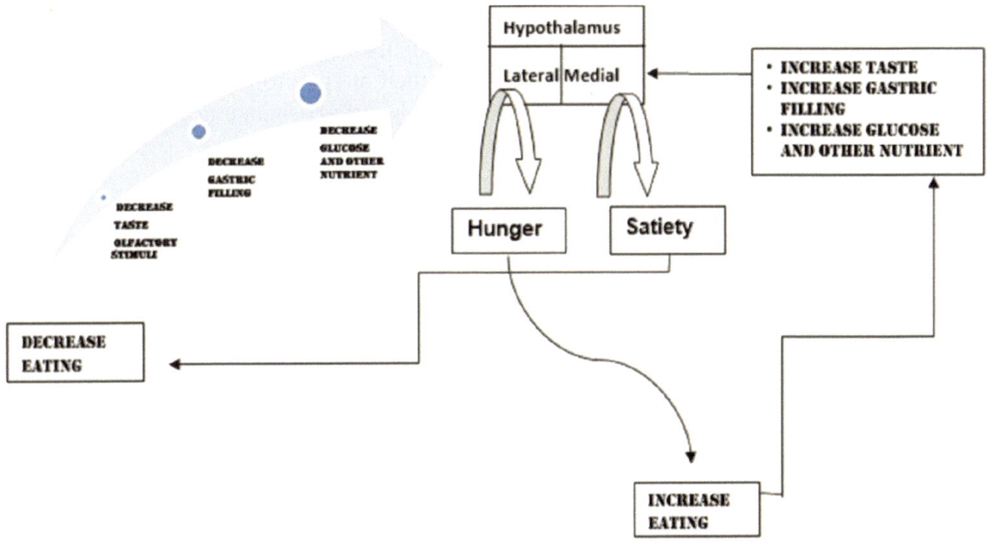

FIGURE 2.1 Interrelationship of various influences in the regulation of appetite.

control such normally rewarding functions as eating, drinking and sexual behaviour, and many more, do not necessarily depend on dopamine or norepinephrine.

2.8.2 Negative Reinforcement Systems

Negatively reinforcing stimuli produce withdrawal, immobility, aversion, escape or similar behaviours. The neural systems that mediate such behaviours occur mostly in the brain stem reticular formation and in the periventricular and periaqueductal areas of the midbrain. The transmitter system most clearly associated with such functions is cholinergic.

2.9 MOTIVATION IN DOGS

Motivation is what influences a dog to behave a certain way. By understanding a dog's motivation, one can change their behaviour. Much of dogs' behaviour is driven by consequences. In other words, what happens when they behave a certain way? Dogs can use their behaviour to get something they want (like food) or to escape something unpleasant (like the vacuum cleaner). Most importantly, a dog decides whether a particular consequence is worth seeking or avoiding. If they don't care about something, it won't motivate them. Emotionally based motivations, like escaping something the dog thinks is scary, are very important to consider. Fear and anxiety can be strong motivators for a dog's behaviour, and often your dog can't control their emotions in the same way that they control their behaviour. For example, they can't choose to not be afraid in the same way that they can choose to work for or ignore a treat. But emotional motivation can change too; with careful training, dogs can learn to feel safe around things that previously scared them.

Some common motivators are food, play and attention; their behaviours can be changed by using these motivators. Every dog needs to eat, so food is one of the "primary" motivators. Many dogs love toys and play. Whether it's chasing a ball, tugging a rope or playing with other dogs, play feels good. A dog's version of play might not always match ours, such as mouthing on our hands or grabbing our clothes. Dogs typically enjoy attention from their favourite people. They may seek out petting or cuddles as well. It's important to note that dogs don't enjoy every type of attention or petting in the same way, so you will need to pay attention to what your dog enjoys. By providing a reward to the dog, one can motivate it and use this to reward any behaviour. Food and treats are a great way to encourage a dog to repeat desirable behaviours. Moreover, the good behaviour needs to be appreciated. If a dog's problematic behaviour is their attempt to get something they want, it can be decreased by removing their reward, for example, ignoring demanding barking (Char, 2024).

2.10 MOTIVATION IN HORSES

Horses aren't motivated by food but rather by the need for comfort and safety. They comfort each other with activities such as mutual grooming and swishing flies away from each other. Horses feel safe when they are together—there is safety in numbers for prey animals. There are four basic drives that motivate horses and other animals to do things. These are fear, aggression, a learned response and instincts. Fear and aggression are often misinterpreted. Did the horse kick because he was fearful, or did he kick because he was aggressive or "bad"? Neurologically, fear and aggression are different emotions which may result in similar behaviours such as kicking or pinning the ears back. It is important to determine which emotion is motivating kicking because punishing the horse for kicking will make a fear-based behaviour worse. If kicking occurs during a training exercise, it is likely to be fear-based. Fear is also most likely motivated if an animal becomes agitated when it is alone, tied up, or held in a squeeze chute. Another factor is genetics. Horses and cattle with a nervous, high-strung temperament are more likely to have fear-motivated behaviour than an animal with a calm, placid temperament. It is unfortunate that some breeders are selecting for "hot-blooded" draft horses. This is likely to result in more problems with fear-motivated behaviour. An animal with a "hot" temperament is more likely to blow up when it is suddenly confronted with a scary novel experience.

REFERENCES

Agelink M.W., Boz C., Ullrich H. and Andrich J. (2002). Relationship between major depression and heart rate variability: clinical consequences and implications for antidepressive treatment. *Psychiatr. Res.* **113**(1–2):139–149.

Bosch J.A., de Geus E.J., Kelder A., Veerman E.C., Hoogstraten J. and Amerongen A.V. (2001). Differential effects of active versus passive coping on secretory immunity. *Psychophysiology.* **38**(5):836–846.

Brugnera A., Zarbo C., Adorni R., Tasca G.A., Rabboni M., Bondi E., Compare A. and Sakatani K. (2017). Cortical and cardiovascular responses to acute stressors and their relations with psychological distress. *Int. J. Psychophysiol.* **114**:38–46.

Char J. (2024). https://www.svpetproject.org/info/display?PageID=20149.

Grandin T. (2002). Understanding motivation will help you solve behavior problems in animals. SAVMA Symposium.

Hori T., Ozaki H., Katayama J. and Suzuki N. (2017a). Editors. *Physiological Psychology and Psychophysiology* II (In Japanese). KitaoojiShobo; Kyoto.

Hori T., Ozaki H., Sakata S. and Yamada F. (2017b). Editors. *Physiological Psychology and Psychophysiology I* (In Japanese). KitaoojiShobo; Kyoto.

Kakkos, N. and Panagiotis, T. (2011). Investigating the link between motivation, work stress and job performance. Evidence from the banking industry. 8th International Conference on Enterprise Systems, Accounting and Logistics (8th ICESAL '11),10–12 July 2011, Thassos, Greece.

Montaquila J.M., Trachik B.J. and Bedwell J.S. (2015). Heart rate variability and vagal tone in schizophrenia: a review. *J. Psychiatr. Res.* **69**:57–66.

Vroom, V.H. (1964). Work and motivation. Wiley.

3 Foraging Behavior in Self-Organization

Madhumita Roy and Santanu Debnath

3.1 INTRODUCTION

Hormones, muscles, and nerves must all function in a specific way for any behavior to manifest. The word "motivation" has been used frequently by ethologists to describe the different underlying components that form the basis of all behavior. The way an animal chooses what to do at any given time is referred to as its motivation (Bateson et al., 1990). An animal has to prioritize one activity over another by weighing a variety of internal and external cues using an internalized decision-making process. The animal's motivation is determined by its internal state, which is the overall outcome of stimuli coming from both within and outside the body. An animal's motivation aids in its decision-making regarding behavior. The control mechanism of higher-level decision-making is largely unknown. For example, an animal that is encouraged to run away and feed simultaneously may exhibit behaviors such as calling out for new members of the group and then feeding as a flock. According to some ecologists, the only way to understand this is to use physiology techniques to examine the creatures and record their nerve impulses, muscle movements, and hormone levels. Some ethologists think that in order to provide a more causal explanation, some characteristics like motivation are necessary for a thorough explanation.

Depending on the time of day, almost all animals exhibit varying behavioral patterns, yet many exhibit a high degree of consistency in their routines. The fact that animals have internal clocks contributes to this daily behavioral shift in some ways. At these moments, their actions are mostly unaffected by what is happening in the outside world. But there are other factors that influence an animal's behavior at different times of the day besides its internal clock. Furthermore, animals are capable of modifying their time budget in response to changing conditions. Unexpectedly, a significant amount of the active portion of a wild animal's day is spent sleeping and seemingly doing nothing (Duncan and Wood-Gush, 1972). This rest interval could be crucial for energy conservation, digestion, or just staying out of the way of predators (Elgar, 1986a, 1986b, 1989). Indeed, one strategy that animals may use to get away from potentially dangerous situations is to sleep in a secluded spot for a portion of the day (Gottmark et al., 1986; Heinrich, 1988; Krebs et al., 1972).

3.2 OPTIMIZATION THEORY

Optimization theory states that there is an ideal solution that yields the optimal outcome in any given circumstance. The theory of optimization may be extremely effectively applied to ecology by calculating the best solution before the situation is analyzed in order to forecast what an organism should be doing. These predictions can then be tested. Put another way, the theory can lead to a testable hypothesis, but in actuality, it is frequently challenging to pinpoint the limitations that the creature lives under. An essential explanation is that animals make choices based on optimizing some factor, like net rate of energy intake, which means that when an animal consumes all of the food in one region, it will receive less food despite continuing to put in the same amount of effort to procure food in the same place over time. It might acquire energy more quickly even though it

DOI: 10.1201/9781003710271-3

would need to expend some energy traveling by foot or by air to reach the new location. In a similar vein, animals engage in a variety of behaviors linked to eating, evading predators, finding cover from the elements, reproducing, and other activities. It is clear that animals are subject to limitations and must make the most of specific circumstances and surroundings that might not be well suited for achieving any other one behavioral objective. The majority of biologists and ecologists have looked at optimization theory in relation to ideal foraging. In this discipline, biologists have applied optimization theory to provide quantitative predictions regarding feeding behavior that can then be verified by experimentation and observation. According to the notion of optimal foraging, an animal should remain in a particular region even if its food supply is gradually being depleted until the net rate of energy intake approaches the environmental average. The animal should then make the decision to move on and look for something else. Making decisions requires a multifaceted and intricate process.

3.3 OPTIMAL FORAGING THEORY

An animal needs food to survive, develop, and procreate. An animal gets the nutrition and energy it needs for growth, healing, and reproduction when it feeds. Observing animals eat is something we have become accustomed to. It is easy to ignore the fact that they feed themselves. Biologists have just recently started to look into the extent to which an animal may be considered to feed or forage optimally—that is, over the last 30 years. Numerous researchers in behavioral ecology have examined the degree to which animals optimize their energy consumption. It will become clear that the optimal foraging hypothesis is predicated on the animal's ability to gauge environmental conditions and its ability to compare the rate of requirement with what it would encounter if it took off for another location and began feeding there. Even though animals are not aware of these facts, optimal foraging theory, actually, at least in some circumstances, captures behavior quite effectively. An animal that is foraging must make a number of decisions, such as where to travel in order to get food. What time of day should it feed itself? Which prey should it choose? How much time should it spend processing the food or prey it has found? Users of the optimal foraging theory model may be able to study each of these foraging decisions. More intricate optimality models take into consideration a variety of variables, such as the time of day and the condition of an animal's food supplies, to account for the actual behavior.

Developing optimality models and putting them to the test can help us understand what principles an animal might use to make decisions. As a result, these models can be used to address concerns of causality pertaining to animal behavior. However, because the rules of decision-making will themselves mirror the process of natural selection, they can also be used to address concerns related to adaptation. Based on the predictions of a basic optimum foraging model, an animal would only optimize its energy intake if it was critical to its life and ability to reproduce. Actually, it would only forage optimally if maximizing energy intake was the only factor affecting survival and success in reproduction; otherwise, the optimization of food intake would unavoidably be compromised by other factors.

Animals are almost never under the influence of a single selection pressure; instead, they are entangled in a complex web of trade-offs between several. As a result, we anticipate that actual animal behavior will represent the best possible balance between various selection forces. To put it another way, we shouldn't expect an animal to forage at its best every moment of its existence because there will be moments (like when it is not starving and is being pursued by a predator) when starvation won't be the biggest threat to its survival and ability to procreate.

However, there will also be instances when finding adequate food quickly and effectively will be crucial. For example, little birds in the winter may need to prioritize finding food or they risk starvation and death. However, in general, we should anticipate that, with regard to the total optimality of the entire animal, each of the many elements of an animal's success (eating, avoiding predators, avoiding dehydration, finding a partner, etc.) will be something less than optimal. Therefore, an animal's behavior may be less than ideal when certain aspects, like feeding, are compared to an optimality model that only represents a single behavior, but optimal in the sense that it has reached the best compromises for maximal lifetime reproductive success. By keeping this distinction clearly in mind, a considerable deal of confusion and debate can be avoided (Laland et al., 1990).

3.3.1 Examples of Optimal Foraging Behavior

Reto Zach (1979) investigated how crows in the northwest fed. The crow eats whelk, clams, or snails. When eating, it scoops up the prey, launches itself into the air, and deposits it upon the stony, hard surface. On the rocks beneath, the mollusk's shell breaks. The crow then descends to the victim and plucks off the flesh that is visible. It is widely known that this behavioral pattern has adaptive benefits due to the crow's inability to break through the mollusks' tough shells. Dropping them on the hard, rocky surface smashes it as a result. This behavior appears to be adaptive, based on observations. However, upon closer examination of the crow's foraging behavior and decision-making, it becomes clear that the voracious bird must choose which mollusk to eat, how high to soar before letting its prey fall to the ground, and how many times to try again if the mollusk shell does not break on its first try. According to the field study, the crows only took up medium- to large-sized whelks, which were between 3.5 and 4.4 cm long. They also dropped their food after flying up to a height of 5 to 6 meters. It has also been observed that despite the necessity of numerous flights, they persisted in trying until the whelk's shell shattered.

Zach investigated whether the crow's choice maximized the amount of food available for consumption per unit of time spent foraging, which would help to explain the behavior of the bird, and whether the following predictions from the optimality theory could be confirmed. 1. After a drop of about 5 meters, large prey would be more likely to be picked up than small ones that break. 2. Breakage rates would be lower for drops under 5 meters. Drops below 5 meters, therefore, shouldn't significantly increase the likelihood of the mollusk shell opening. 3. The likelihood of the whelk shell breaking should be unaffected by the quantity of breaks necessary to shatter it.

A 50-meter pole was placed on a rocky beach, and shells of various sizes were dropped from various heights to test each of the aforementioned theories. Zach also discovered that, in contrast to medium- and small-sized whelks, large whelks required the fewest amount of 5-meter drops. Second, when the height of the drop increased to roughly 5 meters, the likelihood that the whelk shell would break improved significantly. As the height was increased, there was no discernible improvement, but the food was scattered more widely. Third, the quantity of prior whelk drops had no bearing on the quantity of large whelk drops.

Consequently, the crow would not select a large, identically sized, undamaged whelk shell as its next meal because it would require time and effort to locate it. Zach also looked into calculating the food energy contained in a large whelk (2.0 kilocalories) versus the number of calories needed to open it (0.5 kilocalorie), for a net gain of 1.5 kilocalories. On the other hand, if a medium-sized whelk is selected, requiring a greater number of drops, the energy yield would be at least 3.3 kcal. Additionally, it would be considerably less cost-effective to try to open a tiny whelk. The crow's ability to reject all of the large whelks was thus an adaptive trait, and the energy it gained per unit of time improved its foraging fitness.

The marginal value theorem can be used to predict optimal foraging in the European starling (Sturnus vulgaris). Starlings leave their nests and travel to feeding patches in search of larval leatherjackets to bring back to their young. It is necessary to ascertain the optimal load size, or the quantity of prey items the starlings should return with in a single flight. As they can only fit so many leatherjackets in their bills, starlings that forage inside a patch eventually run out of options. The result is an increase in the amount of larvae held in the parent's bill. Thus, the restrictions are the travel time (the total time required to get from the nest to a patch and back) and the shape of the diminishing return curve (Krebs & Davies, 1989).

Kacelnik (1984) aimed to verify the validity of the theory of the species' optimization of net energy gain per unit time. They devised a technique to train the starlings to retrieve mealworms from a fake feeder at different distances from the nest. The scientists gradually increased the time between mealworm drops in order to intentionally establish a fixed curve of diminishing returns for the birds. When mealworms were supplied, the birds gathered them until they reached their "optimal" load and flew for their home. If the starlings were optimizing net energy gain per unit time, a short journey time would imply a small optimal load, whereas a long travel time would indicate a bigger optimal load. Consistent with these assumptions, Kacelnik found that the distance between the fake feeder and the nest increased with the load size. Moreover, there was a good quantitative match between the model's predictions and the observed load sizes. Other models with different currency bases performed worse in forecasting the observed load quantities, such as energy efficiency (i.e., energy gained per energy spent). Kacelnik concluded that in a given length of time, starlings maximize net energy gain. This result was not contradicted by additional testing (Bautista et al., 1998, 2001).

Worker bees provide another illustration of how the marginal value theorem can be used to forecast the best foraging behavior. By foraging from flower to flower and returning with it to the hive, bees collect nectar. The situation may be similar to that of the starlings, but in reality, there are different regulations and a different currency for the bees.

A bee does not experience diminishing returns because of any characteristic of the blooms, such as running out of nectar. The total amount of nectar foraged increases linearly with the length of time spent in a patch. However, the weight of the nectar greatly affects the bee's ability to fly between flowers and back to the hive. Wolf and Schmid-Hempel (1989) experimentally proved that the high cost of heavy nectar reduces the bees' longevity by placing varying weights on their backs. The shorter a worker bee's lifespan, the less time it has overall to contribute to the colony.

Accordingly, the net yield of energy received by the hive follows a curve of decreasing returns when the bee gathers more nectar in a single trip (Krebs & Davies, 1989).

The bees' energy budget is impacted by the cost of rich nectar. Unlike the starlings in the previous example, bees place a higher priority on energy efficiency (energy obtained per energy expended) than net rate of energy gain (net energy gained per time). This is because, as was previously mentioned, the optimal load that results from optimizing net rate of energy gain is too heavy for the bees, reducing both their overall output for the hive and their lifespan.

By maximizing energy efficiency, the bees can prevent consuming too much energy on each journey and survive long enough to maximize their lifetime productivity for their colony (Krebs & Davies, 1989). In a previous study, Schmid-Hempel et al. (1985) showed that there is a substantial correlation between the observed link between load size and flight time and predictions based on maximizing energy efficiency, but a very poor correlation with expectations based on maximizing net rate of energy gain (Figure 3.1).

3.4 CRITICISM OF OPTIMAL FORAGING THEORY

Researchers have determined that predators select prey that offers the most caloric gain relative to the amount of time spent foraging by creating and testing optimality models. However, some scientists have objected to the use of the optimality theory, arguing that animals don't always act in the most efficient manner. However, as we have seen, optimality models are designed to enable testing of the possibility that one has accurately identified the elements that may have contributed to the evolution of an animal's behavior, rather than to make claims about perfection in evolution. The variables included in an optimality model have a significant impact on the ensuing predictions, as we have also seen. Researchers can advance by rejecting an optimality model and developing and testing other hypotheses that take other aspects into account if the model's predictions do not match data.

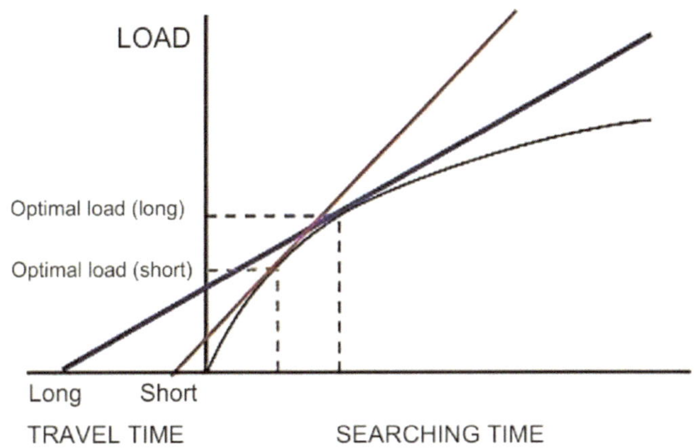

FIGURE 3.1 An increased optimum load is produced by longer trip times if starlings are optimizing their net rate of energy gain (adapted from Krebs and Davies,1989).

A caloric maximization model will, as it should, fail its taste test if ecological factors other than calorie intake influence the foraging behavior. Furthermore, foraging behavior does have ramifications for the majority of foragers that go beyond merely obtaining food. When the risk of unexpected death associated with foraging is significant, we would anticipate that foragers would forgo short-term calorie gain in favor of long-term survival (McFarland, 1973; Maynard, 1978; Milinski, 1978; Meire, 1986; Stephens, 1986; Laland et al., 1990; Shealer, 2002). Garden skinks have been used in experiments to demonstrate this kind of sacrifice. The lizards were raised in enclosures that were split into two microhabitats: open, sunny sections and shady sections with wooden blocks. The enclosures included crickets, which were allowed to freely roam between the different types of environments and served as food for the skinks.

The open spaces in a few of the cages were filled with the scent of lizard-stalking snakes. Compared to skinks kept in enclosures devoid of snake scent, the skinks in these enclosures were much more likely to spend time in the shaded, cover-containing portion of their living space; the control animals moved around more, especially in the open microhabitats, which led to them catching more crickets and growing faster in the final six months of their lives (Ydenberg, 1986; Weimerskirch et al., 1993; While, 2001). An optimality model would be unable to correctly anticipate the behavior of the skink if it ignored the trade-offs between the danger of predation and foraging success.

3.5 OPTIMAL FORAGING AND GAME THEORY

Various examples demonstrate the application of optimality theory to mathematical models of foraging behavior that identify precisely the variables thought to be significant in forming the behavior under study; these models can in some cases allow the production of quantitative predictions that can be verified with real precision against reality, allowing researchers to draw conclusions about the viability of a hypothesis.

Game theory is another related method that may be used to investigate the potential adaptive benefit of feeding behavior. Game theory is particularly helpful in situations when individuals are competing with one another over a valued resource. This method can be used, for instance, in situations where a single species uses two or more different foraging strategies. A few instances of this occurrence are the larger snakes that consume slugs and their cousins who eat tadpoles and fish, as well as the rovers and sitters that are present in populations of fruit fly larvae. There are numerous other documented instances when members of the same species have developed distinct foraging strategies (Milinski, 1978; Meire, 1986; Stephens, 1986; Shealer, 2002). When a species exhibits two or more different foraging phenotypes, the natural question to ask is: Why hasn't the type linked to higher fitness supplanted its competitor throughout the course of evolution? For instance, it seems improbable that hunters and thieves would find precisely the same amount of food in a given length of time spent foraging.

According to a different theory, the main reason why these animals' wide geographic ranges still support both inherited phenotypes is because of the ecological diversity that exists there. However, when two inherited traits coexist within a species at precisely the same location and time, a much larger explanatory issue arises (Krebs et al., 1972; Maynard, 1978; Laland et al., 1990). But as the following example shows, the problem can be countered. Due to genetic variations, the two species of terns—the hunting and thieving terns—have diverse foraging behaviors and might be considered to adopt distinct strategies. In the vernacular of game theory, evolved strategies are behavioral qualities that vary among individuals due to their genetic composition rather than intentionally adopted game plans of the type that humans frequently utilize. We have seen that in the end, only the more effective strategy will endure if it confers greater fitness than the other. However, under certain unique situations, two methods can coexist forever because of the influence of a process known as frequency-dependent selection (Gottmark et al., 1986; Heinrich, 1988). When one phenotypic fitness depends on how frequently it occurs in relation to another phenotype, selection of this type takes place. A phenotype will eventually become more common in the population when the fitness of one type rises with the type's rarity—but only until it reaches the same level of fitness as people using the other tactic. If one type becomes even more prevalent, frequency-dependent selection will then work against it, driving the percentage back toward the equilibrium where both forms have equal fitness. The two phenotypic frequencies will therefore remain constant from generation to generation when this occurs. Few species seem to be flexible in their foraging strategies, since individuals seem to be able to modify their eating strategies in accordance with their capacity to manipulate various food sources. The ability to adapt is given by what game theorists have called a conditional strategy, which is an innate mechanism that allows an individual to change its behavior adaptively in response to the circumstances it faces (Weimerskirch et al., 1993; Ydenberg, 1986; While, 2001).

An immediate explanation for the occurrence of distinct behavioral phenotypes within a population can be found in either genetic or environmental variations. This form of behavioral diversity has been described by ultimate explanations based on game theory as the outcome of either adaptive flexibility or frequency-dependent selection, which allows individuals to modify their behavioral phenotype in response to changing surroundings. A cost-benefit viewpoint is used by both optimality theory and game theory, and it can be utilized to resolve some other perplexing aspects of animal behavior.

REFERENCES

Bateson, P., Mendl, M., and Feaver, J. 1990. Play in the domestic cat is enhanced by rationing of the mother during lactation. *Animal Behaviour* 40: 514–525.

Bautista, L. M., Tinbergen, J. M., and Kacelnik, A. 2001. To walk or to fly? How birds choose among foraging modes. *Proceedings of the National Academy of Sciences of the United States of America* 98(3): 1089–1094. https://doi.org/10.1073/pnas.98.3.1089.

Bautista, L. M., Tinbergen, J. M., Wiersma, P., and Kacelnik, A. 1998. "Optimal foraging and beyond: how starlings cope with changes in food availability." *The American Naturalist* 152(4): 221–238. https://doi.org/10.1086/286189.

Duncan, I. J. H., and Wood-Gush, D. G. M. 1972. Thwarting of feeding behaviour in the domestic fowl. *Animal Behaviour* 20: 444–451.

Elgar, M. 1986a. The establishment of foraging flocks in house sparrows: risk of predation and daily temperature. *Behavioural Ecology and Sociobiology* 19: 433–438.

Elgar, M. 1986b. House sparrows establish foraging flocks by giving chirrup calls if the resources are divisible. *Animal Behaviour* 34: 169–174.
Elgar, M. A. 1989. Predator vigilance and group size among mammals: a critical review of the evidence. *Biological Reviews* 64: 1–34.
Gottmark, F., Winkler, D. W., and Andersson, M. 1986. Flock-feeding on fish schools increases individual success in gulls. *Nature* 319: 589–591.
Heinrich, B. 1988. Winter foraging at carcasses by three sympatric corvids, with emphasis on recruitment by the raven, *Covus corax. Behavioural Ecology and Sociobiology* 23: 141–156.
Kacelnik, A. 1984. Central place foraging in starlings (*Sturnus vulgaris*). I. Patch residence time. *The Journal of Animal Ecology* 53(1): 283–299. https://doi.org/10.2307/4357.
Krebs, J. R., and Davies, N. B. 1989. *An Introduction to Behavioral Ecology*. 4th ed. Oxford: Blackwell Scientific Publications.
Krebs, J. R., Macroberts, M. H., and Cullen, J. M. 1972. Flocking and feeding in the great tit, Parus major - An experimental study. *Ibis* 114: 507–530.
Laland, K. N., and Plotkin, H. C. 1990. Social learning and social transmission of foraging information in Norway rats (*Rattus norvegicus*). *Animal Learning and Behaviour* 18: 246–251.
Maynard Smith, J. 1978. Optimization theory in evolution. *Annual Review of Ecology and Systematics* 9: 31–56.
McFarland, D. J., and Loyd, I. 1973. Time shared feeding and drinking. *Quarterly Journal of Experimental Psychology* 25: 48–61.
Meire, P. M., and Ervynck. A. 1986. Are oystercatchers (*Hemoptopus ostralegus)* selecting the most profitable mussels (*Mytilus edulis*)? *Animal Behaviour* 34: 1427–1435.
Milinski, M., and Heller, R. 1978. Influence of a predator on the optimal foraging behaviour of sticklebacks (*Gasterosteus aculeatus*). *Nature* 275: 642–644.
Schmid-Hempel, P., Kacelnik, A., and Houston, A. I. 1985. Honeybees maximize efficiency by not filling their crop. *Behavioral Ecology and Sociobiology* 17: 61–66. https://doi.org/10.1007/BF00299430.
Shealer, D. A., and Spendelow, J. A. 2002. Individual foraging strategies of kleptoparasitic roseate terns. *Waterbirds* 25: 436–441.
Stephens, D. W., and Krebs, J. R. 1986. *Foraging Theory*. Princeton, NJ: Princeton University Press.
Weimerskirch, H., et al. 1993. Foraging strategy of wandering albatrosses through the breeding season: a study using satellites telemetry. *Auk* 110: 325–342.
While, B., and Gosler, A. G. 2001. The effect of food availability and distance to protective cover on the winter foraging behavior of tits (Aves: *Parus*). *Oecologia* 129: 312–320.
Wolf, T. J., and Schmid-Hempel, P. 1989. Extra loads and foraging life span in honeybee workers. *The Journal of Animal Ecology* 58(3): 943. https://doi.org/10.2307/5134.
Ydenberg, R., and Houston, A. I. 1986. Optimal trade-offs between competing behavioural demands in the great tit. *Animal Behaviour* 34: 1041–1050.
Zach, R. 1979. Shell dropping: decision-making and optimal foraging in northwestern crows. *Behaviour* 68(1–2): 106–117. https://doi.org/10.1163/156853979X00269.

4 Behaviour of Animals during Handling and Restraining

Parag Acharya and Abhijeet Fernandes

4.1 INTRODUCTION

Understanding animal behaviour is pivotal in the realm of animal handling and restraint. Fraser (1960) defines animal behaviour as "the overt and composite functioning of animals individually and collectively ... [and] the means whereby the animal mediates dynamically with its environment, both animate and inanimate." This definition finds resonance in historical accounts, such as the quiet and skilful handling of cattle during the late 1800s, as documented by cowboy Andy Adams. In his diary, he wrote: "Boys, the secret of trailing cattle is to never let your herd know that they are under restraint. Let everything that is done be done voluntarily by the cattle" (Adams, 1903). However, as time progressed, the art of gentle handling seemed to wane, giving way to rougher methods by some modern practitioners (Wyman, 1946; Hough, 1958; Burri, 1968). Contemporary shifts in livestock management, marked by automation and intensification, have further altered the landscape of animal-human interaction. The implications are profound, with increased stress during routine management processes potentially leading to aggressiveness and reduced productivity among livestock. Moreover, stressors can adversely impact physiological processes, affecting immune and reproductive systems. Hence, a nuanced understanding of livestock behaviour is imperative for effective handling practices, with considerations ranging from individual versus group management to the influence of genetic, age, and previous experiences on animal responses. In this context, key behavioural features such as the flocking instinct, visual field, and flight distance emerge as crucial factors to be accounted for in animal handling and restraint protocols, underscoring the interconnectedness of behaviour, stress, and welfare in livestock management.

4.2 IMPACT OF EARLY HANDLING ON ANIMAL BEHAVIOUR AND MANAGEABILITY

Early experiences significantly influence animals' reactions to handling, restraint, and novel situations, with a greater impact on young animals compared to older ones (Hemsworth et al., 1986; Gonyou, 1993). These effects are less easily reversible than later experiences (Scott and Fuller, 1965; Levine, 1968). Cattle breeders in the United States often wonder why European continental cattle genetics may be difficult to handle in the United States but are easier in France. This difference is attributed to early intensive handling of calves in France, making them calm and easy to handle as adults. Research by Fordyce et al. (1985), Boivin et al. (1992), and Mal and McCall (1996) supports the notion that early handling makes cattle and horses easier to manage. While some genetic lines may appear calm in familiar surroundings, they can panic when faced with novelty. Similar findings are reported in sheep by Boissy et al. (1996). In horses, temperament can be masked by learning, with some horses reacting strongly to new environments or stimuli despite seeming calm in familiar settings. Taming and training can reduce reactions to specific handling procedures, but they don't alter innate emotional reactivity. Environmental factors during early life can influence emotional

 DOI: 10.1201/9781003710271-4

reactivity, but highly reactive animals are unlikely to become as calm as genetically calm ones, regardless of early handling efforts.

Research on rats by Denenberg et al. (1967), Levine et al. (1967), Denenberg and Whimbey (1968), and Levine et al. (1979) demonstrates that early treatments lead to long-lasting effects, with younger rats showing excessive emotionality compared to adults in similar conditions (Fuller and Thompson, 1978). Early rearing methods can also alter nervous system reactivity, as observed in rodents' responses to open field testing. Additionally, dogs raised in barren environments exhibit increased fearfulness and reactivity to new stimuli (Melzack, 1954; Melzack and Burns, 1965).

Dellmeier et al. (1985) studied early rearing effects in cattle, finding that calves raised in individual stalls exhibited more energetic behaviour and excessive running compared to those raised in group pens when tested in open field pens. Fordyce (1987) demonstrated that adult cattle handled quietly as calves were calmer during subsequent handling. Boivin et al. (1992) observed that calves reared intensively could be sorted more quickly than those from extensive systems. Additionally, Boissy and Bouissou (1988) found that handling and contact with people near weaning improved cattle's ease of handling.

Lyons (1989) observed that goats raised by humans showed milder responses to environmental changes than those raised by their dams, indicating that early rearing practices lead to lasting temperament variations. Prior experiences also play a crucial role in determining the size of their flight zone and their physiological reactions to handling and restraint (Reid and Mills, 1962; Boissy and Bouissou, 1988; Hastings et al., 1992; Ewbank, 1993; Gonyou, 1993). Training young Bos indicus heifer calves results in calmer adult animals easier to handle (Binstead, 1977; Fordyce et al., 1985; Fordyce, 1987). Similarly, gentle handling sessions reduce escape attempts and charging behaviour in zebu cross-bred calves (Becker and Lobato, 1997). Cattle and deer that are hand-reared tend to have lower cortisol levels during restraint than those with minimal early human contact (Ray et al., 1972; Boandl et al., 1989; Hastings et al., 1992). Training also benefits excitable cattle, with bongo antelope showing minimal cortisol response to injections and blood sampling after training (Phillips et al., 1998).

4.3 EFFECTS OF ENVIRONMENT AND REACTION TO NOVELTY

Individual animals exhibit varying reactions to handling and restraint, with some showing violent responses to novelty. Reid and Mills (1962) propose that animals can be trained to tolerate irregularities in management, reducing such reactions. Exposure to reasonable levels of music or miscellaneous sounds can decrease fear responses to sudden noises. For instance, playing a radio in a pig barn results in milder reactions to sudden sounds like door slamming. Ames (1974) found that playing instrumental music or miscellaneous sounds at 75 dB improved weight gains in sheep, while louder sound had the opposite effect. Grandin (1984, 1987) suggests that compared to animals previously subjected to rough handling, those with experience of gentle handling are calmer and easier to manage in the future.

In tame beef cattle, introducing a novel-coloured ball caused a crouch-flinch reaction in 50% of the animals (Miller et al., 1986). Cattle with prior handling at livestock markets adapted more easily to slaughter plant stockyards due to reduced novelty and fear (Cockram, 1990). Additionally, stress during handling and weaning of calves can be minimised by calmly moving them through chutes and corrals multiple times before veterinary procedures (Humphries, 2006). Zebu cattle in the Philippines are exposed to significant novelty, making them less alarmed by new experiences. To prevent excitement from novelty at feedlots, auctions, or slaughter plants, it's recommended to familiarise cattle with handling by people on foot, horseback, and in vehicles.

Grandin et al. (1987) and Grandin (1984) explored the effects of early environment on the behaviour of pigs reared indoors. They found that environmental enrichment, including assorted objects, straw bedding, and daily petting, resulted in calmer and less reactive pigs. Enriched pigs were also more willing to walk through single-file chutes and startled less when encountering sudden stimuli. Pedersen (1992) and Grandin (1984) noted that pigs raised with enrichment needed less prodding and were more inclined to approach new objects or people compared to those from barren environments. However, Grandin (1987) cautioned against overindulgence, noting that pigs exposed to excessive petting may become indifferent to people and more challenging to manage in chute systems.

4.4 PHYSICAL TRAITS AFFECTING TEMPERAMENT DURING HANDLING

4.4.1 Coat Colour

Dairymen have noted that white Holsteins tend to be more skittish and harder to manage than their more pigmented counterparts. This observation aligns with studies linking coat colour to temperament in animals. Research indicates that genetic selection for physical traits can influence behaviour, and vice versa (Belyaev, 1979). For example, breeding rats for a black non-agouti coat colour resulted in tamer rats (Keeler, 1942). Similarly, foxes with different coat colours exhibited varying temperaments, with tamer behaviours observed as more mutant coat colour genes were introduced (Keeler, 1968, 1975). Tameness in foxes correlated with smaller adrenal glands and increased body weight, suggesting a relationship between coat colour and behaviour.

4.4.2 Hair Whorls

Researchers have found a link between physical traits and temperament. For instance, a horse trainer noted that hair whorls on a horse's forehead are indicative of temperament, a concept recognised in Arabian horses centuries ago. Horses with high forehead whorls or dual spirals tend to be more reactive (Tellington-Jones and Bruns, 1985; Barker, 1990; Friedly, 1990). Body traits and temperament may be connected because they develop simultaneously with the brain.

Many scientists doubted the validity of trainers' observations regarding hair whorls and temperament. To investigate, researchers studied hair whorls in cattle, which have similar patterns on their foreheads. They assessed the temperament of 1500 cattle from different ranches in a Colorado feedlot during vaccination, using a rating scale previously described. The cattle were mainly European X British crosses, with some zebu crosses from Mexico. The study found that cattle with hair whorls above the eyes were significantly more agitated during restraint than those with whorls below the eyes (Grandin et al., 1995). Additionally, these cattle had larger flight zones and were more likely to avoid an approaching person.

Tanner et al. (1994) and Grandin et al. (1995) found that beef cattle had more hair whorls above their eyes compared to Holstein cows from large dairies. Holsteins are generally less reactive to new or unpleasant stimuli and are calmer during handling at commercial feedlots. Voisinet et al. (1997) also discovered that cattle with a calm temperament during restraint in a squeeze chute exhibited the highest weight gains.

The authors suggest that the position of hair whorls on the forehead might be linked to bone structure and temperament. A higher hair whorl could be associated with greater vigilance, sharper senses, and a stronger orienting response, while a lower hair whorl might indicate reduced vigilance and an orienting response. Bone structure might correlate with the overall reactivity of the nervous system and levels of fearfulness. Temperamental traits in animals may be linked to various neurotransmitter mechanisms, suggesting that temperament is influenced by multiple biological processes and genetic factors. In cattle and horses, fine-boned, slender animals with high hair whorls tend to have nervous temperaments, while heavy-boned, muscular animals are generally calmer with low hair whorls. Despite lacking heavy muscling, Holstein dairy cows are heavy-boned and align with the criteria for a calm temperament.

4.5 GENETIC AND BEHAVIOURAL FACTORS INFLUENCING CATTLE TEMPERAMENT AND HANDLING

Cattle breeds exhibit notable differences in temperament, and these genetic variations can significantly influence their stress responses during handling. Animals with more flighty genetics tend to become highly agitated when exposed to sudden novel stimuli, such as encountering a waving flag for the first time, compared to those with calmer temperaments (Grandin and Deesing, 2022). Brahman cattle are generally flightier compared to British breeds like Angus and Hereford, but are also highly inquisitive and likely to follow people or dogs, a trait utilised in Australian herding. With gentle handling, Brahmans can become very docile (Hohenboken, 1986). Conversely, Brahman-cross cattle exhibit more agitation in squeeze chutes than Shorthorns (Fordyce et al., 1988). Temperament in *B. indicus* cattle is heritable (Fordyce et al., 1988).

A fundamental concept is that animals with excitable temperaments require a more gradual introduction to new experiences than those with calm temperaments. Wild species typically exhibit higher fearfulness and stronger reactions to environmental changes than domesticated animals (Price, 1998).

In dairy cattle, heritability estimates for temperament range from 0.40 to 0.53 (O'bleness et al., 1960; Dickson et al., 1970), and in beef cattle from 0.40 to 0.48 (Shrode and Hammack, 1971; Stricklin et al., 1980). Herefords are noted as the most docile British breed, while Galloways are the most excitable, with continental European breeds being more excitable than British breeds. Studies show Angus cows are more temperamental and dominate social ranks despite their smaller size (Tulloh, 1961; Wagnon et al., 1966; Stricklin et al., 1980). Brahman crosses and certain other crossbreeds also show higher excitement levels during restraint compared to Angus (Voisinet et al., 1997). Sires influence temperament scores within breeds.

Fearful species like American bison and antelope can injure themselves when restrained, requiring training for voluntary cooperation (Grandin, 1999). Moving them in small groups through short, single-file races reduces stress. While domestic cattle tolerate this, wild ungulates may become agitated. Some domestic animals habituate to non-painful procedures, while others remain stressed (Lanier et al., 1995). Similar patterns were noted in horses, where reactive breeds like Arabians or Thoroughbreds panic more easily than calmer breeds such as draught horses.

Sires influence cortisol response to stress and learning ability in Holstein calves (Johnston and Buckland, 1976; Arave et al., 1974). Lanier et al. (1995) discovered that while some pigs quickly adapted to swimming and their adrenaline levels returned to baseline, others stayed fearful and maintained high adrenaline levels.

Le Neindre et al. (1996) noted issues when breeds developed for intensive systems are placed on extensive ranges. For instance, bulls can sire daughters that are gentle in intensive settings but aggressive on the range. Excitable cattle often panic in new situations, with some Saler breed of cattle remaining calm with familiar handlers but becoming aggressive in noisy, unfamiliar environments like auctions.

4.6 EFFECTIVE HANDLING TECHNIQUES AND FACILITY DESIGN FOR DIVERSE LIVESTOCK BREEDS

Domestic cattle, sheep, horses, and pigs have been selectively bred for various physical and behavioural traits over thousands of years. For instance, British and European Continental cattle breeds, which are well adapted to cold climates, are primarily found in the northern regions of the United States. In contrast, Brahman cattle, which thrive in warmer climates, are mainly raised in the southern states. These differences in temperament and environmental suitability significantly affect the design of facilities for handling cattle. In the northern United States, placid British cattle can be easily sorted in wide alleys by people on foot or horseback, whereas more excitable Brahman cattle require narrower, single-file races with gates to separate them into pens. Brahmans tend to flock tightly and exhibit shelter-seeking behaviour when agitated, making sorting them in wide alleys challenging. Conversely, Whateley et al. (1974) and Shupe (1978) showed that sheep of all breeds flock tightly, necessitating single-file chutes for sorting. Rambouillet flock tightly and Cheviots act more independently. These design differences have evolved over time as ranchers utilised methods that proved effective for handling different breeds.

Solid barriers and fences are essential for managing extensively reared animals, as they prevent the animals from seeing potential sources of fear or danger. This principle has been observed by mustangers and handlers of various herd animals, including cattle, deer, sheep, and wild horses (Kilgour, 1971; Fowler, 1978; Grandin, 1980). These barriers exploit the natural instinct of herd animals to seek safety behind visually substantial obstacles, reducing their fear and agitation. Studies by Whittington and Chamove (1995) in New Zealand demonstrated that solid barriers in deer pastures reduced vigilance and the tendency to flee from moving stimuli. Additionally, Grandin (1993) found that covering the sides of restraint devices with solid material calmed extensively reared cattle and reduced their agitation. This method also prevents animals from seeing handlers in positions that may trigger panic responses, improving overall handling efficiency and animal welfare.

Experienced dairy cows will walk over drain grates in the milking parlour, while young heifers often hesitate (Kilgour, 1971; Grandin, 1980). Dairy cows can learn that drain grates and other similar objects are harmless. However, animals in unfamiliar environments often balk at walking over puddles or drain grates. Reflections, shadows, and changes in flooring type can cause livestock to stop and refuse to move through handling facilities, particularly in places like auction yards or slaughterhouses where animals lack the opportunity to learn. Leader animals, if given time, will usually navigate around or over these obstacles, encouraging others to follow (Hutson, 1982). Unfortunately, high-speed slaughter plants do not allow enough time for such investigations.

4.7 THE INFLUENCE OF DARKNESS OR LIGHT ON HANDLING

Darkness has paradoxical effects on animals, being both fear-provoking and calming. While animals may be reluctant to enter dark enclosures, they often calm down once inside. This effect is observed in various species, including pigs and cattle. Cattle raised in extensive systems tend to be calmer in restraint devices when their heads are enclosed by a solid barrier. Research indicates that darkness can help reduce stress responses (Douglas et al., 1984). Cattle confined in dark environments exhibit lower cortisol levels compared to those restrained in a standard squeeze chute (Hale et al., 1987). However, unfamiliar darkness may increase stress. Light can be used strategically to attract animals into races and restraint devices, but it must not shine directly into their eyes (Van Putten and Elshof, 1978; Grandin, 1980). Parsons and Helphinstine (1969) used dark box for handling cattle and Littlejohn (1994) and Pollard et al. (1992) for deer, with animals becoming more approachable and less tightly grouped. Darkness can even calm frightened horses, with covering their eyes proving effective in reducing agitation.

4.8 HERDING AND SOCIAL BEHAVIOUR

Genetic factors play a role in the herd behaviour of domestic animals like cattle and sheep, shaped in part by their evolutionary need to defend against predators. African herding animals often exhibit tighter grouping, likely due to higher predator presence such as lions, hyenas, and cheetahs. With eyes on the sides of their heads, herd animals have broad vision, facilitating collective movement and threat detection (Prince, 1977; Hutson, 1993; Matthews, 1993). Panoramic vision allows animals to move cohesively as a herd while continuously monitoring for predators or other threats. This visual capability is linked to an internal state of nervous system arousal that accompanies orienting responses (Davis, 1992; Boissy, 1996; Rogan and LeDoux, 1996). Startled animals orient quickly to novel stimuli, triggering a state of heightened alertness. They may freeze, mill tightly together, or flee when alarmed, resorting to fighting if cornered. During flight, animals such as cattle, deer, and bison use raised tails to signal danger to the herd (Kiley-Worthington, 1976). American bison exhibit stronger herding behaviour compared to domestic cattle. When separated from the herd, bison make intense efforts to rejoin their group (Grandin, 1993), while domestic cattle become challenging to handle when agitated, seeking safety in the middle of the herd. Similar to prey species like elk and deer, domestic herding animals group tightly in response to danger, with smaller species such as sheep forming even tighter groups due to their vulnerability to predators. This behaviour reflects their instinctual need to find safety in numbers.

Social rank within a herd influences an animal's position during movement, with high-ranking animals often avoiding leading to reduce risk. Studies by Hediger (1955) and Hediger and Sircom (1968) support this observation. Social behaviour also affects learning, as demonstrated by Boissy and Le Neindre (1995), who found that cattle learn tasks more quickly in the presence of companions, likely due to reduced fear. Similar principles apply to horse training, where calm companion horses help young horses cope with novel experiences. In Australia, tame "coacher" cattle are used to help gather wild feral cattle (Roche, 1988). Similarly, Fordyce (1987) employed calm steers to train Brahman calves. Historically, excitable leaders were avoided during cattle drives to reduce the risk of stampedes (Harger, 1928).

Fear is a potent stressor for cattle and other grazing, herding animals like horses (Grandin, 1997). It keeps them vigilant to escape predators and can elevate stress hormones more than physical stressors. When cattle are agitated during handling, fear is the main driver. Brain circuits controlling fear-based behaviour have been studied and mapped (Rogan and LeDoux, 1996). Calm animals are easier to manage, while fearful cattle tend to cluster, complicating handling. The key to low-stress cattle handling is maintaining calmness, as frightened cattle take about 20 minutes to settle. Quiet handling, especially during procedures like vaccination, helps cattle resume feeding quickly. Agitated cattle in squeeze chutes tend to gain less weight and are more prone to pre-slaughter stress, leading to tougher meat (Petherick et al., 2002; Vann et al., 2004). King et al. (2005) found that extensively raised yearling steers with excitable temperaments had higher cortisol levels after handling.

Sheep and cattle may inherently fear dogs, with sheep more willing to approach a goat than an unfamiliar human or a quiet dog, likely perceiving the goat as a herd mate (Beausoleil et al., 2005). Grazing animals exhibit a vigilant posture when they perceive threats, such as raising their heads (Welp et al., 2004). The amount of visible white in a bovine's eyes can indicate stress levels, increasing with fear-inducing stimuli and decreasing with tranquillisers like diazepam. Suddenly opening an umbrella frightened cattle, causing a greater visible percentage of the white part of their eyes (Sandem et al., 2004). Additional studies revealed that this percentage increases when a calf is separated from its mother, but decreases with the use of the tranquilliser diazepam (Sandem et al., 2006). Cattle handlers should refine their methods if cattle frequently show the whites of their eyes and maintain a vigilant posture, as these are signs of fear. Calm handling eliminates these stress indicators.

4.9 FLIGHT ZONE

The concept of flight distance, initially formulated for wild ungulates, delineates the space within which an animal allows human proximity before moving away. This principle, articulated by Hediger and Sircom (1968), found further elaboration by Kilgour (1971) and Grandin (1980), particularly concerning domestic sheep and cattle. Understanding the flight zone is crucial, as it dictates the ease with which animals can be manoeuvred during handling. Typically, herd animals will confront a perceived threat when it lies beyond their flight zone but retreat when it encroaches. Notably, flight zone sizes vary, influenced by factors like genetics, experience, and environmental context (Figure 4.1).

Genetic factors play a significant role in determining an animal's flight zone size, with individuals of the same species or breed exhibiting varied distances. Experience and learning also influence

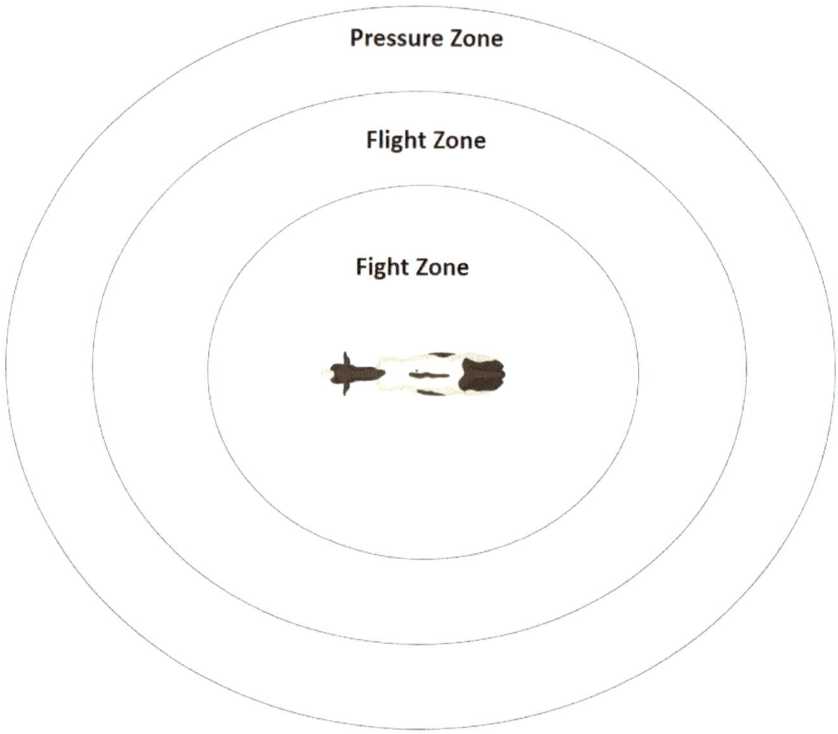

FIGURE 4.1 Depicting fight, flight and pressure zone of an animal.

flight zone size, with animals adjusting their response based on past encounters with danger. Gentle treatment reduces the flight zone in cattle, while abuse enlarges it. Excitement enlarges the flight zone of cattle. Tame dairy cattle might have no flight zone, allowing human contact. This applies to both domestic cattle and wild ungulates. For example, wild cows in Arizona can have a flight distance of 30 meters, while feedlot cattle have a flight distance between 1.5 and 7.61 meters (Grandin, 1980). The size of objects influences flight distance, with cattle staying further away from larger objects (Smith, 1998). Approaching full-face enlarges the flight zone compared to a sideways profile. Hand-reared pronghorn antelope panicked at the introduction of large novel objects but showed no reaction to small ones.

Determining the edge of the flight zone involves slowly approaching a group of cattle until they begin to move away. If the person remains still, the cattle will turn to face them but maintain a distance. Re-entering the edge prompts the animals to move away again. Efficient cattle handling occurs when working on this edge, as noted by Grandin (1980, 1987). The flight zone's boundary isn't distinct and can enlarge if approached quickly or if the cattle are excited. Eye contact with the animals can also expand the flight zone. Handlers should avoid direct eye contact to encourage cattle to pass by and position themselves strategically to prompt forward movement while staying out of the animal's blind spot.

Flight zone principles are versatile tools for handlers, applicable to both individual animals and large groups. When controlling a large herd, handlers utilise similar principles by entering and withdrawing from the collective flight zone. Moving in the opposite direction prompts the entire group to move forward. These techniques are effective for all herd animals. Dogs trained for herding also employ flight zone principles, using similar positions to control livestock movement. Coppinger and Coppinger (1993) suggest that herding dogs exhibit a natural hunting strategy akin to wolves, albeit without attacking or killing due to domestic breeding (Figure 4.2 and Figure 4.3).

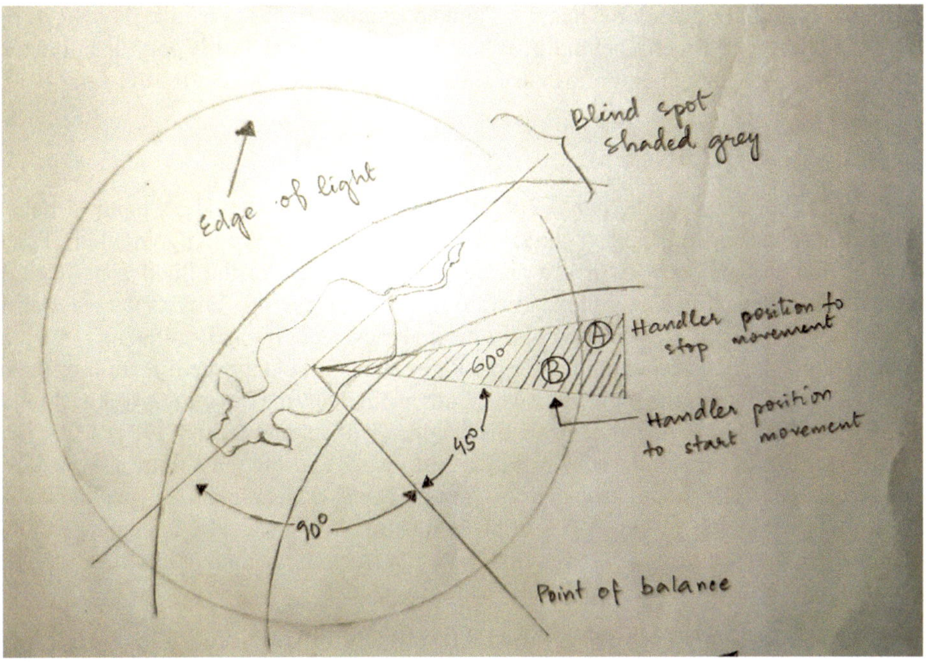

FIGURE 4.2 Flight zone diagram showing optimal handler positions for moving an animal forward.

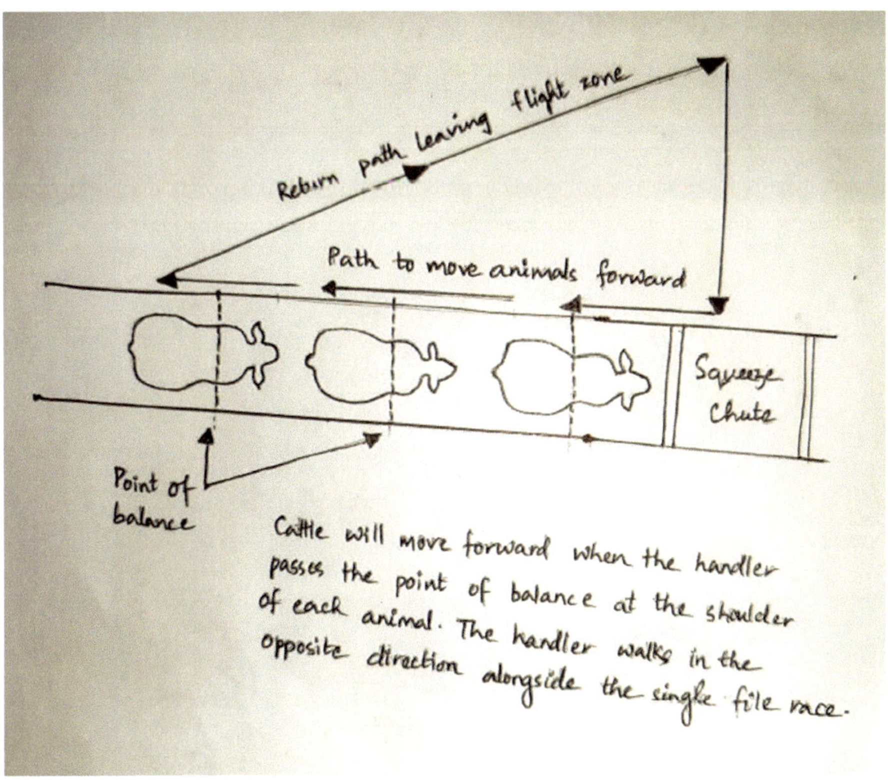

FIGURE 4.3 Handler movement pattern to induce cattle to move forward.

(To move an animal forward, the handler should position themselves behind its point of balance, avoiding the blind spot directly behind it. When close to the animal, this point is at the shoulder. From a greater distance, the point of balance shifts forward to just behind the eye.)

4.10 POINT OF BALANCE

To move an animal forward, the handler should stand behind its shoulder's point of balance. To make the animal back up, the handler should stand in front of this point (Kilgour and Dalton, 1984). To turn the animal, approach its rear at an angle from the edge of the flight zone (Cote, 2003). Additionally, grazing animals will move forward if the handler swiftly crosses the point of balance at the shoulder in the opposite direction of the desired movement (Grandin, 1998a)

To move animals, handlers should enter the flight zone in the opposite direction of the desired movement and stay outside the flight zone when moving in the same direction. Approaching an animal head-on will cause it to turn away from the handler (Kilgour and Dalton, 1984). Calm cattle can be turned using plastic strips on a stick to block their vision, causing them to turn away. Deep penetration of the flight zone should be avoided to prevent panic. If an animal rears up, handlers should back away rather than push it down. The flight zone size is influenced by the approach angle and enclosure size, being smallest at the sides and largest in front and behind (Cote, 2003) (Figure 4.4).

4.11 CATTLE MOVEMENT AND LEADERSHIP IN HANDLING PRACTICES

Cattle's natural tendency to follow can aid in their movement. A good leader is usually a sociable cow and is not the most dominant animal. Historical practices, like old US cattle drives, valued calm leaders over excitable ones (Harger, 1928). In situations where a cattle herd is reluctant to cross a bridge or brook, a calf may be roped and dragged across to encourage the others to follow (Ward,

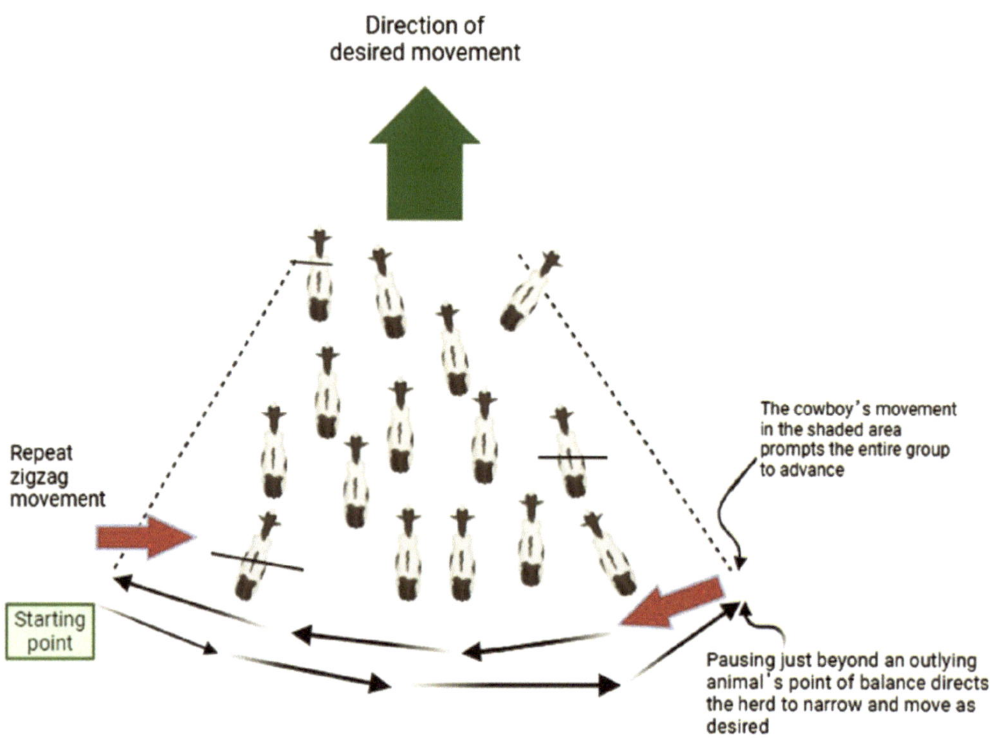

FIGURE 4.4 Handler zigzag pattern for effective cattle bunching.

1958). In Australia, trained "coacher" cattle are used to help gather wild feral cattle (Roche, 1988). Fordyce (1987) suggests using quiet steers to facilitate training for handling procedures. Dumont et al. (2005) recommend observing spontaneous long-distance movements to identify true leaders. Cattle can be trained to come when called, associating a vehicle horn with feed (Hasker and Hirst, 1987). However, caution is needed to prevent them from becoming a nuisance. Intensive grazing systems may stress calves, so handlers should control the transition to new pastures (Savory, 1978; Smith et al., 1986).

4.12 THE IMPACT OF STOCKMANSHIP FOR IMPROVING HANDLING AND WELFARE

Scientific studies demonstrate that good stockmanship significantly benefits animal productivity (Rushen et al., 2015; Hemsworth et al., 2010). Animals that fear handlers and move away from them are less productive. For instance, young pigs subjected to gentle handling or stroking had higher productivity compared to those treated aversively with slaps and shocks, which exhibited lower growth rates (Hemsworth et al., 1991). Similarly, shouting, slapping, and tail twisting in dairy cows resulted in reduced milk yields, while soft talking and stroking improved those (Breuer et al., 2000). Animals like dairy cows, calves, pigs, and sheep can recognise individual people, but rough treatment can cause them to fear all humans (Koba et al., 2001; Tanida et al., 1998; Rybarczyk et al., 2003). Handler attitudes also impact animal productivity (Hemsworth et al., 1991). Positive attitudes, such as people liking animals, correlate with calmer, more productive animals, while negative attitudes have the opposite effect Hemsworth et al., 1991; Breuer et al., 2000; Destrez et al., 2013; Kauppinen et al., 2012; Waiblinger et al., 2002). Additionally, the presence of a familiar stockperson can reduce animal stress, as seen in sheep that remain calmer when isolated with a known person. Qualitative Behavioural Assessment (QBA) is useful for evaluating stockmanship (Ellingsen et al., 2014). A study on human-animal relationships outlined four handler styles: positive, calm/patient, dominating/aggressive, and insecure/nervous. These styles had a significant impact on calves' arousal and mood, highlighting the importance of positive interactions between humans and animals and the necessity for adequate stockperson training.

4.13 METHODS TO ASSESS REACTIONS TO HANDLING

Various behavioural tests are employed by scientists and producers to evaluate animal behaviour during routine handling and restraint. Calmer animals are preferred for breeding due to their easier weight gain and safer handling. Key behavioural tests include chute score, exit speed scoring, pen scoring, vocalisation tests, and aversion tests. The first three assess inherited temperament and the impact of past handling experiences, while vocalisation and aversion tests help researchers evaluate the aversiveness of different treatments.

1. Chute score: Numerical rating scales are used to evaluate cattle reactions during veterinary restraint. Animals are scored while restrained in various devices, such as a single animal scale, squeeze chute, or headgate. Fordyce et al. (1988) utilised a seven-point scale, while Grandin (1994) employed a five-point scale in a manually operated squeeze chute, ranging from calm (1) to berserk frenzy (5). Grandin (1980) used a four-point scale for cattle in a hydraulic squeeze chute, which restricts movement more tightly than a head stanchion. When animals are tightly restrained, exit speed scoring is more effective for distinguishing breed differences Cafe et al. (2011) investigated how persistent assessments of temperament affect productivity, carcass traits, and meat quality, including colour and tenderness. Their study revealed that cattle with more excitable temperaments indicated by higher exit velocities and chute speeds consistently showed lower feed intake and slower growth rates. These cattle produced smaller carcasses with less fat cover, darker longissimus lumborum muscle (LLM) colour, and higher muscle pH.

2. Exit speed: Scoring exit speed from a squeeze chute, measured electronically (Boissy and Bouissou, 1988), correlates with lower weight gain and higher stress levels (King et al., 2006; Muller et al., 2008). It can also be assessed by observing cattle exit gait—walking, trotting, or cantering (Vetters et al., 2013; Baszczak et al., 2006). Beef cattle (Bos indicus) tend to exit faster than Bos taurus cattle, likely reflecting differences in fearfulness (Baszczak et al., 2006; Cafe et al., 2011). However, purebred Bos indicus cattle, such as Nelore, may lie down or freeze when fearful, making exit speed tests less effective for them (Grandin et al., 2014). Purebred *Bos indicus* cattle sometimes freeze instead of having an active response to handling and restraint (Fraser et al., 1960; Grandin et al., 1980).

3. Pen scoring: Cattle are placed in a small enclosure where a person deliberately invades their flight zone to measure either flight distance or the likelihood of charging at the handler (Grignard et al., 2001). Animals that stay farther from the tester have a larger flight distance. Grandin (1997) discovered that temperament evaluations of bulls, conducted at 30-day intervals while restrained in a squeeze chute, yielded consistent scores over time for both the calmest and most excitable animals. To accurately assess temperament, multiple evaluations are necessary, and scores obtained from a squeeze chute provide a more accurate reflection of an animal's true physiological reactivity compared to passive flight zone tests.

4. Vocalisation tests: Sheep, cattle, and pigs exhibit different vocal responses to handling or restraint. Studies indicate that vocalisation in cattle and pigs correlates with higher stress levels (Warriss et al., 1994). Cattle vocalise in response to painful events like branding or excessive restraint pressure (Bourguet et al., 2011; Grandin, 1998b; Grandin, 2001), and reducing such pressure decreases vocalisations (Grandin et al., 2001). Sheep, on the other hand, do not vocalise and remain silent when hurt to avoid attracting predators (Dwyer, 2004) but vocalise loudly when separated from the flock (Rault et al., 2011; Torres-Hernandez and Hohenboken, 1979). Similarly, cattle vocalise loudly when separated from the herd or when cows and calves are separated (Grandin, 2001).

5. Aversion tests: An aversion test assesses animals' perception of handling and restraint by measuring their willingness to re-enter areas where they were previously restrained (Hutson et al., 1985). Sheep, with long memories for negative experiences, may avoid such places even a year later (Hutson, 1985). For instance, sheep took longer to move through a handling race with each trial, especially when restrained in more aversive positions (Hutson and Butler, 1978; Hutson, 1985). Similarly, cows find shouting and hitting highly aversive when moving more slowly through a race leading to a shouting person (Pajor et al., 2003). Aversion tests help identify stressful procedures for animals. Sudden novel or aversive experiences reveal true reactivity more effectively than flight zone tests, which can be influenced by learning.

Aversion testing measures include the time it takes for an animal to move through a single-file race to restraint (Grigor et al., 1998), the force needed to move the animal, and the number of refusals to enter the race from a holding pen. In force studies, the animal's behaviour in the single-file race is classified as: (1) no force, walks in when the gate opens; (2) handler walks past the point of balance to urge movement; (3) taps on the rear; or (4) use of an electric prod (Baszczak et al., 2006)

Another method, the choice test, evaluates the relative aversiveness of two procedures (Rushen et al., 1986) using a Y-shaped race leading to different treatments (Grandin et al., 1986). Animals, such as sheep, tend to avoid paths leading to more aversive treatments like electrical immobilisation, preferring less severe options (Grandin et al., 1986). The severity of treatments affects aversion test results, with severe treatments being reliably avoided. However, for mildly aversive treatments, animals may stick to a known safe path rather than switch, indicating a greater fear of the unknown. These methods help determine how animals perceive handling and restraint procedures: significant reluctance or refusal to re-enter indicates that the procedure is perceived as stressful or aversive.

4.14 CONCLUSION

In conclusion, the interplay of genetics, early handling, environmental factors, and physical traits significantly shapes animal temperament and manageability. Early experiences exert a profound influence, with young animals more susceptible to long-lasting behavioural changes from handling and environmental conditions. This foundational period is critical for developing manageable livestock, as evidenced by calmer behaviours in cattle and horses subjected to early handling. Temperament traits, such as excitability and fearfulness, manifest in various forms across different breeds and species, affecting their reactions to novelty, handling, and environmental changes. Research consistently highlights the importance of gentle and consistent handling in reducing stress and improving manageability. Environmental enrichment and strategic use of light and darkness can further mitigate stress, promoting calmer behaviours during handling. Facility design plays a crucial role in accommodating different temperaments and species-specific behaviours. Understanding the flight zone and point of balance of animals is pivotal for effective livestock management, allowing handlers to manoeuvre animals with minimal stress. Ultimately, effective livestock management requires a holistic approach, integrating knowledge of genetics, early experiences, environmental factors, and specific physical traits. This comprehensive understanding ensures the development of handling practices and facility designs that enhance animal welfare, productivity, and safety for both animals and handlers.

REFERENCES

Adams, A. *Log of a Cowboy.* Houghton Mifflin, New York (reissued in 1964 by Bison Book, University of Nevada Press, Reno, Nevada), (1903).

Ames, David R. "Sound stress and meat animals." In *Proceedings of the International Livestock Environment Symposium. April 17, 18, 19, 1974, University of Nebraska, Lincoln*, pp. 324–330. American Society of Agricultural Engineers, (1974).

Arave, C. W., J. L. Albright, and C. L. Sinclair. "Behavior, milk yield, and leucocytes of dairy cows in reduced space and isolation." *Journal of Dairy science* 57, no. 12 (1974): 1497–1501.

Barker, R. "The mind behind the swirls." *Rocky Mountain Quarter Horse* 28 (1990): 26–27.

Baszczak, J. A., T. Grandin, S. L. Gruber, T. E. Engle, W. J. Platter, S. B. Laudert, A. L. Schroeder, and J. D. Tatum. "Effects of ractopamine supplementation on behavior of British, Continental, and Brahman crossbred steers during routine handling." *Journal of Animal Science* 84, (2006): 3410–3414.

Beausoleil, Ngaio J., Kevin J. Stafford, and David J. Mellor. "Sheep show more aversion to a dog than to a human in an arena test." *Applied Animal Behaviour Science* 91, (2005): 219–232.

Becker, Bettina G., and JoséF P. Lobato. "Effect of gentle handling on the reactivity of zebu crossed calves to humans." *Applied Animal Behaviour Science* 53, (1997): 219–224.

Belyaev, Dmitry K. "Destabilizing selection as a factor in domestication." *Journal of Heredity* 70, (1979): 301–308.

Binstead, M. "Handling cattle." *Queensland Agriculture* 103, (1977): 293–295.

Boandl, K. E., J. E. Wohlt, and R. V. Carsia. "Effects of handling, administration of a local anesthetic, and electrical dehorning on plasma cortisol in Holstein calves." *Journal of Dairy Science* 72, (1989): 2193–2197.

Boissy, A., P. Le Neindre, P. Orgeur, and J. Bouix. "Genetic variability of psychobiological reactivity in lambs reared under open-range management." In *Proceedings of the 30th International Congress of the International Society for Applied Ethology: 14–17 August, 1996, Guelph, Ontario, Canada.* (1996).

Boissy, Alain, and Marie-France Bouissou. "Effects of early handling on heifers' subsequent reactivity to humans and to unfamiliar situations." *Applied Animal Behaviour Science* 20, (1988): 259–273.

Boivin, Xavier, Pierre Le Neindre, J. M. Chupin, J. P. Garel, and Gilbert Trillat. "Influence of breed and early management on ease of handling and open–field behaviour of cattle." *Applied Animal Behaviour Science* 32, (1992): 313–323.

Bourguet, Cécile, Véronique Deiss, Carole Cohen Tannugi, and E. M. Claudia Terlouw. "Behavioural and physiological reactions of cattle in a commercial abattoir: Relationships with organisational aspects of the abattoir and animal characteristics." *Meat Science* 88, (2011): 158–168.

Breuer, Kate, Paul H. Hemsworth, John L. Barnett, Lindsay R. Matthews, and Grahame J. Coleman. "Behavioural response to humans and the productivity of commercial dairy cows." *Applied Animal Behaviour Science* 66, (2000): 273–288.

Burri, R. *The Gaucho.* Crown Publishing, New York, (1968).

Cafe, L. M., Dorothy L. Robinson, D. M. Ferguson, Brooke Louise McIntyre, G. H. Geesink, and P. L. Greenwood. "Cattle temperament: Persistence of assessments and associations with productivity, efficiency, carcass and meat quality traits." *Journal of Animal Science* 89, (2011): 1452–1465.

Cockram, M. S. "Some factors influencing behaviour of cattle in a slaughterhouse lairage." *Animal Science* 50, (1990): 475–481.

Coppinger, L., and Coppinger, R. Dogs for herding livestock. In *Livestock Handling and Transport* (T. Grandin, ed.), pp. 179–196. CAB International, Wallingford, UK, (1993).

Cote, S. *Stockmanship: a powerful tool for grazing management.* USDA Natural Resources Conservation Service, Arco, ID, (2003).

Davis, Michael. "The role of the amygdala in fear and anxiety." *Annual Review of Neuroscience* 15, (1992): 353–375.

Dellmeier, G. R., T. H. Friend, and E. E. Gbur. "Comparison of four methods of calf confinement. II. Behavior." *Journal of Animal Science* 60, (1985): 1102–1109.

Denenberg, Victor H., Joan Thatcher Brumaghim, Gary C. Haltmeyer, and M. X. Zarrow. "Increased adrenocortical activity in the neonatal rat following handling." *Endocrinology* 81, (1967): 1047–1052.

Denenberg, Victor H., and Arthur E. Whimbey. "Experimental programming of life histories: Toward an experimental science of individual differences." *Developmental Psychobiology: The Journal of the International Society for Developmental Psychobiology* 1, (1968): 55–59.

Destrez, Alexandra, Marjorie Coulon, Véronique Deiss, Eric Delval, Alain Boissy, and Xavier Boivin. "The valence of the long-lasting emotional experiences with various handlers modulates discrimination and generalization of individual humans in sheep." *Journal of Animal Science* 91, (2013): 5418–5426.

Dickson, D. P., G. R. Barr, L. P. Johnson, and D. A. Wieckert. "Social dominance and temperament of Holstein cows." *Journal of Dairy Science* 53, (1970): 904–907.

Douglas, A. G., M. J. Darre, and D. M. Kinsman. Sight restriction as a means of reducing stress during slaughter. In *Proc. Euro. Meet. Meat Res. Workers,Bristol, UK, 30th,* Edited by G. John Benson and Bernard E. Rollin. Blackwell Publishing, (1984): 10–11.

Dumont, Bertrand, Alain Boissy, Caroline Achard, A. M. Sibbald, and H. W. Erhard. "Consistency of animal order in spontaneous group movements allows the measurement of leadership in a group of grazing heifers." *Applied Animal Behaviour Science* 95, (2005): 55–66.

Dwyer, C. M. "How has the risk of predation shaped the behavioural responses of sheep to fear and distress?." *Animal Welfare* 13, (2004): 269–281.

Ellingsen, Kristian, Grahame J. Coleman, Vonne Lund, and Cecilie M. Mejdell. "Using qualitative behaviour assessment to explore the link between stockperson behaviour and dairy calf behaviour." *Applied Animal Behaviour Science* 153, (2014): 10–17.

Ewbank, R. Handling cattle in intensive systems. In *Livestock Handling and Transport* (T. Grandin, ed.), CAB International, Wallingford, UK, (1993): 59–73.

Fordyce, G. "Weaner training." *Queensland Agricultural Journal* 113, (1987): 323–324.

Fordyce, G., M. E. Goddard, R. Tyler, G. Williams, and M. A. Toleman. "Temperament and bruising of Bos indicus cross cattle." *Australian Journal of Experimental Agriculture* 25, (1985): 283–288.

Fordyce, G., J. R. Wythes, W. R. Shorthose, D. W. Underwood, and R. K. Shepherd. "Cattle temperaments in extensive beef herds in northern Queensland. 2. Effect of temperament on carcass and meat quality." *Australian Journal of Experimental Agriculture* 28, (1988): 689–693.

Fowler, M. E. *Restraint and handling of wild and domestic animals.* Iowa State University Press, Ames, IA, (1978).

Fraser, A. F. "Spontaneously occurring forms of 'tonic immobility' in farm animals." *Canadian Journal of Comparative Medicine and Veterinary Science* 24, (1960): 330.

Friedly, J. "Dang... there's a trichoglyph on your horse." *Rocky Mountain Quarter Horse* 28 (1990): 26–27.

Fuller, John L., and William Robert Thompson. *Foundations of behavior genetics.* Mosby, St. Louis, MO, (1978).

Gonyou, H. "Behavioural principles of animal handling and transport." In *CAB International* (T. Grandin, ed.), Wallingford, UK, (1993): 11–20.

Grandin, Temple. "Assessment of stress during handling and transport." *Journal of Animal Science* 75, (1997): 249–257.

Grandin, Temple. "Behavioral agitation during handling of cattle is persistent over time." *Applied Animal Behaviour Science* 36, (1993): 1–9.

Grandin, Temple. "Cattle vocalizations are associated with handling and equipment problems at beef slaughter plants." *Applied Animal Behaviour Science* 71, (2001): 191–201.
Grandin, Temple. "Handling methods and facilities to reduce stress on cattle." *Veterinary Clinics of North America: Food Animal Practice* 14, (1998a): 325–341.
Grandin, Temple. "Observations of cattle behavior applied to the design of cattle-handling facilities." *Applied Animal Ethology* 6, (1980): 19–31.
Grandin, Temple. "Reduce stress of handling to improve productivity of livestock." *Veterinary Medicine* 79, (1984): 287–831.
Grandin, Temple. "Safe handling of large animals." *Occupational Medicine: State of the Art Reviews* 14, (1999): 195–212.
Grandin, Temple. "The feasibility of using vocalization scoring as an indicator of poor welfare during cattle slaughter." *Applied Animal Behaviour Science* 56, (1998b): 121–128.
Grandin, Temple, Stanley E. Curtis, Tina M. Widowski, and John C. Thurmon. "Electro-immobilization versus mechanical restraint in an avoid-avoid choice test for ewes." *Journal of Animal Science* 62, (1986): 1469–1480.
Grandin, Temple, and Mark J. Deesing. "Genetics and behavior during handling, restraint, and herding." In *Genetics and the behavior of domestic animals* (T. Grandin, ed.) Academic Press, (2022): 131–181.
Grandin, Temple, Matthew B. Rooney, Megan Phillips, Richard C. Cambre, Nancy A. Irlbeck, and Wendy Graffam. *Conditioning of nyala (Tragelaphus angasi) to blood sampling in a crate with positive reinforcement.* Wiley Subscription Services, Inc., A Wiley Company, New York, (1995): 261–273.
Grigor, P. N., P. J. Goddard, and C. A. Littlewood. "The relative aversiveness to farmed red deer of transport, physical restraint, human proximity and social isolation." *Applied Animal Behaviour Science* 56, (1998): 255–262.
Hale, R. H., T. H. Friend, and A. S. Macaulay. "Effect of method of restraint of cattle on heart rate, cortisol and thyroid hormones." *Journal of Animal Science* 65, (1987): 217.
Harger, C. M. *Frontier days.* Macrae Smith Company, Philadelphia, (1928).
Hasker, P. J. S., and D. Hirst. "Can cattle be mustered by audio conditioning." *Queensland Agriculture Journal* 113, (1987): 325–326.
Hastings, B. E., D. E. Abbott, L. M. George, and S. G. Stadler. "Stress factors influencing plasma cortisol levels and adrenal weights in Chinese water deer (Hydropotes inermis)." *Research in Veterinary Science* 53, (1992): 375–380.
Hediger, Heini. "Studies of the psychology and behavior of captive animals in zoos and circuses." (1955). Butterworths Scientific Publications: London, UK.
Hediger, Heini, and Geoffrey Sircom. "The psychology and behaviour of animals in zoos and circuses." Dover, New York (1968).
Hemsworth, P. H., and J. L. Barnett. "The effects of aversively handling pigs, either individually or in groups, on their behaviour, growth and corticosteroids." *Applied Animal Behaviour Science* 30, (1991): 61–72.
Hemsworth, P. H., J. L. Barnett, C. Hansen, and H. W. Gonyou. "The influence of early contact with humans on subsequent behavioural response of pigs to humans." *Applied Animal Behaviour Science* 15, (1986): 55–63.
Hemsworth, P. H., and G. J. Coleman. "Human livestock interactions: the stockperson and the productivity and welfare of intensively farmed animals." *2nd ed., CABI Publishing, Cambridge, UK* (2010).
Hohenboken, W. D. "Inheritance of behavioural characteristics in livestock. A review." *Animal Breeding Abstracts* 54, (1986): 623–639.
Hough, E. The roundup. In *The Great West* (C. Neider, ed.), Coward-McCann, New York, (1958).
Humphries, E. "Weaning doesn't have to be wearing cattlemen." *Texas and Southwestern Cattle Raisers Association,* (2006): 38–50.
Hutson, G. D. Behavioural principles of sheep handling. In *Livestock handling and transport* (T. Grandin, ed.), CAB International, Wallingford, UK, (1993): 127–146.
Hutson, G. D. "'Flight distance'in Merino sheep." *Animal Science* 35, (1982): 231–235.
Hutson, G. D. "The influence of barley food rewards on sheep movement through a handling system." *Applied Animal Behaviour Science* 14, (1985): 263–273.
Hutson, G. D., and M. L. Butler. "self-feeding sheep race that works." *Journal of Agriculture* 76, (1978): 335–336.
Johnston, J. D., and R. B. Buckland. "Response of male Holstein calves from seven sires to four management stresses as measured by plasma corticoid levels." *Canadian Journal of Animal Science* 56, (1976): 727–732.
Kauppinen, Tiina, Kari Mikko Vesala, and Anna Valros. "Farmer attitude toward improvement of animal welfare is correlated with piglet production parameters." *Livestock Science* 143, (2012): 142–150.

Keeler, Clyde E. "Coat color, physique, and temperament." *Journal of Heredity,* 38 (1968): 82–84.

Keeler, Clyde E. *Genetics of behaviour variations in color phases of the red fox.* The Wild Canids. Van Nostrand Reinhold, New York, (1975): 399–417.

Keeler, Clyde E. "The association of the black (non-agouti) gene with behavior: In the Norway rat." *Journal of Heredity* 33, (1942): 371–384.

Kiley-Worthington, M. "The tail movements of ungulates, canids and felids with particular reference to their causation and function as displays." *Behaviour* 56, (1976): 69–114.

Kilgour, R. "Animal handling in works: pertinent behaviour studies." In *Res Conf Meat Indus New Zeal Pap,* (1971): 9–12.

Kilgour, R., and D. C. Dalton. *Livestock behaviour.* University of New South Wales Press, Sydney, NSW, Australia (1984).

King, D. A., C. E. Schuehle Pfeiffer, R. D. Randel, T. H. Welsh Jr, R. A. Oliphint, B. E. Baird, K. O. Curley Jr, R. C. Vann, D. S. Hale, and J. W. Savell. "Influence of animal temperament and stress responsiveness on the carcass quality and beef tenderness of feedlot cattle." *Meat Science* 74, (2006): 546–556.

King, D. A., C. E. Schuehle Pfeiffer, R. D. Randel, T. H. Welsh Jr, R. A. Oliphint, K. O. Curley Jr, R. C. Vann, and J. W. Savell. "Influence of animal temperament on the tenderness of beef m. longissimus lumborum steaks." *Feeding and Nutrition,* 51st International Congress of Meat Science and Technology August 7–12, Baltimore, MD (2005): 129–133.

Koba, Y., and H. Tanida. "How do miniature pigs discriminate between people?: Discrimination between people wearing coveralls of the same colour." *Applied Animal Behaviour Science* 73, (2001): 45–58.

Lanier, E. K., T. H. Friend, D. M. Bushong, D. A. Knabe, T. H. Champney, and D. G. Lay Jr. "Swim habituation as a model for eustress and distress in the pig." *Journal of Animal Science* 73, (1995): 126.

Le Neindre, Pierre, Gilbert Trillat, Jean Sapa, François Ménissier, J. N. Bonnet, and J. M. Chupin. "Individual differences in docility in Limousin cattle." *Journal of Animal Science* 73, (1995): 2249–2253.

Le Neindre, Pierre, Xavier Boivin, and Alain Boissy. "Handling of extensively kept animals." *Applied Animal Behaviour Science* 49, (1996): 73–81.

Levine, Louis, Joseph Grossfield, and Robert F. Rockwell. "Functional relationships between genotypes and environments in behavior: Effects of different kinds of early social experience on interstrain fighting in male mice." *Journal of Heredity* 70, (1979): 317–320.

Levine, S. (1968). "Influence of infantile stimulation on the response to stress during preweaning development". *Developmental Psychobiology,* 1 (1968): 67–70.

Levine, Seymour, Gary C. Haltmeyer, George G. Karas, and Victor H. Denenberg. "Physiological and behavioral effects of infantile stimulation." *Physiology & Behavior* 2, (1967): 55–59.

Lyons, David M. "Individual differences in temperament of dairy goats and the inhibition of milk ejection." *Applied Animal Behaviour Science* 22, (1989): 269–282.

Mal, M. E., and C. A. McCall. "The influence of handling during different ages on a halter training test in foals." *Applied Animal Behaviour Science* 50, (1996): 115–120.

Matthews, L. R. "Deer handling and transport." In *Livestock Handling and Transport* (T. Grandin, ed.), CAB International, Wallingford, UK, (1993): 253–272.

Melzack, Ronald. "The genesis of emotional behavior: an experimental study of the dog." *Journal of Comparative and Physiological Psychology* 47, (1954): 156–166.

Melzack, R., and S. K. Burns. "Neurophysiological effects of early sensory restriction." *Experimental Neurology* 13, (1965): 163–175.

Miller, C. P., Wood Gush DGM, and P. Martin. "The effect of rearing systems on the responses of calves to novelty." *Biology of Behaviour* 11, (1986): 50–60.

Muller, R., K. S. Schwartzkopf-Genswein, M. A. Shah, and M. A. G. Von Keyserlingk. "Effect of neck injection and handler visibility on behavioral reactivity of beef steers." *Journal of Animal Science* 86, (2008): 1215–1222.

O'bleness, G. V., L. D. Van Vleck, and C. R. Henderson. "Heritabilities of some type appraisal traits and their genetic and phenotypic correlations with production." *Journal of Dairy Science* 43, (1960): 1490–1498.

Pajor, E. A., J. Rushen, and A. M. B. De Passillé. "Dairy cattle's choice of handling treatments in a Y-maze." *Applied Animal Behaviour Science* 80, (2003): 93–107.

Parsons, R. A., and W. N. Helphinstine. *Rambo AI breeding chute for beef cattle.* Agricultural Extension Service, 1969.

Pedersen, Bjarne Kornbek. *Comprehensive evaluation of well-being in pigs: Environmental enrichment and pen space allowance.* University of Illinois at Urbana-Champaign, Champaign, IL, 1992.

Petherick, J. Carol, R. G. Holroyd, V. J. Doogan, and B. K. Venus. "Productivity, carcass and meat quality of lot-fed Bos indicus cross steers grouped according to temperament." *Australian Journal of Experimental Agriculture* 42, (2002): 389–398.

Phillips, Megan, Temple Grandin, Wendy Graffam, Nancy A. Irlbeck, and Richard C. Cambre. *Crate conditioning of bongo (Tragelaphus eurycerus) for veterinary and husbandry procedures at the Denver Zoological Gardens.* Vol. 17, no. 1. Wiley Subscription Services, Inc., A Wiley Company, New York, (1998): 25–32.

Pollard, J. C., and R. P. Littlejohn. "Behavioural effects of light conditions on red deer in a holding pen." *Applied Animal Behaviour Science* 41, (1994): 127–134.

Pollard, J. C., R. P. Littlejohn, P. Johnstone, F. J. Lass, I. D. Carson, and J. M. Suttie. "Behavioural and heart rate responses to velvet antler removal in red deer." *New Zealand Veterinary Journal* 40, (1992): 56–61.

Price, Edward O. "Behavioral genetics and the process of animal domestication." *Genetics and The Behavior of Domestic Animals* (T. Grandin, ed.), Academic Press, New York, (1998): 31–66.

Prince, J. H. "The eye and vision". In *Duke's physiology of domestic animals* (M. J. Swenson, ed.). Cornell University Press, Ithaca, NY, (1977).

Putten, G. van, and W. J. Elshof. "Observations on the effect of transport on the wellbeing and lean quality of slaughter pigs." *Animal Regulation Studies* 1, (1978): 247–271.

Rault, Jean-Loup, Alain Boissy, and Xavier Boivin. "Separation distress in artificially-reared lambs depends on human presence and the number of conspecifics." *Applied Animal Behaviour Science* 132, (2011): 42–50.

Ray, D. E., W. J. Hansen, Brent Theurer, and G. H. Stott. "Physical stress and corticoid levels of strees." In *Proc West Sect Am Soc Anim Sci* 23, (1972): 255–259.

Reid, R. L., and S. C. Mills. "Studies on the carbohydrate metabolism of sheep. XIV. The adrenal response to psychological stress." *Australian Journal of Agricultural Research* 13, (1962): 282–295.

Roche, B. W. "Coacher mustering." *Queensland Agricultural Journal* 114, (1988): 215–216.

Rogan, Michael T., and Joseph E. LeDoux. "Emotion: systems, cells, synaptic plasticity." *Cell* 85, (1996): 469–475.

Rushen, J. "Aversion of sheep for handling treatments: paired-choice studies." *Applied Animal Behaviour Science* 16, (1986): 363–370.

Rushen, Jeffrey, and Anne Marie De Passille. "The importance of good stockmanship and its benefits to animals." *Improving Animal Welfare: A Practical Approach* Ed. 2 (2015): 125–138.

Rybarczyk, Pierre, Jeff Rushen, and Anne Marie de Passillé. "Recognition of people by dairy calves using colour of clothing." *Applied Animal Behaviour Science* 81, (2003): 307–319.

Sandem, A. I., A. M. Janczak, and B. O. Braastad. "A short note on effects of exposure to a novel stimulus (umbrella) on behaviour and percentage of eye-white in cows." *Applied Animal Behaviour Science* 89, (2004): 309–314.

Sandem, A. I., A. M. Janczak, R. Salte, and B. O. Braastad. "The use of diazepam as a pharmacological validation of eye white as an indicator of emotional state in dairy cows." *Applied Animal Behaviour Science* 96, (2006): 177–183.

Savory, Allan. "A holistic approach to range management using short duration grazing." *Proc. Intn'l. Rangel. Congr.* 1, (1978): 5.

Scott, John Paul, and John L. Fuller. *Genetics and the social behavior of the dog.* Vol. 570. University of Chicago Press, Chicago, IL, (1965).

Shrode, R. R., and S. P. Hammack. "Chute behavior of yearling beef cattle." *Journal of Animal Science* 33, (1971): 193.

Shupe, W. L. *Transporting sheep to pastures and markets.* ASAE St. Joseph, Michigan, (1978).

Smith, B. "Moving 'em: A guide to low stress animal handling: The Graziers Hui." Kamuela, Hawaii, (1998).

Smith, Burt, PingSun Leung, and George Love. "Intensive grazing management: forage, animals, men, profits." The Graziers Hui, (1986).

Stricklin, W. R. "Heritability of temperament in beef cattle." *Journal of Animal Science* 51 (1980): 109.

Tanida, H., and Y. Nagano. "The ability of miniature pigs to discriminate between a stranger and their familiar handler." *Applied Animal Behaviour Science* 56, (1998): 149–159.

Tanner, M., T. Grandin, M. Cattell, and M. Deesing. "The relationship between facial hair whorls and milking parlor side preferences." *Journal of Animal Science* 72, (1994): 207.

Tellington-Jones, L., and V. Burns. *The Tellington-Jones equine awareness method.* Breakthrough Publications, Millwood, NY, (1985).

Torres-Hernandez, Glafiro, and William Hohenboken. "An attempt to assess traits of emotionality in crossbred ewes." *Applied Animal Ethology* 5, (1979): 71–83.

Tulloh, N. M. "Behaviour of cattle in yards. II. A study of temperament." *Animal Behaviour* 9, (1961): 25–30.

Vann, R. C., J. C. Paschal, and R. D. Randel. "Relationships between measures of temperament and carcass traits in feedlot steers." *Journal of Dairy Science* 87, (2004): 259.

Vetters, M. D. D., T. E. Engle, J. K. Ahola, and T. Grandin. "Comparison of flight speed and exit score as measurements of temperament in beef cattle." *Journal of Animal Science* 91, (2013): 374–381.

Voisinet, B. D., T. Grandin, J. D. Tatum, S. F. O'connor, and J. J. Struthers. "Feedlot cattle with calm temperaments have higher average daily gains than cattle with excitable temperaments." *Journal of Animal Science* 75, (1997): 892–896.

Wagnon, K. A., R. G. Loy, W. C. Rollins, and F. D. Carroll. "Social dominance in a herd of Angus, Hereford, and Shorthorn cows." *Animal Behaviour* 14, (1966): 474–479.

Waiblinger, Susanne, Christoph Menke, and Grahame Coleman. "The relationship between attitudes, personal characteristics and behaviour of stockpeople and subsequent behaviour and production of dairy cows." *Applied Animal Behaviour Science* 79, (2002): 195–219.

Ward, F. *Cowboy at work.* Hasting House Publishers, New York, 1958.

Warriss, P. D., S. N. Brown, S. J. M. Adams, and I. K. Corlett. "Relationships between subjective and objective assessments of stress at slaughter and meat quality in pigs." *Meat Science* 38, (1994): 329–340.

Welp, T., J. Rushen, D. L. Kramer, M. Festa-Bianchet, and A. M. B. De Passille. "Vigilance as a measure of fear in dairy cattle." *Applied Animal Behaviour Science* 87, (2004): 1–13.

Whateley, J., R. D. C. D. Kilgour, and D. C. Dalton. "Behaviour of hill country sheep breeds during farming routines." *Proceedings of the New Zealand Society of Animal Production* 34, (1974): 28–36.

Whittington, C. J., and A. S. Chamove. "Effects of visual cover on farmed red deer behaviour." *Applied Animal Behaviour Science* 45, (1995): 309–314.

Wyman, Walker Demarquis. *The wild horse of the west.* Caxton Printers, 1946.

5 Learned Behavior

Madhumita Roy

5.1 INTRODUCTION

Learning can be characterized as a behavior change brought about by experience that is comparatively persistent. Nevertheless, there is a problem with this definition of learning: it is frequently challenging to define what exactly qualifies as "relatively permanent." If behavior is viewed as a phenotype, learning therefore falls under the category of phenotypic plasticity in the definition of learning given above, which is "relatively permanent change in behavior as a result of experience." Therefore, learning is regarded as a type of phenotypic plasticity and includes everything from long-lasting morphological alterations to behavioral changes brought about by experience, if behavior is substituted by phenotype (Alcock, 2005). Learning and "phenotypic plasticity," as described by evolutionary ecologists, have an intriguing connection. The observable traits of an organism are referred to as phenotypes, and an organism's capacity to generate distinct phenotypes in response to varying environmental circumstances is known as phenotypic plasticity. *Membranipora membranacea*, a bryozoan, is a good example of phenotypic plasticity. Individuals usually lack the spines that are commonly used as protection against predators, and they live in colonies. *Membranipora membranacea* simply do not generate spines when there are no predators around. On the other hand, individuals develop spines very quickly when they are in the presence of predators. When environmental changes (the arrival of predators) cause such a drastic shift from spineless to spined, it is a case of phenotypic plasticity.

5.2 HOW ANIMALS LEARN

What processes underlie learning? Animal behaviorists have been captivated by this question. Heyes (1994) proposed three typically arranged experiences that have the potential to be educational. These are the following: one single stimulus, two stimuli in combination, and a response reinforcer round out the list.

a) Learning from a single stimulus experience: The simplest type of learning involves a single stimulus, which can take any shape. For instance, if a color stick is placed in a rat cage several times a day, the rats will likely notice any disruptions. They would look that way with their heads turned. Over time, if the rat becomes increasingly focused on the blue stick, it indicates that it has become sensitized, or unresponsive to the stimuli. Conversely, it is considered that habituation has occurred if the rats start to pay less and less attention to the stick. Two straightforward methods of learning from a single stimulus are sensitization and habituation.

5.2.1 Consequences of Habituation or Sensitization in Learning

It becomes exceedingly challenging to study some types of behavior in animals if they become acclimated to stimuli quickly, especially in lab settings. The example of antipredatory behavior serves as a clear demonstration of this. Assuming that two aquariums are taken and positioned next to each other. A predatory fish is kept in one, while some prey fish are kept in the other; predator and prey may engage in visual exchanges in this situation, but the predator is unable to hurt the animal.

DOI: 10.1201/9781003710271-5

The potential prey's life is spared by such an experimental approach. But it does set up a situation where the prey eventually becomes accustomed to the predator after realizing that it is not a real threat. Therefore, when conducting studies, ethologists must take considerable care to ensure that habituation has not happened in their study models. A single stimulus that causes sensitization or habituation may also have additional learning effects. For instance, i) an animal may become habituated to a stimulus and then become resistant to subsequent attempts to acquaint it with another event. It would be more challenging for the rats in the blue stick experiment, for instance, to learn later on that the blue stick indicates the coming of food or any other event because they had grown accustomed to it. ii) Should there have been sensitivity to a single trigger, later, it can be used to associate subsequent occurrences.

b) Stimulus-stimulus (Pavlovian conditioning): Here, the blue stick is not the only stimulus provided; rather, it is combined with another stimulus, such as the smell of a cat, which the rats are genetically frightened of. In this experiment, a cat's odor is sprayed at one cage corner five seconds after the blue stick is introduced. The rat then gains the ability to link stimulus two (cat odor) with stimulus one (blue stick). In response, it hides as soon as the blue stick shows up and before the cat's scent is released. We refer to this type of controlled experiment as classical or Pavlovian conditioning. Ivan Pavlov created Pavlovian conditioning for the first time in the late 1800s. Pavlovian conditioning experiments comprise two stimuli—conditioned and unconditioned. A stimulus that does not initially evoke a specific reaction but does so when it is linked to an unconditioned stimulus is known as a conditioned stimulus (CS). The conditioned stimulus is thought to be the blue stick. In the absence of training, an unconditioned stimulus (US) is one that produces a strong response. The cat odor is therefore an unconditioned sensation. The term "conditioned response" refers to the rat's ability to conceal when it only sees the blue stick (CR). Positive and negative relationships can also be distinguished from one another. Positive associations are those in which the location of the blue stick in the first event (Pavlovian conditioning experiment above) predicts the occurrence of the second event (cat odor). It is a negative association if the first event indicates that the second event won't happen. Negative interactions result in inhibitory conditioning, whereas positive ones result in excitatory conditioning. It is common to distinguish between pleasurable and unpleasant stimuli when studying literature. An appetitive stimulus is any stimulus that is regarded as rewarding, pleasurable, or positive. It consists of things like food, a secure place to live, possible partners, etc. An aversive stimulus is any stimulus that is connected to a negative experience.
c) Response-reinforcer: Instrumental conditioning happens when an animal's response is in some way reinforced. Another name for it is goal-oriented or operant learning (Skinner, 1966). A typical illustration of instrumental learning is a rat pressing a lever to cause food to fall into its cage. In order to elicit a reaction (food falling into the cage), the rat in this instance rapidly associates itself with an action (pressing a bar). In 1911, while researching instrumental learning, Edward Thorndike proposed the law of effect. According to the law of effect, there will be a stronger association between a stimulus and a reaction if the response to a stimulus is followed by a pleasant or fulfilling occurrence. Conversely, the association will be diminished if the response is followed by an unfavorable or hostile incident. One disadvantage of instrumental learning is that it fails to take into account the behavior's status as a continuous, freely varying variable.

The fundamental difference between instrumental learning and Pavlovian conditioning is that the latter requires an animal to respond or act in order for the conditioning process to result in learning. Nonetheless, there is still some disagreement over the relative benefits of instrumental and Pavlovian learning methods.

5.3 TYPES OF LEARNED BEHAVIOR

Learned behavior can be distinguished into eight types.

a) Habituation: When a stimulus is repeatedly presented without an incentive or consequence, the subsequent gradual reduction in behavioral reactivity is known as habituation. Here, the animal gains the ability to block out unimportant information. A few instances of habituation are as follows: (i) Scarecrows placed in crop fields to scare birds work well for a limited period of time. The birds quickly become accustomed to these benign scarecrows. (ii) When a cloud shadow passes overhead, juvenile birds have a diminishing escape response. (iii) Because of the disturbance caused by passing trains, fish in a body of water adjacent to a railroad track gradually exhibit a decline in behavioral reaction.

b) Imprinting: One cannot define imprinting precisely. It is, instead, the process of forcing a young animal to exhibit a consistent behavior pattern by subjecting it to certain stimuli at a pivotal juncture in its development. The term "imprinting" describes a variety of behavioral alterations that occur when a young animal forms an attachment to a potential mate or "mother figure." Through his research with geese, Konrad Lorenz (1937) was the first to introduce imprinting when he managed to get gosling broods to follow him and regard him as their mother figure.

   After hatching or birth, imprinting typically occurs and frequently results in a strongly permanent bond that is challenging to modify. These kinds of lessons are distinct from those found in other forms, irreversible, and limited to the short "critical period" immediately following hatching. Before anything else is learned, imprinting takes place. It has been shown to be particularly helpful for the study of the neurological underpinnings of learning and memory because it is frequently a very distinct and recognizable occurrence. Imprinting can fall into two types:

   i) Filial imprinting: We refer to the reaction to the mother figure as filial imprinting. There is a wide variety of objects that can cause young birds to approach and bond. Imprinting stimuli can be smelled, heard, or seen (Hampton et al., 1995). It appears that the variety of visual stimuli has no bounds. Similar to flashing lights, movements can draw attention. Young birds will be drawn to a stationary object as long as it stands out against the background. Many young birds are reported to be attracted to auditory stimulation. For instance, sound plays a crucial role in encouraging mallard ducklings to follow their mother. Wood ducks make their nests in tree holes. The little ones approach the mother despite the dangers when they hear her call from the water outside the nest hole; the children are encouraged to come closer to the mother despite their incomplete impression of her.

   ii) Sexual imprinting: Lorenz claims that a young duck or goose's early experiences influence their choice of sexual partner as they get older. When it comes to sexual imprinting, children "imprint" on what makes an acceptable partner by watching adults in their community. Similar to filial imprinting, the variety of objects that might elicit attachment is essentially limitless in sexual imprinting as well (Hamilton, 1981). Bengalese and zebra finches were used in a comprehensive sexual imprinting experiment. Immelmann (1972) permitted the Bengalese finch parent to raise the entire brood after he nested a single zebra finch egg amid several Bengalese finch eggs. The male zebra finch that was fostered was later kept apart until it reached sexual maturity. Subsequently, the adult male zebra finch was positioned in the center of a cage with three sections. Bengalese and zebra finches, both females, were housed in compartments on either side. The outcome that was seen was quite intriguing. The Bengalese finch female was the object of the male zebra finch's courting. When the male Bengalese finch approached the female, she would typically display avoidance and remained neutral. Conversely, the female zebra finch instinctively replied by making all of the typical, distinct greeting calls and perched as close to the male as the partition permitted. Thus, imprinting is distinct from other learning methods due to three reasons: It only occurs at a limited, sensitive time in the early stages of life; it is quite stable and frequently lasts the entirety of the animal's life; and it affects the adult, social, and sexual behavior of the animal.

c) Conditional reflex: A response is referred to as a conditioned reflex when it is altered by prior experiences. The brain controls the coordination of these reflexes. I. P. Pavlov's (1901) research on dogs helps to understand the mechanism of conditional reflex. A dog that was ravenous was put on a platform and held down with a harness. After that, the dog was fed in powder form on a regular basis. A bell or other external stimulation was used to signal the beginning of the food delivery. The dog started leaking lip fluid and secreting saliva in response to the bell by itself.

Our everyday lives are filled with numerous examples of conditioned reflexes. When a finger is touched by an empty hot utensil, it burns and is quickly put down (basic reflex action). However, when touched, an equally hot utensil full of food hurts the fingers just as much and must be put down promptly but gently to prevent spills (conditioned reaction). The variations in reaction intensity indicate the role of memory and conditioning, which is followed by the brain's conscious choice.

d) Trial-and-error learning: Through this process, animals pick up knowledge about daily activities. They experiment with several options and eventually, through failure and success, they learn how to address the issue. Trial and error learning is demonstrated in B.F. Skinner's (1966) experiment in which a rat was forced to push a lever, which caused the animal to drop food. In a different experiment, Skinner trained pigeons to press a lever to open a box's door and retrieve food grains. The experiment was conducted numerous times in both situations. After progressively pressing the bar for fewer and fewer time intervals, the animal did so quickly.

e) Latent learning: Thorpe et al. (1983) asserts that learning occurs even in situations in which there is no related reward or punishment. The connection of various stimuli without any immediate reward or penalty is known as latent learning. Throughout their lives, animals are exposed to novel experiences and are always learning new things that they may use later in life—hence the phrase latent. Metzgar (1967) gave a wonderful example of latent learning using the deer mouse (Peromyscus leucopus). A group of deer mice was housed in an enclosed space with natural elements such as plants, hay, logs, twigs, and so on to give them experience. A second set of mice was housed in a lab cage away from the natural objects mentioned above. After that, a predatory owl and the mice from each group were put within the confined space. Merely two out of twenty mice who had visited the hall before were caught by the owl, according to Metzgar's observations. Eleven out of the twenty mice that had never visited the habitat before were captured. This indicates that the experienced deer mouse was able to evade the predator (owl) thanks to latent learning.

f) Insight: The capacity to react appropriately to a circumstance that one encounters for the first time in life and that differs greatly from any prior experience is known as insight (Wrangham et al., 1994; Kohler, 1927) demonstrated understanding by using chimpanzees in his experiment. Kohler placed a chimpanzee in a room with several boxes strewn all across the floor. A few bananas that were too high for the chimpanzee to reach were suspended from the ceiling by him. The chimpanzee first attempted to jump in order to get to the bananas. Upon not making it, he took a seat and pondered. After a while, he stood up and stacked the wooden boxes on top of one another before climbing up to get the bananas. This answer, which came to mind after some apparent contemplation, is referred to as insight.

g) Reasoning: The capacity to impromptu integrate two or more disparate experiences to create a new experience that is successful in achieving a specific goal is known as reasoning. The mental process of reasoning is highly evolved from behavior and is arguably unique to humans, dolphins, apes, and killer whales. But there's still a lot we don't know about the brain processes involved in reasoning.

h) Cognition: Animals that possess the ability to perceive, interpret, and make decisions about their actions are said to possess cognition. Foraging, mate selection, and many other activities are influenced by cognitive processes like perception, learning, memory, and decision-making. It is impossible to directly observe cognition. It is an advancement over insight in which one obtains knowledge and applies it to a variety of situations. Considered the founder of the contemporary cognitive approach to animal behavior is Tolman (1886–1959). He developed the concept of cognitive maps, which are mental representations of the outside world that can be created via exploratory behavior. Animals learn to associate relevant spatial or causal characteristics of their environment with one another through the acquisition of such cognitive maps.

5.4 DIFFERENCES BETWEEN INSTINCTIVE AND LEARNED BEHAVIOR

The differences between instinctive and learned behavior are:

1) Natural selection gradually alters instinctive behavior to make it as environment-adaptive as feasible. It is inherited by subsequent generations via the DNA and provides a kind of species memory. When an animal exhibits learned behavior, it lacks innate situational responsiveness but yet possesses the capacity to adapt its behavior based on past experiences. The next generation does not inherit such experiences.
2) Adaptive behavior is guaranteed by both instinct and learning; the former is achieved by selection acting throughout a species' history, while the latter occurs within an individual's lifetime. An animal with instinct is endowed with a repertoire of adaptive behaviors that appear prefabricated during its first behavior. For animals with short lifespans and little to no parental care, this is undoubtedly advantageous.
3) Rigid stereotypical reactions and movement patterns define behavior. Conversely, learning leads to adaptable response patterns. Learning is a process that is varied and requires time. The onset of a stimulus soon elicits instinctive actions.
4) Animal nervous and muscular systems' development is correlated with their instinctive behavior. For instance, when still in the nest, young birds are frequently observed vigorously flipping their wings. Though innate, sexual behavior is correlated with gonad growth. It is only carried out when the blood contains a specific amount of sex hormones. With practice, or learning, behavior performance improves. For instance, the combination between instinctive and learned behavior is seen in the development of pecking in just-hatched chicks. Chicks that have just hatched possess an innate inclination to peck at items without discrimination. With repeated tries and errors, they eventually become better. Learning, then, improves instincts. As a result, these two processes cannot be separated. The genes represent innate traits. Through learning, they are refined or perfected for more precise changes.

It is believed that learning is subject to a complicated genetic influence; however, it is debatable if natural selection favors impairment. It is preferred. Then, in what situation? According to psychologists, behavioral ecologists, and ethologists, learning should take precedence over hereditary transmission of permanent features when an animal's natural surroundings shift frequently but not excessively. When the environment doesn't vary much, genetic transmission is used to pass on

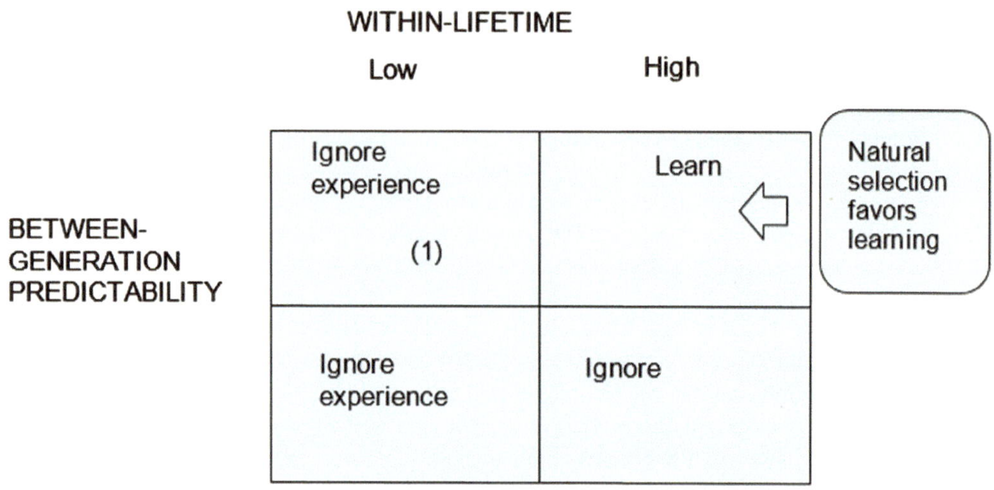

FIGURE 5.1 Model showing evolutionary tradeoffs and environmental stability based on Stephens (1993)

knowledge because it minimizes learning costs and ensures that the environment that offspring interact with is comparable to that of their parents. But since things change all the time, there's nothing worth learning because everything that is taught is already fully irrelevant in this next situation. Therefore, genetic transmission of a fixed reaction is preferred over learned responses, as the past has no predictive value. The environment that constantly learns is preferred above the genetic transmission of a set reaction, somewhere in the middle, because learning is worth the price. The environment in this scenario is sufficient enough to support transmission, but not as steady as to favor transmission. Although he acknowledged the aforementioned reality, David Stephens (1993) questioned the notion of environmental stability, asserting the necessity to distinguish between two forms of stability. Environmental predictability has been divided into two categories by Stephens: (i) predictability within an individual's generation (lifetime) and (ii) predictability between an offspring's and parent's environment (Figure 5.1).

Stephens proposed a model of the evolution of learning. Stephens' model demonstrates that learning is more advantageous when an individual's lifetime predictability is strong, but environmental predictability over generations is low (box 3). In all other scenarios (boxes 1–2 and 4), fixed genetic transmission is preferred since learning has a cost. Box 3 favors learning because once an organism learns a behavior it can repeat it for the rest of its life.

BIBLIOGRAPHY

Alcock, J. 2005. *Animal Behaviour: An Evolutionary Approach.* Sunderland, Mass.: Sinauer Associates.

Altman, S. A. 1962. A field study of the sociobiology of rhesus monkeys, *Macaca mulatta. Annals of the New York Academy of Science* 102: 338–435.

Axelrod, R. D., Hamilton, W. D. 1981. The evolution of cooperation. *Science* 211: 1390–1396.

Baldwin, J. D., Baldwin J. L. 1973. The role of play in social organisation: comparative observations of squirrel monkeys (*Samiri*). *Primates* 14: 369–381.

Bateson, P. P. G. 1976. How do sensitive periods arise and what are they for? *Animal Behaviour* 27: 470–486.

Boesch, C. 1994. Cooperative hunting in wild chimpanzees. *Animal Behaviour* 48: 653–667.

Bolhuis, J. J. 1995. Development of perceptual mechanisms in birds: predisposition and imprinting. In *Neuroethological Studies of Cognitive and Perceptual Processes,* ed. C. F. Moss and S. J. Shettleworth, pp. 150–176. Boulder, CO: Westview Press.

Bolles, R. C. 1973. The comparative psychology of learning: the selective association principle and some problems with "general" laws of learning. In *Perspectives in Animal Behaviour*, ed. Bermant, G. Glenview, IL: Scott, Foresman and Company.

Brown, R. E., and Eklund, A. 1994. Kin recognition and the major histocompatibility complex: an integrative review. *American Naturalist* 143: 435–461.

Caro, T. M. 1994. *Cheetahs of the Serengeti Plains: Group Living in Asocial Species.* Chicago, IL: The University of Chicago Press.

Curio, E. 1996. Conservation needs ethology. *Trends in Ecology and Evolution* 11: 260–263.

Dawkins, M. S. 1995. *Unravelling Animal Behavior*, 2nd edition. Harlow, UK: Longman.

Duffy, S. 1996. Eusociality in a coral-reef shrimp. *Nature* 381: 512–514.

Dunbar, R. I. M. 1988. *Primate Social System.* London: Croom Helm.

Gouzoules, S., and Gouzoules, H. 1987. Kinship. In *Primate Scocieties*, ed. B. B. Smuts, D. L. Chenny, R. M. Seyfarth, R. W. Wragham, and T. T. Strauhsaker. Chicago, IL: University of Chicago Press.

Grafen, A. 1984. Natural selection, kin selection and group selection. In *Behavioural Ecology: An Evolutionary Approach*, 2nd edition, ed. J. R. Krebs and N. B. Davies, pp. 62–84. Oxford: Blackwell Scientific Publication.

Hampton, N. G., Bolhuis, J. J., and Horn, G. 1995. Induction and development of a filial predisposition in the chick. *Behaviour* 132: 451–477.

Heyes, C. M. 1994. Social learning in animals: categories and mechanisms. *Biological Reviews: Cambridge Philosophical Society* 69(2): 207–231.

Hinde, R. A., and Fisher, J. 1952. Further observations on the opening of milk bottles by birds. *British Birds* 44: 393–396.

Immelmann, K. 1972. Sexual and other long-term aspects of imprinting in birds and other species. *Adv. Study Behav.* 4: 147–174.

Jarvis, J. U. M., Oriain, J. M., Bennett,N. C., and Sherman, P. W. 1994. Mammalian eusociality: a family affair. *Trends in Ecology and Evolution* 9(2): 47–51.

Kenward, R. E. 1978. Hawks and doves: factors affecting success and selection in goshawk attacks on wood-pigeons. *Journal of Animal Ecology* 47: 449–460.

Kohler, W. 1927. *The Mentality of Apes.* New York: Kegan Paul, Trench, Trubner & Co. Ltd.

Krebs, J. R., and Davies, N. P. 1993. *An Introduction to Behavioural Ecology*, 3rd edition. Oxford, UK: Blackwell Scientific Publication.

Kruijt, J. P., and Meeuwisse, G. B. 1991. Sexual preferences of male zebra finches: effects of early and adult experience. *Animal Behaviour* 42: 91–102.

Lorenz, K. (1937). Über die Bildung des Instinktbegriffes. *Naturwiss* 25: 289–300, 307–308 and 324–331.

Metzgar, L. H. (1971). Behavioral population regulation in the woodmouse, Peromyscus leucopus. *Amer. Midland Nat.* 86: 434–448.

Maynard Smith, J. 1964. Group selection and kin selection. *Nature* 201: 1145–1147.

Packer, C. 1977. Reciprocal altruism in *Papio anubis. Nature* 265: 441–443.

Pavlov, I. P. (1901). *The work of the digestive glands.* London: Griffin.

Pearce, J. M. 1996. *Animal Learning and Cognition: An Introduction*, 2nd edition. Hove, UK: Psychology Press.

Roper, T. J. 1983. Learning as a biological phenomenon. In *Animal Behaviour vol 3. Genes, Development and Learning*, ed. P. J. V. Slater and T. R. Halliday, pp. 178–212. Oxford: Blackwell Scientific Publication.

Skinner, B. F. 1966. Operant behavior. In *Operant Behaviour*, ed. W. Honing. New York: Appleton-Century-Crofts.

Slagsvold, T., and B. T. Hansen. 2001. Sexual imprinting and the origin of obligate brood parasitism in birds. *American Naturalist* 158: 354–367.

Sonnemann, P., and Sjolander, S. 1977. Effect of cross fostering on the sexual imprinting of the female zebra finch *Taeniopygia guttata. Journal of Psychology* 45: 337–348.

Stephens, D. (1993). Learning and behavioral ecology: Incomplete information and environmental predictability. *Insect Learning.*

Ten Cate, C. 1982. Behavioural differences between zebra finches and Bengalese finch (foster) parents raising zebra finch offspring. *Behaviour* 81: 152–172.

Thorpe, S. J., Rolls, E. T., and Maddison, S. 1983. The orbitofrontal cortex: neuronal activity in the behaving monkey. *Exp. Brain. Res.* 49: 93–115.

Van Shaik, C. P. 1983. Why are diurnal primates living in groups? *Behaviour* 87: 120–144.

Wrangham, R. W., McGrew, W. C., De Waal, F. B. M., and Heltne, P. G. 1994. *Chimpanzee Cultures.* Cambridge, MA: Harvard University Press.

6 Animal Cognition

Elohozino Oghale Benneth, Isaac Babatunde Fasipe, Oluwashina Owolabi and Emmanuel Onche

6.1 INTRODUCTION

6.1.1 What Is Animal Cognition?

Animal cognition encompasses various mental processes, including perception, attention, memory formation, awareness, reasoning, and consciousness, alongside executive functions like learning and problem-solving, all vital for information processing. It involves a spectrum of cognitive abilities essential for animals to navigate their environments effectively and adapt to diverse challenges they encounter (Brown et al., 2011). The study of animals' capacity to receive, analyze, and respond to information has captivated numerous scientists, emerging as a compelling area of research which explores how animals perceive, interpret, and engage with their surroundings, therefore offering insights into their cognitive abilities and adaptive behaviors (Bush et al., 2000). While some early studies argue that animals lack cognitive abilities and their behavior is merely influenced by human manipulation (Ferguson, 1954), others contend that similar to humans, many animals possess distinct mechanisms for processing information, communicating it within their species, and making decisions based on it (Panksepp, 2011). As a result, numerous studies have arisen with the aim of assessing the degree to which animals can process information and subsequently translate it into actions, thereby leading to investigations seeking to not only elucidate the cognitive capacities of animals but to also understand the mechanisms underlying their behavioral responses to stimuli (Bräuer et al., 2020). From this perspective, the potential for animals to engage in sophisticated cognitive processes is well highlighted, thereby suggesting a richer understanding of their behaviors and interactions within their environments (Panksepp, 2011). According to Tang et al. (2001), the introduction of novel objects as a source of environmental enrichment for animals not only improves their interactive abilities but also their cognitive abilities (Figure 6.1).

Therefore, cognition as defined by Shettleworth (2009) is the ability of animals to process information via their senses about the world they live in, which includes their abilities to discern and learn, their ability to retain such information, and, most importantly, their ability to decide on how to act on it. Over time, robust debates have emerged concerning the existence of consciousness in animals, including discussions about which species might possess such a state of mind, probing fundamental questions about the nature of consciousness across the animal kingdom and expanding our understanding of cognitive experiences beyond humans (Berridge & Winkielman, 2003; Krupenye & Call, 2019). Even though there are questions regarding the possibility that animals experience consciousness which cannot be proffered with specific answers, certain species display behaviors possessing functional parallels characteristic of human conscious cognitive abilities in studies of metacognition and blind sight (Mendl & Paul, 2004). According to Burgdorf & Panksepp (2006), there is a prevalent tendency to make numerous assumptions regarding the cognitive complexity of animals. They argue that as the likelihood of animals possessing intricate cognitive abilities increases, so does the probability of them experiencing consciousness, highlighting the interconnectedness between cognitive capacity and consciousness in animals. Bräuer et al. (2020) propose that for an animal to undergo conscious emotional experiences, it must align with the animal's awareness of being able to feel those emotions. They suggest that the capacity for an animal to

DOI: 10.1201/9781003710271-6

FIGURE 6.1 Animal cognition – primate processing information about its environment.

acknowledge and interpret its emotions plays a pivotal role in the conscious experience of emotions. As the years have progressed, numerous studies have suggested that animals possess cognitive abilities that enable them to anticipate or predict future events, allowing them to establish temporal and consequential relationships. These cognitive abilities facilitate the anticipation of events, while emotional reactions aid in the management of immediate situations, collectively contributing to animals' adaptive responses to their environments (Flavell, 1979; Mendl & Paul, 2020).

6.1.2 Brief Overview of the Significance of Studying Animal Cognition

Animal cognition has received a lot of attention through the years and has led to several researchers making notable discoveries on why studying the field is a necessity for the non-human and human race. Pepperberg (2000) puts forward that animal cognition research serves as a bridge that offers essential connections between crucial natural behaviors which are shaped by evolution, theories as well as empirical evidence from cognitive development across species. The author proposes that this interdisciplinary approach assists in comprehending both the similarities and differences in cognitive processes among animal species, elucidating the evolutionary origins of diverse cognitive characteristics, thereby allowing researchers to glean insights into the mechanisms that drive cognitive development across various species and enhancing our understanding of animal cognition. As a result, this linkage not only enhances our understanding of the evolutionary continuity of cognitive abilities but also provides a framework for exploring the adaptive significance of cognitive processes across diverse ecological contexts.

According to Watanabe and Huber (2006), the investigation of animal cognition reveals the complex inner workings of non-human mental processes, providing deep understandings of intelligence, perception, and behavior across different species. They suggest that exploring the cognitive abilities of animals extends beyond mere curiosity but holds significant implications for disciplines like evolutionary biology, psychology, as well as artificial intelligence, and offers valuable insights into the evolution of intelligence, the influence of perception on behavior, as well as how these understandings can drive progress across various domains. Furthermore, findings derived from the study of animal cognition carry considerable weight for both the welfare of animals and conservation endeavors. Mendl et al. (2009) suggest that by understanding the cognitive capacities of diverse species in a comprehensive manner, their welfare needs and viewpoints can be better assessed, thereby guiding more conscientious approaches to captivity, rehabilitation, and conservation. This can further lead to the creation of strategies to mitigate human-animal conflicts and promote peaceful coexistence obtained through a nuanced grasp of animal cognition information.

Research on animal cognition has contributed significantly to our understanding of collective cognitive abilities in group-living animals. Couzin (2009) suggests that studying collective behavior in animals can offer valuable insights into cognitive abilities. The author emphasizes the parallels between findings in animal cognition and neuronal processes, pointing out that through the exploration of how animals interact within groups, researchers can uncover shared cognitive mechanisms that underpin coordinated behaviors, which provides insight into the complex dynamics of social organization and decision-making in animal societies. This perspective not only enriches our understanding of animal cognition but also offers a framework for exploring the cognitive foundations of collective behavior in various contexts, from social insects to human communities.

Moreover, Chieffi et al. (2017) believe that the study of animal cognition offers a distinctive perspective into human cognition, providing a lens through which researchers can identify shared cognitive processes across species. The authors suggest that this approach aids in understanding the parallels and divergences between the cognitive processes of humans and other species, illuminating diverse cognitive traits and allowing researchers to gain insights into the mechanisms driving cognitive development across species, enhancing our understanding of both animal and human minds. Through comparative analysis, scientists reveal fundamental cognitive principles, enriching disciplines such as psychology and neuroscience, and gaining valuable insights into diverse areas such as language acquisition, social cognition, memory, and decision-making, thus deepening our understanding of the human mind's workings.

Additionally, in fields such as animal training, animal-assisted therapy, and even robotics, studying animal cognition plays a significant role in practical applications. From the perspective of Rowell et al. (2021), studying animal cognition enhances training methods for working animals and informs biomimicry, where insights into how animals solve problems drive the development of innovative technologies. Therefore, through understanding how animals perceive and interact with their environment, researchers can create sustainable solutions inspired by nature's efficiency and adaptability. In essence, studying animal cognition transcends academic boundaries and offers profound insights into the natural world, informing ethics and conservation, advancing our grasp of human cognition, as well as igniting innovation. Exploration into the cognitive realms of other species not only expands our knowledge but also fosters empathy and appreciation for the diversity of cognitive life on Earth.

6.1.3 Importance of Understanding Cognitive Processes in Animals across Different Environments

Understanding cognitive processes in animals across different environments is crucial for several reasons, spanning scientific research, conservation efforts, and our broader comprehension of the natural world. Firstly, understanding cognitive processes in animals helps us grasp the diversity of intelligence across species (Thorndike, 2000). Examining how animals perceive, acquire knowledge, and tackle challenges across diverse habitats provides valuable insights into the development of cognitive skills, challenging human-centric notions of intelligence and showcasing the distinct adaptations various species have evolved to ensure their survival and success in their respective environments (Rowell et al., 2021). Secondly, insights into animal cognition can inform conservation strategies. Conservationists can design more effective habitat management plans and wildlife corridors by understanding how animals perceive and interact with their environment, exemplified through knowing how certain species navigate complex landscapes or find food sources can aid in mitigating human-wildlife conflicts and preserving biodiversity hotspots (Nathan et al., 2022).

Moreover, studying cognitive processes in animals enhances our understanding of animal welfare. Understanding the intricate cognitive abilities of animals informs ethical considerations about how they should be treated both in captivity and in their natural habitats, prompting the creation of enrichment initiatives tailored to meet the cognitive requirements of animals in settings like zoos, sanctuaries, or rehabilitation centers, thereby enhancing their mental welfare. (Hemsworth et al., 2015). In addition, seeing that nature's evolutionary solutions to intricate challenges can inspire innovative technologies, understanding animal cognition holds practical significance for disciplines like robotics and artificial intelligence. Biomimicry has spurred advancements across various fields, from robotics to material science, involving emulating biological systems for engineering solutions (Crosby, 2020). Finally, recognizing and comprehending the cognitive processes of animals nurtures a stronger bond between humans and the natural realm, encouraging empathy and respect for diverse species by acknowledging their individuality and complexity, resulting in such awareness being essential for advancing conservation initiatives and fostering sustainable cohabitation with wildlife (Andrews, 2020).

In conclusion, insights on the cognitive processes in animals across diverse habitats are essential for the progression of scientific understanding, the enhancement of conservation initiatives, the safeguarding of animal well-being, the stimulation of technological advancements, and the cultivation of a deeper respect for the variety of life on our planet.

6.2 AQUATIC ANIMALS

6.2.1 Overview of Cognitive Abilities Observed in Aquatic Animals

Aquatic animals exhibit a diverse array of cognitive abilities shaped by their ecological niche, which encompasses sensory perception, learning and memory, problem-solving, and social cognition (Brown et al., 2011). In aquatic life, sensory systems adapted to underwater environments enable detection of prey, predators, and mates; learning and memory facilitate navigation, foraging, and social interactions; problem-solving skills are vital for adapting to environmental challenges, such as finding shelter or manipulating objects; and social cognition, which is evident in complex social structures and communication, underpins cooperative behaviors and group cohesion (Brown, 2015). These cognitive abilities vary across species, influenced by factors like habitat complexity, social structure, and evolutionary history. Morgan (2022) is of the opinion that studying cognitive abilities in aquatic animals provides insights into the evolutionary origins of intelligence and the diversity of cognitive solutions to ecological challenges in underwater environments. For the purpose of this study, we will be looking at aquatic animal groups such as marine mammals, fish species, crustaceans, cephalopods, and mollusks.

6.2.2 Marine Mammals

Marine mammals including dolphins, whales, seals, and sea lions exhibit a range of cognitive mechanisms tailored to their oceanic habitats and have evolved sophisticated cognitive abilities to navigate complex underwater environments, adapt to dynamic ecological challenges, and communicate with conspecifics (Reidenberg, 2007).

The communication systems of marine mammals such as dolphins and whales, which have been extensively researched, are multifaceted and comprise various vocalizations including clicks, whistles, and pulsed sounds that convey deep information about social hierarchy, prey presence, as well as navigation, such that their intricate communication significantly reflects a level of social acumen and cognitive processing (Bauer et al., 2020). Many marine mammals, such as dolphins and certain species of whales, utilize echolocation as a sensory mechanism, which involves emitting high-frequency sound pulses and interpreting the returning echoes to create a detailed mental map of their surroundings for navigating and locating prey in their aquatic habitats (Erbe et al., 2016). Echolocation requires precise coordination of sensory input and motor control, indicating advanced cognitive processing. Marine mammals also demonstrate remarkable problem-solving skills and

FIGURE 6.2 Cognitive skills for navigation – female pregnant gray whale migrating for delivery.

innovative behaviors. For instance, dolphins have been observed using tools, such as sponges or marine debris, to protect their rostrums while foraging on the seafloor. Such behaviors suggest flexible cognitive abilities and the capacity to adapt to novel challenges in their environment (Bauer et al., 2020) (Figure 6.2).

Certain marine mammals like dolphins display sophisticated social cognition through intricate social structures and behaviors including cooperative hunting and cultural transmission of learned behaviors within groups, revealing advanced social intelligence. These dynamics entail individual recognition, social bond maintenance, and group activity coordination, highlighting the animals' high cognitive sophistication (Frick et al., 2017). This group of aquatic animals exhibits remarkable memory and learning skills crucial for thriving in their vast oceanic environments and possesses the capacity to retain information over extended periods that enable them to navigate unique ocean landscapes and respond effectively to changing environmental conditions (Davis & Thomson, 2019). Their ability to recall intricate migration routes and master novel foraging strategies underscores their profound long-term memory and adaptive behavior. Such cognitive prowess not only ensures their survival but also highlights the intricate intelligence adapted to their marine lifestyles (Erbe et al., 2016).

In conclusion, comprehending the cognitive mechanisms of marine mammals underscores their adaptation to ocean life thereby revealing a diverse range of abilities influenced by their ecological niche and evolutionary past, offering insight into their intelligence and emphasizing the significance of cognitive adaptations for thriving in underwater ecosystems (Frick et al., 2017).

6.2.3 Fish Species

Fish species exhibit a wide array of cognitive mechanisms that defy conventional notions of their intelligence because they lack the neocortex which is typically linked with advanced cognition in mammals, yet their cognitive prowess reflects a sophisticated adaptation to evolutionary pressures and environmental needs (Brown et al., 2011) (Figure 6.3).

Many fish species demonstrate remarkable spatial cognition that is crucial for maneuvering and finding food in intricate underwater habitats, utilizing cognitive maps to remember and navigate spatial layouts, benefiting activities like defending territory, evading predators, and optimizing foraging techniques (Lucon-Xiccato & Bisazza, 2017). Fish exhibit impressive memory and learning capabilities that are vital for adjusting to environmental changes and establishing connections

FIGURE 6.3 Problem-solving abilities in fish – Stickleback male building nesting area.

between stimuli and consequences, as evidenced by their ability to recall spatial food locations, identify and evade predators, and engage in observational learning through conspecific observation and imitation, according to research findings (Brown, 2015). Despite the common perception of fish as solitary beings, numerous species exhibit intricate social interactions necessitating advanced social cognition, manifesting in the establishment of elaborate social hierarchies, individual conspecific recognition, and participation in cooperative activities like group hunting and parental nurturing (Pouca & Brown, 2017). Social cognition in fish therefore encompasses recognition, communication, and cooperation which significantly contribute to their survival and reproductive success. Furthermore, fish species display problem-solving abilities and inventive behaviors across different situations, showing cognitive flexibility and creativity by manipulating objects for food access, navigating obstacles, and adapting to environmental changes (Lucon-Xiccato et al., 2020).

While fish may not communicate vocally like mammals, they utilize a varied range of visual, olfactory, and tactile cues to convey information to peers, essential for mate selection, territorial defense, and group coordination, emphasizing the significance of cognitive processes in social behavior (Escobar-Camacho & Carleton, 2015). Overall, fish species exhibit a diverse range of cognitive mechanisms suited to their aquatic habitats, and comprehending these abilities illuminates both their intelligence and the intricacies of cognition across various taxa.

6.2.4 Crustaceans and Cephalopods

6.2.4.1 Crustaceans (e.g. Crabs, Shrimp, Lobsters)

Despite possessing neural structures that appear simpler compared to mammals, crustaceans showcase a diverse array of cognitive mechanisms finely tuned to their ecological niche, evidencing remarkable cognitive prowess (Ramm & Scholtz, 2017). Crustaceans like crabs and lobsters demonstrate problem-solving abilities in convoluted surroundings by recalling and adapting from previous encounters, showcasing cognitive flexibility, as seen in hermit crabs' capacity to evaluate and select shells based on their appropriateness and size (Birch et al., 2021). While not as sophisticated as in

FIGURE 6.4 Camouflage tactics in crustaceans – mud crab avoiding being preyed upon through camouflaging.

mammals, crustaceans communicate through various means, including chemical signals and visual cues. Social crustaceans, like certain species of shrimp, engage in signaling behaviors to establish dominance hierarchies and coordinate group activities (Ramm & Scholtz, 2017). Crustaceans possess memory skills vital for their survival, recalling food locations, shelter, and dangers, facilitating foraging and predator evasion; certain species like crayfish can navigate mazes, highlighting their spatial memory prowess (Birch et al., 2021) (Figure 6.4).

6.2.4.2 Cephalopods (e.g. Octopuses, Squid, Cuttlefish)

Cephalopods are renowned for their complex nervous systems and advanced cognitive abilities, making them some of the most intelligent invertebrates. Octopuses, known for their exceptional problem-solving skills and remarkable tool usage, demonstrate cognitive flexibility by manipulating objects, solving puzzles with their dexterous arms, and exhibiting behaviors like opening jars, unscrewing lids, and using coconut shells for shelter, especially when in captivity (Mather & Kuba, 2013). According to Hanlon and Messenger (2018), cephalopods employ a variety of communication methods, including visual signals, body patterns, and chromatophore displays; for instance, octopuses utilize color changes and postures to communicate aggression, mating readiness, and camouflage tactics, while cuttlefish utilize intricate color patterns in courtship rituals and territorial conflicts. The authors believe that cephalopods display exceptional learning capabilities, enabling them to adeptly adjust to diverse surroundings, tackle new challenges, and acquire knowledge through both observation and direct experience, with octopuses demonstrating the ability to recall escape routes and distinguish between individual humans, showcasing their capacity for long-term memory and cognitive recognition.

In conclusion, crustaceans and cephalopods possess unique cognitive mechanisms honed by millions of years of evolution, enabling them to thrive in diverse aquatic environments. Understanding these cognitive adaptations enhances our appreciation for the intelligence and complexity of these fascinating marine invertebrates.

6.2.5 Challenges and Opportunities in Studying Cognition Underwater

Studying animal cognition underwater poses distinctive challenges and prospects, with the aquatic realm boasting a diverse array of species behaviors, enticing cognitive researchers, yet navigating logistical, technical, and methodological obstacles (Pouca & Brown, 2017).

Reidenberg (2007) believes that the main hurdle lies in the constrained visibility and accessibility for direct observation underwater, thereby requiring specialized equipment as well as skilled divers, thus inflating the complexity and expenses of research initiatives, which is compounded by environmental factors like currents, temperature shifts, and turbidity, all capable of influencing experimental results and data analysis. The author further reveals that designing experiments for aquatic species must consider their sensory adaptations, including electroreception, echolocation, and hydrodynamic sensing, which differ from those of terrestrial animals, to ensure a meaningful understanding of how underwater animals perceive and interact with their environment.

Despite these challenges, studying animal cognition underwater also offers unique opportunities. The underwater realm offers a diverse environment for researching sensory ecology, social behavior, and cognitive abilities in a wide array of marine life, including fish, cephalopods, and marine mammals, with advancements in technology such as underwater cameras among others expanding studies in underwater ecosystems (Davis & Thomson, 2019). Furthermore, studying cognition in underwater animals offers insights into broader evolutionary inquiries concerning intelligence's origins and diversity, as researchers compare cognitive skills among aquatic and land-dwelling species to reveal the adaptive importance of cognition in diverse ecological settings (Morgan, 2022).

In summary, while studying animal cognition underwater poses numerous challenges, it also offers exciting opportunities to explore the cognitive abilities of aquatic species and deepen our understanding of the evolution of intelligence.

6.3 TERRESTRIAL ANIMALS

6.3.1 Cognitive Abilities across Terrestrial Animals

Terrestrial animals, comprising mammals, birds, reptiles, amphibians, insects, and arachnids, inhabit land and exhibit diverse cognitive capacities shaped by evolutionary adaptations, encompassing perception, learning, memory, problem-solving, and communication, influenced by factors such as brain size and social structure (Allen & Trestman, 2017). While certain terrestrial animals showcase impressive problem-solving abilities and intricate social behaviors, some depend heavily on instinct; mammals such as primates and cetaceans, along with birds, frequently display advanced cognitive skills, whereas certain invertebrates like insects also demonstrate unexpected intelligence (Bräuer et al., 2020). Examining the various cognitive capacities of land-dwelling animals provides valuable understanding into their behaviors and mental processes; hence, this study focuses on categorizing terrestrial animals based on their habitats, including wildlife, livestock, and companion animals.

6.3.2 Wildlife

Land-dwelling wild animals utilize a diverse range of cognitive processes evolved for adaptation, survival, and reproduction in terrestrial habitats, allowing them to effectively navigate landscapes, interact with peers and other species, and optimize resource utilization (Allen & Trestman, 2017). Cognition is therefore the core factor responsible for animal foraging, mating, and defending against predators in the wild (Morand-Ferron et al., 2016)

Perception, crucial for cognitive processing, enables animals to interpret environmental sensory input; land-dwelling wildlife exhibit sharp sensory skills tailored to detect vital cues like food, predators, or mates, varying among species due to adaptations to ecological niches and lifestyles

(Gaynor et al., 2019). Mendl & Paul (2020) emphasize how learning is essential for animals to adapt their behavior and make decisions, utilizing methods like classical conditioning, operant conditioning, and observational learning as a means of gathering information about their surroundings, refining survival tactics, identifying peers, and evading dangers. Also, wild animals often learn from social relationships with mates and not solely from their memory capacity, which has been displayed by vervet monkeys demonstrating associative learning (Robira et al., 2023) (Figure 6.5).

A crucial cognitive function that empowers wild animals to store and recall information is memory, and it includes spatial, episodic, and social aspects which facilitate behaviors like navigation and social interactions, encompassing abilities to remember food locations, past predator encounters, and familiar faces (Morand-Ferron et al., 2016). Wild western gorillas display spatial and heuristic cognitive abilities in foraging, as observed from their likelihood to revisit familiar sites with long-term foraging potential instead of persistently searching out new locations, which reflects a complex understanding of food resources available (Robira et al., 2023). According to Marks (2016), many terrestrial animals exhibit problem-solving abilities, especially when faced with ecological dilemmas that are intricate and employ creative skillfulness, like tool usage to navigate complex habitats for food acquisition and adept evasion of man-made obstacles, often relying on trial-and-error learning, spatial cognition, and adaptable decision-making.

Several studies have relayed how essential communication is for social species in the wild because it enables them to exchange information through various modes like vocalizations, visual displays, chemical signals, and tactile cues, which relay details about territory, social hierarchy, and reproductive status, ultimately engendering coordination within groups, aiding mate selection, and resolving conflicts (Snijders & Naguib, 2017). Overall, studying the cognitive mechanisms of terrestrial wildlife reveals the complex fusion of environmental demands and evolutionary influences that offer researchers deep insights into wild animals' perceptions of their natural habitats, memory functions, learning abilities, communication methods, and problem-solving skills.

FIGURE 6.5 Perception in wildlife – prey detection in wild tiger.

6.3.3 Livestock Animals

Livestock animals display cognitive processes involving several mental abilities that enable them to effectively navigate their surroundings, interact with conspecifics, and adapt to various environmental situations. Recent studies have highlighted the complex cognitive abilities of livestock animals, which are shaped by social dynamics, environmental factors, and evolutionary pressures, thereby challenging earlier conceptions about their cognitive limitations (Nawroth et al., 2019) (Figure 6.6).

The cognitive processes of livestock animals are dependent on knowledge acquisition, which is achieved through learning methods that are associative, observational, and operant (Mendl & Paul, 2020). These methods play a vital role in their survival such that, through associative learning, livestock animals collate information about their habitats and immediate surroundings, distinguishing food sources as well as potential dangers, while observational learning enables them to gain insights from social interactive behaviors among their mates. Lastly, operant learning helps them modify their behaviors by fortifying beneficial actions and avoiding unfavorable ones using lessons from past encounters, thereby promoting the display of novel behaviors (Thorndike, 2000). As seen from the studies of Krupenye and Call (2019), memory plays a significant role in livestock cognitive processes due to their possession of both short-term and long-term memory capacities, enabling them to recall past occurrences and modify their behavior accordingly, which is essential for spatial navigation tasks, remembering feeding locations, as well as recognizing conspecifics. Herrero et al. (2015) emphasize how livestock animals make use of memory to retain details about their environments and past activities, thereby enhancing their ability to make informed decisions and adapt to evolving circumstances. Overall, memory is crucial in the cognitive processes of livestock animals and enables them to thrive in various environments, effectively fulfilling their behavioral needs.

Social cognition is another crucial aspect of livestock cognition, particularly in species with complex social structures like cattle, sheep, and pigs. According to Nawroth et al. (2016), livestock animals with complex social structures display varied social behaviors, which include communication through vocalizations and olfactory cues, cooperation, dominance hierarchies, as well as body language. Therefore, comprehending the social dynamics within these livestock animal groups allows them to coordinate group activities, form alliances, and navigate social hierarchies. The authors further highlight the presence of problem-solving abilities within this group of animals, although varying to certain degrees across species. From their findings, animals such as goats and pigs have exhibited problem-solving abilities in experimental setups that are impressive, involving spatial reasoning and tool use tasks. These cognitive processes highlight their adaptive capacities to novel challenges, resolving environmental problems through innovative means.

FIGURE 6.6 Spatial cognition in livestock – pig digging up hidden food from mates.

Understanding the cognitive functions of livestock animals, including learning, memory, social cognition, and problem-solving, enhances our recognition of their intelligence and informs welfare practices and management strategies to promote their well-being (Mendl & Paul, 2020).

6.3.4 Companion Animals

Companion animals, including dogs, cats, and rabbits, possess cognitive mechanisms majorly shaped by domestication, evolution, as well as individual experiences, which encompass varied processes that allow them to perceive, comprehend, and adapt in their environment, thereby reflecting a combination of inborn abilities and learned behaviors (Vitztum & Urbanik, 2016). With regards to animal intelligence, Thorndike (2000) explains that these animals display a wide range of learning and memory capabilities, where operant and classical conditioning are basic cognitive processes. Through repeated associations and rewards, dogs, for example, can be taught to carry out tasks effectively or follow instructions. Cats and rabbits, similar to dogs, have been observed to modify their behaviors by utilizing past experiences to avoid negative outcomes and maximize positive ones, demonstrating learning through mechanisms such as associative learning and operant conditioning, where they adapt their actions based on previous encounters with rewards or punishments, thereby showcasing a capacity for behavioral flexibility and cognitive sophistication that allows them to navigate their environments more effectively and efficiently (Thomas, 2022) (Figure 6.7).

According to Jardat & Lansade (2022), companion animals are very much reliant on their sensory systems for daily navigation and task completion, as evidenced in dogs, which use their acute sense of smell in detecting and interpreting varied scents vital for survival and communication, while the visual excellence, particularly in low light, for cats aids in spatial navigation and hunting. Problem-solving skills in companion animals are apparent in their navigation of complex environments and acquisition of rewards, exemplified by dogs solving puzzles to earn treats, thereby indicating an understanding of cause and effect, as seen in their ability to manipulate objects, remember past interactions, and apply learned strategies to new situations, which collectively demonstrate their cognitive

FIGURE 6.7 Problem-solving in dogs – dog herding flock of sheep.

flexibility and adaptive behaviors (Bräuer et al., 2020). Although this is not commonly observed in cats, some have been identified to effectively manipulate objects or use tools in achieving an expected outcome, depicting a level of flexibility in their cognitive processes (Thorndike, 2000). Companion animals also exhibit advanced social cognition, as observed with dogs excellently interpreting human gesticulations as well as emotions due to favored domesticated traits that enhance communication, while irrespective of the independence of cats, they also form strong bonds with humans and other animals, displaying comprehension of social cues (Vitztum & Urbanik, 2016).

Conclusively, cognitive mechanisms in terrestrial companion animals reflect a complex interplay of sensory processing, learning, problem-solving, and social interaction, enabling them to thrive in human environments and form meaningful relationships with their human companions.

6.3.5 Challenges and Opportunities in Studying Cognition on Land

Studying cognition in land animals presents a range of challenges and opportunities that are essential to understanding their behaviors and mental processes. According to Pritchard et al. (2016), the diversity of terrestrial animals causes the formulation of standardized testing methods of cognition to be complicated due to the unique behavioral and sensory traits each species possesses, which necessitates customized approaches for accurate cognitive assessment. Kays et al. (2015) believe that interpreting these animal behaviors is demanding due to the absence of a shared language, thereby necessitating researchers to infer cognitive processes from observable actions, which can lead to false interpretations, further influenced by factors like habitat, social dynamics, and human interaction. Furthermore, ethical guidelines that must be observed to prevent harm or unnecessary stress to the animals are prerequisites for ensuring their well-being during such cognitive research, and can complicate the experimental design, limiting the scope of certain studies (Mendl & Paul, 2004).

Andrews (2020), on the other hand, is of the opinion that studying a variety of species offers insights into the adaptive significance of different cognitive traits and the evolution of cognitive abilities, causing this comparative approach to reveal fundamental principles of cognition that can be applicable across species. Additionally, the comprehension of cognition in land animals can improve animal welfare and promote conservation efforts through well-informed and improved management practices, as well as enrichment strategies that take care of their cognitive needs (Shettleworth, 2009). Advances in technology, such as automated behavior tracking and neuroimaging, enable non-invasive studies of animal cognition that are more precise, while interdisciplinary collaborations among biology, computer science, and psychology engender more innovative research approaches that are comprehensive (Bräuer et al., 2020). Furthermore, Pritchard et al. (2016) emphasizes a higher possibility of noting more precise cognitive behaviors of animals when researchers study them in their natural habitats, as well as access to a wider range of animal species for natural behaviors, and opportunities for specifically investigating certain species of interest.

In summary, while the study of cognition in land animals is fraught with challenges like species diversity, ethical concerns, communication barriers, and environmental variables, it also offers significant opportunities for scientific discovery and practical applications in conservation and animal welfare.

6.4 AERIAL ANIMALS

6.4.1 Overview of Cognitive Abilities in Animals' Aerial Environments

Animals in aerial environments like birds display extraordinary cognitive abilities, with species like parrots and crows demonstrating tool use and problem-solving skills that can be compared to primates, such as New Caledonian crows creating and using tools to access food, reflecting advanced reasoning and planning (Hunt & Gray, 2004). The studies of Hagstrum (2013) reveals how pigeons have been shown to possess navigational skills that are sophisticated through the use of visual landmarks, Earth's magnetic field, as well as olfactory cues for migration across vast distances. Additionally, African gray parrots exhibit mimicry and understanding of human language, which is not only exceptional but indicates high levels of memory and social cognition (Pepperberg, 2000). Supported by brain structures that are well developed, like the avian pallium, which is approximately similar to the neocortex of mammals, these cognitive capabilities permit intricate learning processes and behaviors (Güntürkün, 2005). Therefore, aerial animals' cognitive abilities reflect adaptations that are significantly evolutionary to their environments.

6.4.2 Cognitive Analysis of Aerial Animals That Inhabit the Skies

6.4.2.1 Birds

Birds have consistently been the focus of cognitive research due to their impressive problem-solving skills, sophisticated social cognition, and advanced spatial memory, as these abilities not only underscore their intricate neural mechanisms – comparable to those observed in mammals, including primates – but also play a crucial role in their survival, thereby highlighting the evolutionary significance and adaptability of their cognitive processes (Taylor et al., 2012). Birds exhibit exceptional problem-solving abilities, such as seen in New Caledonian crows (Corvus moneduloides), where these remarkable birds demonstrate advanced cognitive skills by crafting tools from sticks and leaves, which they use adeptly to extract insects from the bark of trees, a behavior meticulously documented by Hunt and Gray (2004). In recent studies by Taylor et al. (2012), it was observed that crows can solve compound tasks indicating a high level of cognitive planning, such as the use of a series of tools to achieve a goal using the tools in a specific order, which demonstrates their advanced problem-solving abilities and highlights the sophisticated nature of their cognitive processes (Figure 6.8).

Research on avian cognitive abilities, particularly spatial memory, elucidates the intricate nature of their cognition, showcasing the remarkable complexity inherent in species that engage in food-caching behaviors, as these birds demonstrate highly developed spatial memory capabilities and possess a larger hippocampus in comparison to non-caching avian counterparts, thereby emphasizing the evolutionary significance of cognitive adaptations in facilitating survival strategies such as cache retrieval and navigation within complex environments (Biegler et al., 2001). For instance, Clark's nutcrackers (Nucifraga columbiana) can recall the locations of thousands of food caches for months, a remarkable memory capacity facilitated by a highly developed hippocampus, supporting the notion that environmental demands can drive cognitive specialization (Lucas et al., 2004; Balda & Kamil, 1998).

FIGURE 6.8 Problem-solving skills – crows extracting insects from the bark of trees using leaves.

Hobson and DeDeo (2015) emphasize the advanced social cognition evident in birds, elucidating their capacity to comprehend and respond to complex group dynamics, alongside showcasing sophisticated social learning and communication skills, as demonstrated by pigeons (Columba livia) adeptly learning to associate cues with social hierarchy, thus employing this understanding as a mechanism for behavior modification. Furthermore, ravens (Corvus corax) demonstrate advanced social cognition by strategically deceiving their peers to secure food benefits, reflecting their ability to manipulate social interactions for personal gain (Bugnyar & Heinrich, 2005), which underscores their remarkable understanding of social dynamics and highlights the complexity of their cognitive processes, as they not only comprehend the intentions and perspectives of others but also possess the capacity to employ deceptive tactics in a manner suggestive of a sophisticated understanding of reciprocity and social hierarchies within their group dynamics, thus showcasing an intricate level of cognitive flexibility and awareness that extends beyond mere instinctual behavior. Intelligence therefore can develop from diverse neural architectures as indicated from the complex cognitive abilities observed in birds revealing that despite avian brains differing from mammalian brains, it is capable of supporting such functions through evolutionary convergence (Emery & Clayton, 2004). This convergence suggests that similar environmental pressures can lead to the evolution of similar cognitive abilities across diverse species.

6.4.2.2 Bats

Bats are a varied and widespread group of mammals that possess advanced cognitive skills in problem-solving, spatial memory, and social cognition, which are essential for locating food, navigating complex environments, and social interaction. For example, Thiele and Winter (2005) discovered that Phyllostomid bats utilize echolocation for prey detection as well as to differentiate between objects and obstacles, linking specific acoustic cues with food rewards and solving complex tasks thereby demonstrating significant cognitive adaptability. Furthermore, the authors reveal that bats demonstrate exceptional spatial memory and also utilize echolocation and visual landmarks which is essential for navigating their extensive home ranges and roosting sites to recall feeding locations and roost efficiently over long distances and time intervals. Additionally, Pollock et al. (2016) firmly point out that the hippocampus plays a significant role in the spatial memory of bats, and similar to other mammals it is important in their navigation and memory retention, underscoring the conserved nature of hippocampal function across diverse taxa and highlighting the importance of spatial cognition in the survival and ecological niche exploitation of these nocturnal flying mammals (Figure 6.9).

FIGURE 6.9 Echolocation – prey detection in brown bats.

Thiele and Winter (2005) are opined that bats, in addition to their remarkable echolocation abilities, are widely recognized for their intricate social structures and advanced social cognition, which they convey through a rich repertoire of vocalizations, encompassing various facets such as social identification, emotional states, and the communication of vital information regarding food sources and other pertinent aspects of their environment. Studies by Carter and Wilkinson (2013) reveal that vampire bats and Mexican free-tailed bats have unveiled their ability to identify individuals, organize group activities, and participate in reciprocal altruism, highlighting their intricate social abilities and capacity for cooperative behaviors modified by past interactions. Overall, bats exhibit a range of advanced cognitive abilities that support their ecological success. These abilities in problem-solving, spatial memory, and social cognition highlight the complex and adaptive nature of their behavior, driven by both environmental demands and social interactions (Figure 6.10).

6.4.2.3 Insects

Despite their diminutive stature and apparently rudimentary nervous systems, insects showcase an impressive spectrum of cognitive skills, spanning from problem-solving to spatial navigation, all vital for their survival and reproduction. Delving into the complexities of insect cognition not only illuminates the evolutionary trajectory of intelligence but also offers valuable perspectives for crafting streamlined algorithms for robotics and artificial intelligence (Menzel, 2014).

Research demonstrates that insects, such as honeybees, have advanced associative learning skills, as evidenced by their ability to link specific odors with rewards, exemplifying their aptitude in olfactory learning. Additionally, insects possess both short-term and long-term memory, essential for activities like foraging and identifying mates, as highlighted by Dukas (2008). Insects demonstrate exceptional spatial cognition, captivating researchers with their navigational prowess, particularly evident in ants that utilize visual, celestial, and path integration cues to adeptly maneuver through intricate landscapes and return to their nests with pinpoint accuracy, as evidenced by studies on desert ants, which have unveiled their reliance on panoramic skyline views for navigation (Giurfa, 2015).

Social insects like ants and bees demonstrate sophisticated cognitive skills in social organization and cooperation, utilizing intricate communication systems such as pheromones and behaviors like waggle dancing to coordinate activities such as foraging and nest construction, and recent studies emphasize individual variation in insect cognition, suggesting that within a colony, certain bees may specialize in tasks according to their cognitive abilities, thus improving the efficiency of

FIGURE 6.10 Social cognition in insects – a worker bee leading a group of other workers to find nectar.

colony-level activities such as foraging and brood care (Menzel, 2014). In summary, insects possess a diverse range of cognitive abilities, including learning and memory, spatial cognition, and social organization. Understanding these abilities not only provides insights into the evolution of intelligence but also has practical applications in fields such as robotics and agriculture.

6.4.5 Cognitive Adaptations for Flight and Navigation

6.4.5.1 Insects

Over the last two decades to the present, scientific research has unveiled insights into the mechanisms underlying insect flight and navigation, shedding light on the sophisticated cognitive abilities of these small creatures and revealing how insects have evolved remarkable cognitive adaptations for flight and navigation, enabling them to navigate complex environments with precision and efficiency, encompassing various sensory modalities and behavioral strategies that reflect the diverse ecological niches they inhabit (Perry et al., 2017).

One key aspect of insect flight is their reliance on visual cues for navigation, as research by Menzel (2014) demonstrated that honeybees utilize optic flow patterns to estimate their ground speed and maintain stable flight, and this reliance on visual feedback allows insects to adjust their flight trajectory in response to environmental changes; moreover, studies by Giurfa (2015) revealed the role of polarized light cues in insect navigation, particularly in tasks such as orientation and homing. Beyond visual cues, insects also utilize olfactory and windborne cues for navigation exemplified by the studies of Perry et al. (2017) on moths which exhibit an extraordinary ability to track pheromone plumes released by potential mates over long distances. According to the authors, male moths possess the ability to detect and follow female sex pheromones from several kilometers away even with the faintest of traces, due to this ability being facilitated by their antennae which are highly sensitive and equipped with olfactory receptors that are numerous and uniquely designed for detecting specific pheromones. They further reveal the flight of the moth towards the pheromone source involves a series of movements that are zigzag usually identified as casting and allow them to adjust their direction based on modifications in the concentration of the chemical signal as well as to stay within the plume. This olfactory guidance system enables moths to locate suitable mating partners even in vast and complex environments.

Furthermore, recent studies by Thorndike (2000) have highlighted the role of cognitive maps in insect navigation, showing that by integrating visual landmarks with internal representations of space, insects such as ants can navigate back to their nest with remarkable accuracy, and have elucidated how the neural circuits of fruit flies are responsible for flight control and decision-making. This cognitive mapping ability allows insects to flexibly navigate their surroundings and adapt to changes in their environment, demonstrating how both sensory adaptations and neural mechanisms influence insect flight and navigation. Therefore, understanding these neural mechanisms provides insights into the computational strategies employed by insects to process sensory information and execute precise flight maneuvers.

In conclusion, the cognitive adaptations for flight and navigation in insects represent a fascinating area of study that has seen significant advances in recent years. By integrating sensory cues, neural mechanisms, and behavioral strategies, insects have evolved sophisticated mechanisms for navigating their environments with precision and efficiency. For instance, the assessment of birds to determine if they know their location (Taylor et al., 2012), the ability of pigeons to locate their home using information from their environment, pigeon-dependent discoveries such as the role of the sun compass, the utilization of magnetic and olfactory data, and the use of differing mechanisms inside and outside familiar areas would not have been possible if studies on pigeons were solely done in the laboratory (Hobson & DeDeo, 2015).

6.4.5.2 Birds

Birds exhibit remarkable cognitive adaptations for flight and navigation, which are supported by specialized neural structures and cognitive mechanisms and are essential for their survival, enabling efficient foraging, migration, and predator avoidance exemplified by a significant adaptation which is the highly developed avian hippocampus, a brain region crucial for spatial memory and navigation (Herold et al., 2015). The hippocampus in birds is proportionately larger than in mammals, reflecting its importance in spatial cognition. This enlargement is particularly evident in species that engage in long-distance migration or complex foraging behaviors, such as homing pigeons (Columba livia) and Clark's nutcrackers (Nucifraga columbiana), which store food in numerous locations (Smulders, 2018).

Another critical adaptation is the use of Earth's magnetic field for navigation. Birds possess magnetoreceptors, likely located in the retina, that detect geomagnetic information, enabling them to navigate accurately over vast distances (Mouritsen, 2018). This magnetic sense is complemented by other navigational cues, such as the position of the sun and stars, olfactory landmarks, and visual landmarks, forming a multimodal navigational system (Holland, 2014). Cognitive flexibility and problem-solving skills are also highly developed in birds, particularly in corvids and parrots. These species exhibit advanced tool use, problem-solving, and even elements of planning for future needs, which indicate a high level of cognitive sophistication (Pepperberg, 2017). Such cognitive abilities are underpinned by an expanded pallium, the avian equivalent of the mammalian cerebral cortex, which supports complex behaviors (Jarvis et al., 2013).

The interplay between neural specializations and cognitive abilities in birds highlights their evolutionary adaptations for flight and navigation. These adaptations are not merely physical but are deeply rooted in their neuroanatomy and cognitive processes, showcasing a remarkable convergence of form and function. The ongoing study of avian cognition continues to provide insights into the neural underpinnings of navigation and spatial memory, illustrating the sophisticated nature of bird brains in relation to their ecological demands (Güntürkün & Bugnyar, 2016).

6.5 COMPARATIVE ANALYSIS OF COGNITIVE MECHANISMS IN ANIMALS

Comparative analysis of cognitive mechanisms across animals is a multidisciplinary field within cognitive science that seeks to understand the similarities and differences in cognitive processes across different species. This approach allows researchers to uncover fundamental principles underlying cognition and behavior, elucidating the evolutionary origins of cognitive abilities. Through careful observation, experimentation, and theoretical modeling, scientists aim to unravel the complex interplay between genetic, neural, and environmental factors that shape cognition.

One key aspect of comparative analysis is investigating cognitive abilities such as perception, memory, problem-solving, and social cognition across a wide range of species, from invertebrates like bees and octopuses to mammals like rats, primates, and humans. For instance, studies on animal cognition have revealed remarkable problem-solving skills in species like the New Caledonian crow (Corvus moneduloides), which uses tools to extract food from hard-to-reach places (Hunt & Gray, 2004). Similarly, studies examining spatial memory in rodents have offered valuable insights into the brain mechanisms that support navigation and the formation of memories, expanding our understanding of how these cognitive processes are neurologically encoded (Vorhees & Williams, 2014).

Comparative analyses also shed light on the evolutionary trajectories of cognitive abilities; for example, investigations into social cognition have highlighted the role of complex social structures in shaping cognitive skills, as studies on primates, such as chimpanzees and bonobos, have revealed sophisticated social behaviors, including cooperation, empathy, and perspective-taking (Silk, 2007). These findings suggest that social complexity may have driven the evolution of advanced cognitive abilities in certain species.

Furthermore, advances in neuroimaging techniques have enabled researchers to explore the neural underpinnings of cognitive processes across species. Comparative neuroanatomy studies have identified conserved brain regions involved in functions like decision-making, attention, and emotion regulation (Thorndike, 2000). For instance, the prefrontal cortex, a brain region associated with executive functions in humans, exhibits similar functional properties in non-human primates (Silk, 2007).

In summary, comparative analysis of cognitive mechanisms across animals offers valuable insights into the diversity and universality of cognitive processes. By elucidating the evolutionary origins and neural substrates of cognition, this interdisciplinary approach deepens our understanding of the nature of mind and intelligence in both humans and other animals.

6.6 CONCLUSION

In the past two decades, the study of animal cognition has experienced remarkable growth, yet several gaps persist in our understanding. Despite significant advancements, there remain areas where research is nascent or underexplored, while others present opportunities for future breakthroughs. This conclusion aims to synthesize the existing literature, highlight gaps, and recommend directions for future investigation.

One notable area of progress lies in our understanding of cognitive abilities across various species. Comparative studies have illuminated the cognitive capacities of non-human animals, revealing complex problem-solving skills, social learning mechanisms, and even rudimentary forms of self-awareness. For instance, the mirror self-recognition test conducted on elephants by Plotnik et al. (2006) demonstrated a level of self-awareness previously thought exclusive to humans and great apes. However, despite these advancements, several gaps persist. One such gap pertains to the study of cognition in lesser-studied taxa. While much research has focused on mammals and birds, other taxa such as reptiles, amphibians, and invertebrates remain relatively understudied. Investigating cognition across a broader taxonomic range can provide insights into the evolution of cognitive abilities and enhance our understanding of the diversity of cognitive mechanisms. Furthermore, there is a need for interdisciplinary approaches to studying animal cognition. Incorporating methodologies from neuroscience, psychology, ethology, and computational modeling can provide a more comprehensive understanding of cognitive processes. For instance, recent advances in neuroimaging techniques have allowed researchers to explore the neural correlates of cognitive functions in non-human animals, shedding light on the underlying mechanisms.

Another area that warrants attention is the role of ecological and environmental factors in shaping cognitive abilities. While laboratory studies provide valuable insights, they often overlook the complexities of natural environments. Investigating how cognitive abilities evolve in response to ecological challenges and social dynamics can provide a more nuanced understanding of animal cognition. Looking ahead, there are several promising avenues for future research. One such avenue involves the application of advanced quantitative methods, such as network analysis and machine learning, to analyze complex cognitive datasets. These approaches can uncover hidden patterns and relationships within data, facilitating a deeper understanding of cognitive processes. Additionally, there is a growing interest in studying the impact of anthropogenic activities on animal cognition. Human-induced environmental changes, such as habitat destruction and climate change, pose significant challenges to animal populations. Understanding how these changes affect cognitive abilities can inform conservation efforts and aid in the development of strategies to mitigate their impact.

In conclusion, while significant progress has been made in understanding animal cognition, there are still many unanswered questions and untapped opportunities. By addressing existing gaps and embracing interdisciplinary approaches, researchers can continue to unravel the mysteries of the animal mind, ultimately leading to a more profound appreciation of the cognitive diversity of life on Earth.

REFERENCES

Allen, C., & Trestman, M. (2017). Animal consciousness. In *The Blackwell companion to consciousness* (pp. 63–76). Wiley-Blackwell.

Andrews, K. (2020). *The animal mind: an introduction to the philosophy of animal cognition*. Routledge.

Balda, R. P., & Kamil, A. C. (1998). The ecology and evolution of spatial memory in corvids of the southwestern USA: The perplexing difficulty of identifying adaptation (pp. 29–64). Academic Press.

Bauer, G. B., Cook, P. F., & Harley, H. E. (2020). The relevance of ecological transitions to intelligence in marine mammals. *Frontiers in Psychology, 11*, 557228.

Berridge, K., & Winkielman, P. (2003). What is an unconscious emotion? (The case for unconscious" liking"). *Cognition & Emotion, 17*(2), 181–211.

Biegler, R., McGregor, A., Krebs, J. R., & Healy, S. D. (2001). A larger hippocampus is associated with longer-lasting spatial memory. *Proceedings of the National Academy of Sciences, 98*(12), 6941–6944.

Birch, J., Burn, C., Schnell, A., Browning, H., & Crump, A. (2021). Review of the evidence of sentience in cephalopod molluscs and decapod crustaceans. LSE Consulting. LSE Enterprise Ltd. The London School of Economics and Political Science. Available at: https://www.lse.ac.uk/News/News-Assets/PDFs/2021/Sentience-in-Cephalopod-Molluscs-and-Decapod-Crustaceans-Final-Report-November-2021.pdf.

Bräuer, J., Hanus, D., Pika, S., Gray, R., & Uomini, N. (2020). Old and new approaches to animal cognition: there is not "one cognition." *Journal of Intelligence, 8*(3), 28.

Brown, C. (2015). Fish intelligence, sentience and ethics. *Animal Cognition, 18*(1), 1–17.

Brown, C., Laland, K., & Krause, J. (2011). *Fish cognition and behavior*. John Wiley & Sons.

Bugnyar, T., & Heinrich, B. (2005). Ravens, Corvus corax, differentiate between knowledgeable and ignorant competitors. *Proceedings of Biological Sciences, 272*(1573), 1641–1646.

Burgdorf, J., & Panksepp, J. (2006). The neurobiology of positive emotions. *Neuroscience & Biobehavioral Reviews, 30*(2), 173–187.

Bush, G., Luu, P., & Posner, M. I. (2000). Cognitive and emotional influences in anterior cingulate cortex. *Trends in Cognitive Sciences, 4*(6), 215–222

Carter, G. G., & Wilkinson, G. S. (2013). Cooperation and conflict in the social lives of bats. In *Bat evolution, ecology, and conservation* (pp. 225–242). Springer.

Chieffi, S., Messina, G., Villano, I., Messina, A., Valenzano, A., Moscatelli, F., ... & Monda, M. (2017). Neuroprotective effects of physical activity: evidence from human and animal studies. *Frontiers in Neurology, 8*, 188.

Couzin, I. D. (2009). Collective cognition in animal groups. *Trends in Cognitive Sciences, 13*(1), 36–43.

Crosby, M. (2020). Building thinking machines by solving animal cognition tasks. *Minds and Machines, 30*(4), 589–615.

Davis, R. W., & Thomson, D. H. (2019). Marine mammals. In *Sensory systems, adaptations for an aquatic life*. (pp. 177–217). Springer.

Dukas, R. (2008). Evolutionary biology of insect learning. *Annual Review of Entomology, 53*, 145–160.

Emery, N. J., & Clayton, N. S. (2004). The mentality of crows: Convergent evolution of intelligence in corvids and apes. *Science, 306*(5703), 1903–1907.

Erbe, C., Reichmuth, C., Cunningham, K., Lucke, K., & Dooling, R. (2016). Communication masking in marine mammals: A review and research strategy. *Marine Pollution Bulletin, 103*(1–2), 15–38.

Escobar-Camacho, D., & Carleton, K. L. (2015). Sensory modalities in cichlid fish behavior. *Current Opinion in Behavioral Sciences, 6*, 115–124.

Ferguson, G. A. (1954). On learning and human ability. *Canadian Journal of Psychology/Revue canadienne de psychologie, 8*(2), 95.

Flavell, J. H. (1979). Metacognition and cognitive monitoring: A new area of cognitive–developmental inquiry. *American Psychologist, 34*(10), 906

Frick, E. E., de Vere, A. J., & Kuczaj, S. A. (2017). What do we want to know about personality in marine mammals? In *Personality in nonhuman animals* (pp. 237–253). Springer International Publishing/Springer Nature.

Gaynor, K. M., Brown, J. S., Middleton, A. D., Power, M. E., & Brashares, J. S. (2019). Landscapes of fear: spatial patterns of risk perception and response. *Trends in Ecology & Evolution, 34*(4), 355–368.

Giurfa, M. (2015). Learning and cognition in insects. *Wiley Interdisciplinary Reviews: Cognitive Science, 6*(4), 383–395.

Güntürkün, O. (2005). The avian 'prefrontal cortex' and cognition. *Current Opinion in Neurobiology, 15*(6), 686–693.

Güntürkün, O., & Bugnyar, T. (2016). Cognition without cortex. *Trends in Cognitive Sciences, 20*(4), 291–303.

Hagstrum, J. T. (2013). Atmospheric propagation modeling indicates homing pigeons use loft-specific infrasonic 'map' cues. *The Journal of Experimental Biology, 216*(4), 687–699.

Hanlon, R. T., & Messenger, J. B. (2018). *Cephalopod behaviour.* Cambridge University Press.

Hemsworth, P. H., Mellor, D. J., Cronin, G. M., & Tilbrook, A. J. (2015). Scientific assessment of animal welfare. *New Zealand Veterinary Journal, 63*(1), 24–30.

Herold, C., Coppola, V. J., & Bingman, V. P. (2015). The evolution of the hippocampus in reptiles and birds. *Journal of Comparative Neurology, 524*(3), 496–517.

Herrero, M., Wirsenius, S., Henderson, B., Rigolot, C., Thornton, P., Havlík, P., ... & Gerber, P. J. (2015). Livestock and the environment: what have we learned in the past decade? *Annual Review of Environment and Resources, 40*, 177–202.

Hobson, E. A., & DeDeo, S. (2015). Social feedback and the emergence of rank in animal society. *PLoS Computational Biology, 11*(9), e1004411.

Holland, R. A. (2014). True navigation in birds: from quantum physics to global migration. *Journal of Zoology, 293*(1), 1–15.

Hunt, G. R., & Gray, R. D. (2004). The crafting of hook tools by wild New Caledonian crows. *Proceedings of the Royal Society B: Biological Sciences, 271*(Suppl_3), S88–S90.

Jardat, P., & Lansade, L. (2022). Cognition and the human–animal relationship: a review of the sociocognitive skills of domestic mammals toward humans. *Animal Cognition, 25*(2), 369–384.

Jarvis, E. D., et al. (2013). Whole-brain maps of the highly modular bird song system: a link between perception and action. *Proceedings of the National Academy of Sciences, 110*(28), 11570–11575.

Kays, R., Crofoot, M. C., Jetz, W., & Wikelski, M. (2015). Terrestrial animal tracking as an eye on life and planet. *Science, 348*(6240), aaa2478.

Krupenye, C., & Call, J. (2019). Theory of mind in animals: Current and future directions. *Wiley Interdisciplinary Reviews: Cognitive Science, 10*(6), e1503.

Lucas, J. R., Brodin, A., de Kort, S. R., & Clayton, N. S. (2004). Does hippocampal size correlate with the degree of caching specialization? A comparative study of the passerine birds. *Proceedings of the Royal Society of London. Series B*, 271, 2423–2429.

Lucon-Xiccato, T., & Bisazza, A. (2017). Individual differences in cognition among teleost fishes. *Behavioural Processes, 141*, 184–195.

Lucon-Xiccato, T., Gatto, E., & Bisazza, A. (2020). Male and female guppies differ in problem-solving abilities. *Current Zoology, 66*(1), 83–90.

Marks, J. (2016). A tale of ex-apes. The evolution of wisdom, 1st Edition, University of California Press Oakland, California, pp.152.

Mather, J. A., & Kuba, M. J. (2013). The cephalopod specialties: complex nervous system, learning, and cognition. *Canadian Journal of Zoology, 91*(6), 431–449.

Mendl, M., Burman, O. H., Parker, R. M., & Paul, E. S. (2009). Cognitive bias as an indicator of animal emotion and welfare: emerging evidence and underlying mechanisms. *Applied Animal Behaviour Science, 118*(3–4), 161–181.

Mendl, M., & Paul, E. S. (2004). Consciousness, emotion and animal welfare: insights from cognitive science. *Animal Welfare, 13*(S1), S17–S25.

Mendl, M., & Paul, E. S. (2020). Animal affect and decision-making. *Neuroscience & Biobehavioral Reviews, 112*, 144–163.

Menzel, R. (2014). Learning, memory, and "cognition" in honey bees. In *Neurobiology of comparative cognition* (pp. 255–310). Psychology Press.

Morand-Ferron, J., Cole, E. F., & Quinn, J. L. (2016). Studying the evolutionary ecology of cognition in the wild: a review of practical and conceptual challenges. *Biological Reviews, 91*(2), 367–389.

Morgan, C. L. (2022). *Animal life and intelligence.* DigiCat.

Mouritsen, H. (2018). Long-distance navigation and magnetoreception in migratory animals. *Nature, 558*(7708), 50–59.

Nathan, R., Monk, C. T., Arlinghaus, R., Adam, T., Alós, J., Assaf, M., ... & Jarić, I. (2022). Big-data approaches lead to an increased understanding of the ecology of animal movement. *Science, 375*(6582), eabg1780.

Nawroth, C., Baciadonna, L., & McElligott, A. G. (2016). Goats learn socially from humans in a spatial problem-solving task. *Animal Behaviour, 121*, 123–129.

Nawroth, C., Langbein, J., Coulon, M., Gabor, V., Oesterwind, S., Benz-Schwarzburg, J., & Von Borell, E. (2019). Farm animal cognition—linking behavior, welfare and ethics. *Frontiers in Veterinary Science, 6*, 24.

Panksepp, J. (2011). Cross-species affective neuroscience decoding of the primal affective experiences of humans and related animals. *PLOS One, 6*(9), e21236

Pepperberg, I. M. (2000). *The Alex studies: cognitive and communicative abilities of grey parrots.* Harvard University Press.

Pepperberg, I. M. (2017). *Alex & me: How a scientist and a parrot uncovered a hidden world of animal intelligence—and formed a deep bond in the process.* Harper Perennial.

Perry, C. J., Barron, A. B., & Chittka, L. (2017). The frontiers of insect cognition. *Current Opinion in Behavioral Sciences, 16*, 111–118.

Plotnik, J. M., De Waal, F. B., & Reiss, D. (2006). Self-recognition in an Asian elephant. *Proceedings of the National Academy of Sciences, 103*(45), 17053–17057.

Pollock, T., Moreno, C. R., Sánchez, L., Ceballos-Vasquez, A., Faure, P. A., & Mora, E. C. (2016). Wound healing in the flight membranes of wild big brown bats. *The Journal of Wildlife Management, 80*(1), 19–26.

Pouca, C. V., & Brown, C. (2017). Contemporary topics in fish cognition and behaviour. *Current Opinion in Behavioral Sciences, 16*, 46–52.

Pritchard, D. J., Hurly, T. A., Tello-Ramos, M. C., & Healy, S. D. (2016). Why study cognition in the wild (and how to test it)? *Journal of the Experimental Analysis of Behavior, 105*(1), 41–55.

Ramm, T., & Scholtz, G. (2017). No sight, no smell?–Brain anatomy of two amphipod crustaceans with different lifestyles. *Arthropod Structure & Development, 46*(4), 537–551.

Reidenberg, J. S. (2007). Anatomical adaptations of aquatic mammals. *The Anatomical Record: Advances in Integrative Anatomy and Evolutionary Biology, 290*(6), 507–513.

Robira, B., Benhamou, S., Obeki Bayanga, E., Breuer, T., & Masi, S. (2023). How do primates decide where to feed? Insights from wild western gorillas. *Animal Behaviour, 204*, 25–36.

Rowell, M. K., Pillay, N., & Rymer, T. L. (2021). Problem solving in animals: proposal for an ontogenetic perspective. *Animals, 11*(3), 866.

Shettleworth, S. J. (2009). *Cognition, evolution, and behavior.* Oxford University Press.

Silk, J. B. (2007). The adaptive value of sociality in mammalian groups. *Philosophical Transactions of the Royal Society B: Biological Sciences, 362*(1480), 539–559.

Smulders, T. V. (2018). The avian hippocampal formation and the stress response. *Brain, Behavior and Evolution, 92*(1–2), 10–20.

Snijders, L., & Naguib, M. (2017). Communication in animal social networks: a missing link?. In *Advances in the study of behavior* (Vol. 49, pp. 297–359). Academic Press.

Tang, Y. P., Wang, H., Feng, R., Kyin, M., & Tsien, J. Z. (2001). Differential effects of enrichment on learning and memory function in NR2B transgenic mice. *Neuropharmacology, 41*(6), 779–790.

Taylor, A. H., Hunt, G. R., Holzhaider, J. C., & Gray, R. D. (2012). Spontaneous metatool use by New Caledonian crows. *Current Biology, 17*(17), 1504–1507.

Thiele, J., & Winter, Y. (2005). Hierarchical strategy for relocating food targets in flower bats: spatial memory versus cue-directed search. *Animal Behaviour, 69*(2), 315-327.

Thomas, R. K. (2022). Learning to learn. In *Encyclopedia of animal cognition and behavior* (pp. 3932–3939). Cham: Springer International Publishing.

Thorndike, E. (2000). *Animal intelligence: Experimental studies,* 318pp. Routledge. Taylor and Francis group, ISBN 9780765804822.

Vitztum, C., & Urbanik, J. (2016). Assessing the dog: A theoretical analysis of the companion animal's actions in human-animal interactions. *Society & Animals, 24*(2), 172–185.

Vorhees, C. V., & Williams, M. T. (2014). Assessing spatial learning and memory in rodents. *ILAR Journal, 55*(2), 310–332.

Watanabe, S., & Huber, L. (2006). Animal logics: decisions in the absence of human language. *Animal Cognition, 9*, 235–245.

7 Animal Migration

Shashi Pradhan, Riya Mathur, R. S. Jatav and Arpana Raikwar

7.1 INTRODUCTION

Animal migration is the study of paramount importance for conservation efforts, as it provides valuable insights into the behaviour and needs of various species. By understanding the patterns and drivers of migration, researchers can identify critical habitats, migration corridors, and potential threats to migrating animals. With this knowledge, some strategies have been found for the conservation of migratory species and the landscapes upon which they depend.

Animal migration represents the regular, seasonal movement of individuals or entire populations from one habitat to another. This phenomenon is often driven by factors such as food availability, climate change, breeding requirements, and the quest for suitable habitats. The migration may encompass vast distances, sometimes spanning continents, and can involve a diverse array of species, including mammals, birds, fish, insects, and even some reptiles. Animal migration is a fascinating phenomenon observed in various species across the globe. Migration refers to the seasonal movement of animals from one region to another, often driven by factors such as food availability, breeding grounds, or climate change.

Ecological balance—The significance of animal migration transcends individual species and ecosystems, exerting a profound influence on the overall ecological balance. By redistributing nutrients, seeds, and biomass, migratory species contribute to the maintenance of diverse habitats, thus supporting the health and stability of ecosystems. The movements of migratory animals often play a pivotal role in regulating populations of other species, thereby preventing overgrazing and promoting the growth of vegetation.

The mechanisms behind animal migration are multifaceted, involving a complex interplay of genetic, physiological, and environmental factors. Navigational abilities, often guided by celestial cues, landmarks, and Earth's magnetic field, enable animals to undertake arduous journeys with remarkable precision. However, in the face of mounting challenges such as habitat loss, climate change, and human-induced obstacles like barriers and pollution, many migratory species are confronted with perilous threats to their survival.

Conservation imperative—As human activities continue to exert unprecedented pressures on natural habitats, the need to conserve and protect migratory species has never been more urgent. Collaborative conservation initiatives, encompassing measures such as habitat preservation, the establishment of wildlife corridors, and international cooperation to address transboundary issues, are instrumental in ensuring the continuity of these extraordinary natural phenomena. Additionally, raising awareness about the ecological significance of animal migration and promoting sustainable practices are integral to fostering a harmonious coexistence between human development and wildlife preservation.

In conclusion, the migration of animals stands as a compelling testament to the remarkable adaptability and resilience of the natural world. By understanding and safeguarding these intricate journeys, we embrace the responsibility to preserve the rich tapestry of life that defines our planet. The migration of animals is a phenomenon that has captivated and puzzled scientists and enthusiasts for centuries. This natural behaviour encompasses the movement of animals from one region to another, often over vast distances, and is driven by a variety of factors such as seasonal changes, food availability, breeding patterns, and environmental conditions.

DOI: 10.1201/9781003710271-7

Furthermore, the impact of human activities on animal migration cannot be understated. Habitat loss, climate change, pollution, and barriers such as roads and infrastructure pose significant challenges to migratory species, disrupting their traditional routes and leading to population declines. Addressing these threats requires collaboration across borders, as many migratory animals traverse international boundaries during their journeys.

7.2 ANIMAL MOVEMENT

The evolution of animal migration is driven by natural selection, fitness benefits, and essentially the contribution of an individual to the gene pool of two variables: mainly survival and reproductive success. There is a great deal of research on the cost and benefits of migratory versus resident species across a range of animals. In ancient migrations, some now-extinct creatures experienced highly migratory journeys analogous to those made by extant species. Dinosaurs themselves made migratory journeys that have been debated as explaining the European distribution of prosauropod Plateosaurus.

Some types of *taxonomic patterns of migration* are ubiquitous within all the major vertebrate and invertebrate groups. Migratory behaviour is a commonly evolved response to periodic and predictable shifts in environmental conditions, but there seems to be no clear taxonomic pattern. The migration of animals is a remarkable and complex phenomenon that merits our attention and protection.

Annual movements of species such as zebras that migrate in vast numbers represent the most famous ungulate migration. The ungulates among mammals perform the most visible and well-documented migrations. The sea-dwelling cetaceans also make great seasonal migrations. The record for the longest mammalian migration is currently held by a female humpback whale, Megaptera novaeangliae, which migrated over 9,800 km from Brazil to Madagascar. Bat migration study is a new and exciting taxonomic frontier. Bats are mostly nocturnal flyers and relatively small, which makes them difficult to tag and track. They can make incredible migrations, in terms of both the numbers of animals that migrate and the distance travelled. Many other species of bats also migrate, but migration is still thought to be a relatively rare strategy, with estimations currently at around 3% of extant bat species.

7.2.1 Bird Movement

Birds are the most studied and arguably the most iconic migrants, with a spectacularly diverse array of migratory behaviours.They must move about from one place to another in search of food. Birds, with their fantastic ability to fly and their unparalleled mobility, have explored almost the entire globe searching for food, territory, nesting sites, and other necessities for survival. When they move to warmer regions from the deep chill of winter, and in some countries around the equator, the summer or dry season reduces the plants, insects, and worms, and the temperature gets very high. To escape this, birds must migrate to a more favourable place. For avian migration, knowledge of causes and consequences is increasing at a rapid pace, fueled by technical progression and innovation (Chapman et al., 2012). One way in which bird migration is diverse is in respect to scale. The birds may travel from 500 km to 17,000 km either at a stretch or by making stopovers, depending upon the distance, taking a few hours to days to months. Birds mainly cover short distances; for example blue bird and wood cock migrate from southern Canada to North America, covering only a few hundred kilometers. The longest and most remarkable of all migrations is that of the Arctic tern; its breeding range is circumpolar. Other extreme migratory behaviours include the high altitude journeys of bar-tailed geese and Anser indicus over the Himalayas. In the Himalayas, migrants fly at a height between 250 and 500 meters and night migrants fly even lower. Similarly, warblers fly at a height of 6,400 meters, godwits at 6,100 meters, and sandpipers at 4,000 meters. At the other end of the spectrum, birds can migrate over very short distances, latitudinally or altitudinally, requiring no species adaptations to survive the migratory journey. The speed of migratory birds is measured using Doppler radar; speeds range from 32 to 64 km/h in small songbirds, and from 40 to 96 km/h in larger birds. Research into bird migration has provided some of the most powerful insights into the evolution of this fascinating phenomenon. Some advantages of migration are securing a better climate and locating a habitat that provides greater variety in birds' diets, promoting the geographic dispersal of birds, reducing parasites and infections of microorganisms, and providing birds with longer working hours (Table 7.1).

7.2.2 Fish Migration

Fish migration is a mysterious event of nature and extremely widespread among fishes, and occurs in a diverse array of forms and for a number of functions, including feeding, breeding, and refuging from predators. Nearly all fish move in shoals (groups of fish with similar size and biological conditions) generally with no leader. Homing is a complex, instinctively and physiologically dictated behaviour of migratory fish that compels the fish to return to its original locality. Migratory behaviour in this group is often classified into typologies based on the habitats between which individuals migrate. Oceanodromous concerns migrations that occur solely in the marine environment, migrating only within the sea, such as Atlantic herring, tuna, and sardine mackerel. Conversely, migrations limited to freshwater are known as potamodromous and are widespread across many groups and include seasonal migrations from lakes into streams or lateral migrations of fish into floodplains, which are well described from tropical and subtropical regions with dry and wet seasons, such as carps, trouts, white sucker, and brook lamprey. Further, many fish species perform a diel horizontal migration, where they migrate from the littoral zone of lakes into offshore areas at dusk and then return to the littoral at dawn. Diadromous fish migrate between freshwater and seawater—anadromous, catadromous, and amphidromous. Migration in fish has traditionally received less research attention than in other taxa such as birds, most likely due to the relative difficulty in tracking individual movements over potentially great distances beneath the water. Some factors influencing migration are olfactory hypothesis (strong sense of smell), internal biological clock, sun compass mechanism, groundwater seepage hypothesis, pheromones, celestial reference points, gradient theory, and use of Earth's magnetic field (Table 7.2).

7.2.3 Amphibian migration

Amphibian migrations are mostly breeding migrations to suitable ponds and wetlands from overwintering sites. They generally take place over a limited spatial scale: mean estimates of the distance

TABLE 7.1
Bird Species, Breeding Area, Wintering Area, and Distance Apart

Species	Breeding Area	Wintering Area	Distance Apart in Miles	Km
Artic tern	Canada/Artic	Antartica	11000	17600
Pectoral sandpiper	Artic	Argentina	9000	14400
Bobolink	Canada	South America	8000	12800
American golden plover	Artic tundra	Argentina	10000	
Willow warbler	Siberia	Africa	8000	16000
Sooty shearwater	New Zealand	North Atlantic	6000–8000	12800
Ruft	Europe	India, Srilanka	3000–6000	4800–8000
Snow goose	Alaska	California,Asia	2000–3000	3200–4800
Pintail	California, Alaska	South America	1000–4000	1600–6400

TABLE 7.2
Types of Migratory Fish and Examples

Types of Migratory Fish	Examples
Diadromous fish	Pacific salmon, Hilsa, American eel
Potamodromous fish	Carp, Brook lamprey, white fishes
Oceanodromous fish	Atlantic herring, sardines, tuna

traveled from the pond edge for 32 species were between 142 and 289 m, although longer distance migrations have been recorded for amphibians. Long-distance migrations for amphibians would be extremely local movements for other migratory taxa. Many amphibians are prone to desiccation, and their movements are usually constrained by weather, with migrations being triggered in many cases by nocturnal rainfall. Sex differences in migratory behaviour are also very common in this group, with males typically arriving earlier to maximise reproductive opportunities at the breeding grounds.

7.2.4 Reptile migration

Reptile migration is short distance and largely limited to oceanic migration. Lizards, for the most part, adopt hibernation as an alternative strategy to overcome challenges associated with predictable seasonal changes to the environment. Sea turtles are the best documented migration of reptiles. In many cases, turtles migrate between foraging and nesting grounds, often over exceptional distances. Leatherback turtles, Dermochelys coriacea, can range over thousands of kilometers to forage upon jellyfish in distant waters before returning to nest on their natal beaches, whereas terrestrial migrations in reptiles are typically on a much smaller scale.

The study of migration in **molluscs** is still in its early stage, but it's speculated to be widespread. In cephalopods, migration occurs at all phases of their life cycle, from passive drifting of eggs and larvae to diel vertical migrations (DVM), to adult migrations of animals that are long distance (from thousands to kilometers) for feeding and spawning grounds.

7.2.5 Arthropods Migration

Arthropods involve many crustaceans making migratory journeys in the traditional sense. For example, many zooplankton have been observed making daily migrations vertically in the water column. Other crustaceans that migrate include Christmas Island red crabs, Gecarcoidea natalis, which make a highly synchronised mass breeding migration to the coast from the inland forest habitats where they forage for the rest of the year. Males migrate in a first wave, with females following shortly after. Social migrations are also common among lobsters. The spiny lobster, Panulirus argus, makes autumnal migrations into the deep water, potentially to avoid seasonal storms, and sometimes thousands of individuals line up to form long queues during migratory movements. The function of queue-forming is speculated to be defensive, as lobsters must travel across shelterless substrate during migration. Individuals in queues can reduce their predation risk via increased vigilance, cooperative defense, and risk dilution.

7.2.6 Insect Migration

Insect migration is also thought to be prevalent. Many arthropods such as insects are very short-lived, and hence in most cases traditional definitions of seasonal migration cannot be applied to these organisms. If we relax the requirement of a return journey in seasonal migration, billions of insects make migratory journeys each year. Studies of noctuid moth migration have highlighted that wind is an important factor in their migratory movements. Silver Y moths, Autographa gamma, migrate at high altitude (425 m) and are able to exert some degree of control over their flight direction by both choosing favourable winds and also adjusting their direction using an internal compass. Migration has adaptive benefits in this species, with migrants having increased reproductive success. Some insects (lepidopterans) have evolved a fascinating form of migratory behaviour, which encompasses multiple generations, such that the individuals that begin the migratory journeys do not complete them. Instead, their descendants complete different stages of the migratory journey (which can take up to six generations). We will return to this phenomenon, known as population patterns of migration.

7.3 FACTORS INITIATING MIGRATION

Many migrations are annual events, and hence their timing is in relation to the annual cycle. A possible cue is the cyclical change in day length that occurs in non-tropical regions. Migratory behaviour (often measured as an increase in nocturnal restlessness) has been initiated in winter in a number of birds by increasing their day lengths. Day-length changes cannot act as a cue for migrants wintering in equatorial regions. Here there is evidence that there are endogenous circannual rhythms in some birds; willow warblers and garden warblers showed nocturnal restlessness at the appropriate times when kept for up to 33 months at constant day length. It seems likely that hormonal changes are involved in the links between the timing event and the change in behaviour. Factors initiating less regular movements are likely to be environmental. The irruptions (irregular mass movements out of their normal range) of crossbills and some other northern forest birds seem to be related to the size of the seed crop. Many insect migrations are also related to environmental changes. Crowding may also be important here and may materialize at an earlier stage in the developmental process so as to produce specifically migratory individuals.

7.3.1 Population Pattern of Migration

Population pattern of migration is great diversity between species in migratory behaviour. Also between individuals within a species, patterns of migratory bchaviour arc usually not uniform. Variation between and within populations has been widely observed, and such intraspecies diversity of movement strategies is fascinating both from an evolutionary perspective and from a more applied point of view (i.e., how population variation in migratory behaviour would influence a species's response to global change). Insights into the degree of within-species variation have been assisted by the advent of new tagging and tracking technologies that allow us to follow the migrations and fate of individual animals.

7.3.2 Between Population Variation

Between population variation in migratory propensity and destination is for species with broad distributions; different populations can experience radically different environments, thereby selective pressures. This can result in population differences in life-history strategies such as migratory tendency, and so comparing populations can illuminate some of the adaptations linked to the evolution of migratory behaviour. Monarch butterflies, Danaus plexippus, have a wide distribution and include populations that differ in migratory behaviour, with some resident and some migratory. Morphological trait analysis has shown that migratory butterflies have large and elongated wings compared with individuals from resident populations, which suggests that wing shape may be an important adaptation to promote successful migratory behaviour. Comparisons between salmonid populations (sea migratory versus resident) have revealed the many life-history adaptations (such as fast growth rate and high investment in reproductive output) that can balance some of the putative costs of sea migration, such as increased predation pressure. Similarly, many passerine birds with a broad distribution have both migratory and resident populations, for example European blackbirds, Turdus merula, and blackcaps, Sylvia atricapilla. In these cases, the migratory strategy of different populations is linked with latitude, with northerly populations tending toward obligate migration, central populations being partially migratory, and southern populations being fully sedentary, highlighting the importance of winter climate (Semmens et al., 2007).

Within a species, populations can differ in their movements and distribution during the migratory journey, which can produce interesting patterns of migration. There are at least three main patterns of latitudinal population movements: the first is known as chain migration, and occurs when populations maintain their relative latitudinal position following migration. In this scenario, the more northerly populations in the breeding season remain more northerly in the non-breeding season following migration. This kind of migration pattern has been documented in many birds, including species of waders, passerines, and gulls. More recently, the post-breeding season movements of northern gannets, Morus bassanus, a large pelagic seabird, were tracked using geolocator tags and revealed evidence of chain migration in this species that is quite common among a variety of migratory bird species. Here (as the name suggests), migratory populations 'leapfrog' other populations during the migratory period, leading to a switch in the relative latitudinal position of the populations between breeding and non-breeding seasons. Analysis of 946 ring recoveries of bar-tailed godwits, Limosa lapponica, showed clear evidence of a leapfrog migration pattern.

Finally, telescopic migration occurs when two populations coexist during one season and migrate to different latitudes during the migratory period. Populations may also differ in their migratory trajectories and destinations. An example of this is the between-population variation found in Scandinavian willow warblers, Phylloscopus trochilus. More northerly populations spend the winter in East and South Africa while more southerly populations make a radically different journey to tropical West Africa. The boundary between populations that migrate to alternative locations is known as a migratory divide. In many cases, such as in the willow warbler example, the reason why we see variation in flyways between different populations is still a mystery. One hypothesis here is that the migratory divide is a zone of secondary contact following post-glacial reinvasion of northern European habitats by populations of warblers with distinct migratory pathways (directions and destinations) from different refugia, i.e., there are historical reasons for the divergent migration patterns between these populations.

7.3.3 Intra-Population Variation

In intra-population variation in migratory behaviour, many animals show great variation in migratory strategies among populations, but perhaps even more surprising is that individuals from the same population can radically differ from one another in their migratory behaviour. One of the most extreme forms of intra-population variation is known as partial migration, where populations are composed of both migratory and entirely resident individuals. Partial migration is extremely common and highly taxonomically diverse. We can distinguish between three types of seasonal partial migration: (1) migrants and residents are together during the breeding season and overwinter in separate locations; (2) migrants and residents breed separately and overwinter together; and (3) animals migrate to spawn or breed, but in some years skip spawning and hence remain resident. Partial migration also occurs at timescales shorter than seasonally—for example, this variation has been documented among tiny zooplankton that migrate (or not) up and down in the water column each day.

Perhaps the most intriguing question when faced with the phenomenon of partial migration is why some individuals stay and some go? Many hypotheses have been proposed to try and answer this question. Here, we present some of the ideas that have the best evidential support for a more detailed and broader examination of theories of partial migration.

7.3.4 Partial Migration Competition

Partial migration competition for limited resources during winter may be a key process in partial migration in some cases. For many species, food availability is reduced during the winter, and hence there is increased competition between individuals for access to the remaining resources. Some evidence (mostly from birds) suggests that poor competitors are forced to migrate to gain access to food during the winter months. The idea that during seasons when resources are limited (whether these are food or nest sites), poor competitors are forced to migrate is known as the competitive release hypothesis. Individual animals may also differ physiologically in their ability to cope with the environmental challenges of certain seasons. The fasting endurance and thermal tolerance hypotheses postulate that individuals that cannot cope with either reduced food or changes in temperature adopt a migratory strategy, while those that can remain resident (Hawkes and Balachandran, 2013). Another idea that was developed with birds in mind is known as the arrival time hypothesis. This is based upon the well-documented observation that, in many migratory species, those that arrive early have the best chance of securing a good breeding site. As long as it is possible to survive an entire winter at the breeding site, and the prior residency effect is strong enough, some individuals may take an early arrival strategy to the extreme and never leave the breeding grounds. Furthermore, an idea which has some support from fishes, zooplankton, and some ungulates is that the costs and benefits of a migratory versus resident strategy are dependent on the relative strength of trade-offs between predation risk and growth (the p/g hypothesis). Individuals can adopt either a high risk, high gain strategy (e.g., residency in cyprinid fishes, where both growth and predation risk are high) or a low risk, low gain strategy (e.g., migration in cyprinid fishes to low risk and food-poor streams during winter).

Differential migration: Partial migration is an extreme form of within-population migratory variation, where individuals either migrate or do not. Yet animals can differ from one another in many migratory traits than just the simple propensity to migrate, for example in their migratory timing or destination. This kind of within-population variation is known as differential migration, and is exceptionally widespread among migratory species. Explanations for differential migration are similar to those for partial migration. For example, the arrival time hypothesis has been cited as a potential reason for males migrating earlier than females during breeding migration.

7.3.5 Daily Migration

Daily migration habitats can change in various ecological properties over very short timescales. Differences in light intensity between the day and night can shift the cost-benefit ratio within a 24-hour period, and some animals have evolved behaviours to track these regular changes by making daily migrations between different habitats. It has been documented that arthropods, arachnids, such as spiders, make predictable daily movements that can be classified as diurnal migrations. Some of the earliest work on DVM was with zooplankton and identified sunlight as a key driver of such migratory movements. Sunlight can act as both a proximate cue (to induce negative phototaxis) and an ultimate factor (beneficial in terms of avoiding damaging UV radiation). The ultimate causes driving the evolution of DVM have since remained a hotly debated subject, with the majority of hypotheses focusing on bioenergetic efficiency, feeding opportunities, UV radiation, and predator avoidance. Of these hypotheses, predator avoidance has received the most attention and is now widely accepted as a powerful driving force of DVM in both invertebrates and fish, and also potentially in dinoflagellates. However, DVM has also been reported in lakes lacking predators, and the magnitude of Daphnia vertical migration is often most pronounced in very transparent lakes, suggesting that ultraviolet radiation (UV) may indeed be key.

7.3.6 Temporal Patterns of Migrating

Temporal patterns of migrating animals inhabiting the natural world are subject to environmental conditions that fluctuate over a wide temporal range. Currently, a high quality habitat may be suboptimal, poor, or even uninhabitable within months, days, or hours. However, many of these changes are cyclic and occur at predictable time intervals, and organisms can adapt to these changes by migrating away during periods of low quality, returning once conditions have ameliorated, and in this way increase their fitness. A frequent observation is therefore that animals move within relatively narrow time windows and in step with circannual (seasonal) and circadian (daily) rhythms. Although temporal patterns of movement are diverse among life forms, there are many common themes that apply across temporal scales.

7.3.7 Temporal distortion

Temporal distortion: Daily rhythms in movements and activity triggered by light cycles are common in many animals, but such movements require regular light/dark fluctuations and are lost when such stimuli disappear. Further, a latitudinal experiment showed that freshwater zooplankton perform strong diel migrations in Spain, where the day:night length is 15:9 h, but that this behaviour eventually ceases as the day length increases with latitude, and completely disappears close to the polar circle, where the daylight is almost constant in summer. Similarly, DVM in fishes has been shown to cease when the difference between night and day is reduced, for example under ice and midnight-sun conditions. However, some evidence shows that marine zooplankton in the Arctic winter continue to perform synchronised DVM in the polar night, which is potentially regulated by diel variations in solar and lunar illumination.

7.3.8 Seasonal migration

Seasonal migration is the quality of habitats also changes over longer, seasonal timescales. Seasonality in some form is ubiquitous across the globe, irrespective of geographical location. In temperate northern and southern hemisphere environments, there are considered to be four seasons (spring, summer, autumn, winter), while in tropical habitats there are only two (dry and wet season). Animals have evolved mechanisms including hibernation and migration to cope with differences between the seasons in habitat quality and resource availability. Of the various kinds of migration, seasonal migration has historically received the most attention and, in a sense, constitutes the form that is classical and typically involves repeat journeys over consecutive years for individuals (Gross et al., 1988).

With the changing seasons, many environmental and ecological variables also fluctuate, affecting the habitat's quality for many species. In many cases, animals are thought to migrate simply to maximise resource intake by following seasonal shifts in food/water availability, as is probably the case with many of the great African ungulate migrations. Similarly, seasonal changes in the intensity and frequency of storms can affect feeding opportunities for tropical birds at higher altitudes, leading to migration downhill to drier sites. Alternatively, migration can be influenced by seasonal changes in the predation risk/growth potential trade-off (p/g). In cyprinid fishes that inhabit lakes in the summer, food availability dwindles in the autumn, and individuals migrate to streams to avoid predators during the food-poor winter months. Migration can also be seasonal, as it is strongly tied to breeding, which is seasonal in many species. In breeding migrations, the optimal conditions for offspring differ from those of adults, and hence individuals migrate to spawning or breeding grounds to mate. This has been documented in a wide range of organisms, from amphibians to fishes to insects. Seasonal migration should evolve when the fitness benefits of seasonal movement outweigh the costs of year-round residency.

Quantifying the costs and benefits of seasonal (and often long-distance) migration is not trivial, however. A powerful study of migratory ground-nesting shorebirds provided strong evidence of a reproductive benefit to their Arctic migration. Monitoring the 'survival' of >1,500 artificial nests along a latitudinal gradient of approximately 3,350 km, researchers showed that by migrating to the Arctic, individual shorebirds had a twofold increase in nest success. Until recently, the seasonal journeys of billions of insects to utilise temperate areas during summer were thought to be an evolutionary dead end, a sink where the progeny of migrants seldom returned (the 'Pied Piper' effect). However, evidence now shows that annual migrations in the noctuid moth Autographia gamma (Silver Y) can lead to a fourfold increase in reproductive success.

7.3.9 Timing of the Migratory Journey

Timing of the migratory journey is a great diversity of patterns of migratory timing in the animal kingdom. Migratory timing may be highly synchronised, such as is the case with the mass migration of Christmas Island red crabs. The degree of synchrony can vary depending on season; for example, the spring migrations of the European roe deer, Capreolus capreolus, are generally synchronised among individuals, whereas the timing of the autumn migration from summer to winter habitats varies considerably between regions, individuals, and even sexes. Also in other species, such as migratory wood thrushes, Hylocichla mustelina, tracking has revealed surprising consistency in timing in spring, but not in autumn. Alternatively, it can be passive and driven by exogenous cues shared by many individuals of the same species, such as photoperiod. There is also variability in the diurnal timing of migratory journeys. A number of bird species migrate at night (nocturnal migrants) and use stars to navigate during their migratory sojourns. Others migrate solely during daytime, and still more during both day and night. Migrating by night may have a number of benefits, including increased foraging opportunities during the day, predator avoidance during the journey, and atmospheric structure (i.e., a relative reduction in turbulence and strong winds during night. However, potential costs to nocturnal migration can arise due to sleep loss (Chapman et al., 2011).

7.3.10 Restoration of Lost Migration

Several different approaches have been used to restore migration pathways for populations that have lost them. These include the restoration of damaged habitat along migration routes, removal of migration barriers, and provision of passages that circumvent obstacles to migration.

7.3.10.1 Habitat restoration

If damage to critical habitat is the key factor underlying migration loss, efforts to restore or alter affected habitat might be sufficient to promote the re-establishment of lost migrations (although formal assessments of restoration effectiveness are limited. Results from recent studies show that it is possible to restore degraded habitat to a condition that produces increased use by wildlife that previously avoided such areas. It has been argued that habitat restoration could be an effective approach for restoring lost migration routes among reindeer in Norway. However, this approach would require the rehabilitation of very large areas, which could be prohibitively expensive and logistically infeasible.

7.3.10.2 Barrier removal

In addition to the incremental degradation of important habitats, the migrations of both aquatic and terrestrial organisms are extremely vulnerable to anthropogenic barriers developed along migratory routes. Freshwater systems are particularly vulnerable to such blockages, as almost three-quarters of North American and European rivers have regulated flows and more than 800000 dams divert

Earth's waterways. Along with other aquatic barriers (e.g., culverts and road crossings), dams are major obstacles to migration and consequently are major contributors to species loss and population decline among freshwater organisms. Dam removal is generally advocated for a variety of environmental restoration-based reasons, including the restoration of aquatic migration routes. The limited evidence that does exist indicates that dam removals can be successful in restoring lost migrations. Positive outcomes appear to be dependent on the attributes of the species targeted for restoration. For example, if a barrier has been in place over the long term and a species is dependent on natal stream imprinting or conspecific pheromone cues to migrate, natural re-establishment of migration subsequent to barrier removal may be slow or non-existent. Potential negative impacts associated with dam removal must also be considered (e.g., removal of barriers between native and hatchery-raised populations or between native and invasive species). While there appear to be few terrestrial analogues for the absolute barriers to migration that can affect aquatic migrants, one example is the 'veterinary fences' constructed in Botswana in the 1950s to control the transmission of disease to cattle. These fences span several hundred kilometers, block the migration of wildebeest, zebras (Equus quagga), and giraffes (Giraffa camelopardalis), and produce high mortalities. While restoration efforts are not yet widespread, the removals of two such fences have been followed by population increases among affected migrants.

7.3.10.3 Barrier circumvention

In recent years, the negative impacts of dams on freshwater migrants have been mitigated by increased efforts to include fish passages in dam construction protocols. Similar efforts have been made in terrestrial environments, where human-constructed linear features such as roads and railways represent the most common obstacle for land-bound migrants. While not impenetrable, these features can disrupt migratory behaviour and reduce the size of migratory populations via an increase in collision-related mortalities. Several different forms of wildlife passage have been built over or under these barriers as a means of reducing vehicular mortalities (of wildlife and humans) and facilitating animal movement between bisected areas. Assessments of the effectiveness of these structures have revealed a mixed record of success in promoting movement and reducing vehicular mortalities. For example, railway underpasses were successful at reducing Tibetan antelope (Pantholops hodgsonii) collision mortalities in China, but highway underpasses had little effect on moose collision mortalities in Sweden. Effectiveness appears to be species-dependent and is affected by factors such as passage type and dimensions; habitat suitability in the areas surrounding the crossing structure; and the presence of fencing around the linear feature.

7.3.10.4 Re-introduction and translocation of migratory individuals

For several decades, re-introductions (via the release of translocated or captive-bred individuals) have served as a strategy for re-establishing populations in parts of their historic range where they are in decline or have been extirpated. The outcomes of these re-introduction efforts have been mixed, although it seems that the probability of success is higher if translocated individuals are obtained from a wild population (vs. captive born); relatively large numbers of animals are released (i.e., $n > 100$); and the threat that resulted in the original decline is eliminated. For re-introductions to serve as an effective strategy for re-establishing lost migrations, they must not only succeed at creating self-sustaining populations in areas where migratory species have been lost, but they must also result in the return of migratory behaviour. Whether or not re-introduction efforts manage to accomplish these goals depends on whether the approach chosen is appropriate for the circumstances of the specific situation. For example, while relatively few studies have explicitly evaluated this, success at restoring migratory behaviour may depend on the extent to which that behaviour is learned (e.g., transmitted from parents to offspring) or innate, with the advantage possibly going to those organisms with a strong innate component.

7.3.10.5 Variation between media

While migrations through two-dimensional environments (e.g., terrestrial migrations and those that occur in river systems) are more vulnerable to physical obstruction, they are also easier to protect than migrations through three-dimensional environments (e.g., aerial and many marine migrations). Similar corridors—for instance of preserved habitat or no-exploitation zones—may have to be much broader (in terms of space, the numbers of habitat types that require conservation action, and the size of human use areas that are impacted) to effectively conserve wide-ranging aerial or oceanic animals if they are unable to traverse migration routes that are very narrow or focused at particular vulnerable points. In an avian migrant's route, key migratory stopovers and intermediate habitats will need to be conserved, similar to those for terrestrial animals. The effectiveness of conservation efforts may also vary depending upon the jurisdiction that the migratory journey covers. For example, for most migratory birds, the vulnerable portions of their migration routes are over land and, therefore, conservation efforts need only deal with limited and clear national jurisdictional issues (although these may not be trivial problems to overcome). However, for oceanic seabirds such as albatrosses and petrels, much of their vulnerable migration takes place over the high seas, which have no national jurisdiction and no natural conservation or management bodies to regulate the activities of national fleets of merchant or fishing vessels operating in these waters. In such cases, effective conservation often requires the development of specific international agreements, which can be very difficult both to create and to enforce. As these birds use few intermediate habitats on their migrations, it may be sufficient to conserve only the migration start- and end-points.

7.4 CONCLUSIONS

The extinction or local extirpation of any organism is lamentable from a biodiversity conservation perspective and often from an economic perspective (e.g., due to the loss of exploitation opportunities and/or ecosystem services). However, the loss of a migratory species, or the loss of migratory behaviour from a population, can have additional consequences. These include: the loss of energetic and material linkages between spatially distinct ecosystems; the loss of a behavioural 'spectacle' that is of fundamental evolutionary and ecological interest; and the loss of a phenomenon with aesthetic appeal that can generate tangible socio-economic benefits. The negative impacts associated with migration loss should be of particular concern, as recent studies indicate that migration is a phenomenon at risk, due to local extirpation and population declines among migratory species. The extrinsic threats that migrants and migration face are numerous; however, most can be assigned to one of four general categories, each of which is anthropogenic in nature. These four processes also have a negative impact on non-migratory species. However, habitat loss and degradation, over-exploitation, and interrupted migration have all been associated with relatively unique consequences for migratory species, and they have played a major role in observed migration losses. The impacts of global climate change on migratory species are just beginning to be documented. Yet researchers have already identified ways in which climate change is likely to have differential impacts on migrant compared with non-migratory species (e.g., the decoupling of migration timing from resource availability). Each of the major threats that migratory organisms face is associated with its own general pattern of impacts. Yet determining the extent to which a particular migratory species or population is vulnerable to these impacts requires a detailed evaluation of their migratory attributes. This should include careful consideration of characteristics such as migration distance and behavioural plasticity, and the determination of whether migrants possess these, and, if so, identifying which stage(s) of the migratory journey they apply to. A detailed understanding of the migratory behaviour of a given organism is not just crucial for assessing its susceptibility to particular threats; it is also critical for developing successful conservation and management strategies. A variety of different approaches (e.g., protected areas, barrier removal) are available for conserving, restoring, and managing migratory populations, but the probability that each will be effective is highly dependent on both the migratory attributes of a targeted population and the extrinsic threats that they face. Finally, while some migrants might have characteristics that render them easier to protect or harvest than non-migratory organisms, the difficulties associated with conserving or exploiting migratory species (vs. more sedentary ones) appear to outweigh the advantages. Specifically, the multiple scales at which migrants move and select habitats, and the long-distance movements that they undertake, are likely to expose them to a broader range of threats than those encountered by their non-migratory counterparts.

Additionally, many migrants travel through and take up temporary residency in multiple jurisdictions, which makes it very difficult to apply conservation measures consistently or, for exploited species, to ensure that harvesting is sustainable across a migrant's entire range. The numerous regulations and multilateral agreements devoted to migratory species are testimony to the difficulties associated with the conservation and management of such wide-ranging organisms. These institutional frameworks are not just indicators of the challenges associated with managing migrants, but they are often a critical component of sustainable exploitation and conservation success. This is because they formalise the inter-jurisdictional cooperation that is essential for the long-term persistence of both migrations and migratory populations. Existing legislation and conventions concerning migratory species are not free of shortcomings. However, they can serve as valuable starting points for developing regulations and agreements that promote cohesive or complementary conservation and management approaches throughout a migrant's range. While the development and consistent application of effective conservation strategies for migrants can be extremely difficult, recent increases in migration loss demonstrate the importance of facing and overcoming these difficulties so that further losses are avoided and so that effective strategies for the conservation of migratory organisms become entrenched in management practice.

BIBLIOGRAPHY

Chapman, B. B., Brönmark, C., Nilsson, J.-Å., and Hans-son, L.-A. (2011). The ecology and evolution of partial migration. *Aquatic Ecology*, 120(12), 1764–1775.

Chapman, B. B, Hulthen, K., Brodersen, J., et al. (2012). Partial migration in fishes: causes and consequences. *Journal of Fish Biology*, 81(2), 456–478.

Gross, M. R., Coleman, R. M., and McDowall, R. M. (1988). Aquatic productivity and the evolution of diadromous fish migration. *Science*, 239(4845), 1291–1293.

Hawkes, L., and Balachandran, S. (2013). The paradox of extreme high-altitude migration in bar-headed geese Anser indicus. *Proceedings of the Royal Society B: Biological Sciences*, 280(1750), 20122114.

Semmens, J. M, Pecl, G. T., and Gillanders, B. M. (2007). Approaches to migration patterns. *Reviews in Fish Biology and Fisheries*, 17, 401–423.

8 Social Biology and Behavior in Animals

Elohozino Oghale Benneth, Isaac Babatunde Fasipe, Oluwashina Owolabi and Emmanuel Onche

8.1 INTRODUCTION

Social biology examines the development and function of animal social behaviors and structures, highlighting their crucial impact on survival, reproductive success, foraging strategies, mating systems, and group defense mechanisms, which makes adequate comprehension of these dynamics vital for this field of study. Recent research by Smith and Holekamp (2023) highlights the intricacy of animal societies, showcasing advanced communication methods, cooperative behaviors, and complex social hierarchies. For example, primate research reveals that social bonds significantly influence individual health and longevity, suggesting strong social ties yield tangible fitness benefits (Silk, 2007).

Similarly, the studies of Wilson and Hölldobler (2005) on eusocial insects like bees and ants demonstrate that division of labor and cooperative breeding are crucial for colony success, thus revealing how these findings emphasize the adaptive significance of social behavior across diverse species. Furthermore, innovative methodologies like bio-logging and network analysis offer deeper insights into animals' social lives in natural environments through the exploration of social behavior across various animal categories, allowing humans to gain valuable insights (Sasaki & Biro, 2017). This study aims to clarify the universal and unique aspects of sociality, providing a comprehensive understanding of animal societies' biological foundations.

8.2 OVERVIEW OF SOCIOBIOLOGY AND ITS SIGNIFICANCE IN UNDERSTANDING ANIMAL BEHAVIOR

Sociobiology explores the biological basis of social behavior across species, integrating biology, ethology, anthropology, and evolutionary psychology to explain how genetic and environmental factors influence social behaviors (Wilson, 1975). This interdisciplinary approach revolutionizes our understanding of animal behavior, offering insights into behavioral evolution and persistence within populations.

The basis of social biology lies in studying altruism and cooperation, where individuals sacrifice their well-being for others and seem paradoxical from a survivalist viewpoint, explaining such behaviors through kin selection and inclusive fitness which suggest helping relatives ensures shared gene propagation as seen from recent research by Nowak and Highfield (2021). Their study supports this theory, demonstrating altruistic behaviors among closely related individuals, like eusocial insects. According to Cornwallis et al. (2022), social biology which is a critical area of study, also investigates mating systems and sexual selection by examining how evolutionary pressures influence reproductive behaviors and mate choices exemplified by elaborate displays and ornaments in birds like the peacock's tail which are understood as sexually selected traits. The recent studies of these authors reveal how environmental changes and social structures impact mating behaviors and success.

DOI: 10.1201/9781003710271-8

Parental investment and care, which are significant aspects of social biology as explained by Trivers' theory, predict that the sex investing more in offspring will be more selective in mate choice, while the less-investing sex competes for access. Clutton-Brock (2002) reveals that this theory is substantiated by numerous studies across species, highlighting the evolutionary basis of diverse parenting strategies, with recent evidence showing how environmental changes lead to shifts in parental investment strategies, impacting population social structure. Social hierarchies and dominance interactions are another key area of study, as discovered in many animal societies, where hierarchical structures dictate resource access and mating opportunities, influencing individual fitness and social stability, exemplified by Sapolsky's (2018) research on primates. The author reveals that social rank affects stress levels, health, and longevity, underscoring social status's biological significance and examining how changes in group dynamics alter hierarchies.

Social biology significantly advances our understanding of cooperation beyond kinship through reciprocal altruism, whereby individuals feel obligated to help non-relatives with the expectation of future reciprocation, indicated through several studies on various species providing empirical support for this theory and illustrating complex social interactions and memory use in cooperative relationships (Carter & Wilkinson, 2017). According to Sapolsky (2018), social biology's significance extends beyond understanding animal behavior by also offering insights into human social structures and behaviors, providing a framework for understanding contemporary human issues like social inequality, cooperation, and conflict by examining evolutionary underpinnings. It bridges biological and social sciences, fostering a comprehensive understanding of the interplay between genetics, environment, and behavior.

In conclusion, social biology plays a crucial role in understanding social behaviors across species and integrates evolutionary theory with behavioral studies to illuminate the adaptive significance of social interactions, thereby enhancing our comprehension of animal and human societies.

8.3 IMPORTANCE OF STUDYING SOCIAL BEHAVIOR ACROSS DIFFERENT ANIMAL CATEGORIES

Studying social behavior across various animal categories is crucial for several reasons, with each contributing to our broader understanding of biology, ecology, and human behavior, as well as encompassing animal interactions within their species, including cooperation, competition, mating, and communication (Kappeler et al., 2022).

Firstly, understanding social behavior can provide insights into the evolutionary processes that shape species, noting that these behaviors often evolve because they confer survival and reproductive advantages, which for instance is observed through cooperative hunting in lions (Panthera leo) and wolves (Canis lupus), leading to more efficient predation and increased survival rates (Dunston et al., 2016). Similarly, the role of social bonds and alliances in survival and reproductive success is highlighted through complex social structures in primates, such as those seen in chimpanzees (Pan troglodytes) (Goodall, 1986). Secondly, studying social behavior helps us understand ecological dynamics and species interactions within ecosystems, as observed from social animals that often play pivotal roles in their environments. For instance, the social structure of bees (Apis spp.) and their roles in pollination are vital for the health of many ecosystems and agriculture (Gill et al., 2016). Social behaviors also impact population dynamics, resource distribution, and community structure, ultimately shaping ecosystem stability and resilience.

Furthermore, studying animal social behavior can offer parallels for understanding human social evolution and behavior. Similar social behaviors such as altruism, aggression, and communication, which are observed in animals, have counterparts in human societies, as revealed by research on social learning and cultural transmission in animals like crows (Corvus spp.) and dolphins (Tursiops truncatus), illuminating the origins of human culture and intelligence (Hunt, 2020; Krützen et al., 2005). Additionally, McComb et al. (2011) are opined that social structures and behaviors, which must be considered when developing conservation strategies, can inform conservation efforts through investigations, as exemplified in elephant herds (Loxodonta africana), where the matriarchal leadership is crucial for the group's survival and knowledge transfer, highlighting the need for strategies that preserve social groups and not just individuals. As seen in recent studies by Silk et al. (2009) which underscore the importance of this field, it demonstrates that social bonds in animals like baboons (Papio spp.) significantly impact their longevity and offspring survival, emphasizing that social behavior is integral to the life history strategies of many species and underscoring the need for its consideration in biological research and conservation policies.

In summary, studying social behavior across various animal categories yields vital insights into evolutionary biology, ecology, human anthropology, and conservation. It reveals the interconnectedness of social systems and ecological health, providing critical knowledge for advancing both scientific understanding and informing practical conservation efforts in preserving biodiversity.

8.4 WILDLIFE

8.4.1 Social Structures in Wild Animals

Wild animals rely on social structures for survival and reproduction, often organized in a hierarchical system as seen from various species like wolf packs and elephant herds, showcasing diverse social dynamics through their structure (Pacheco & Madden, 2021). Wolf packs are a classic example of a social hierarchy led by an alpha male and female playing crucial roles, where the alpha pair is responsible for mating, leading hunts, and defending territory while subordinate wolves including beta and omega individuals have specific roles that support pack cohesion and efficiency, and the pack's hierarchy resolves conflicts, allocates resources, and maintains social order. On the other hand, elephant herds are matriarchal and led by the oldest female with valuable experience, where the matriarch's leadership is vital for survival, guiding the herd in finding water, food, and avoiding predators, as explained by Karenina et al. (2018). According to the authors, the matriarchal leader's guidance ensures the group's well-being and stability, providing a sense of security with young elephants learning essential social and survival skills within this structure, emphasizing cooperation and care. These hierarchical social structures are dynamic, changing due to factors like dominant individual deaths or environmental shifts, which makes such adaptability crucial for species resilience and survival, according to Peterson et al. (2021).

8.4.2 Roles and Hierarchies within Groups

Wild animals' societal structures are intricate systems comprising roles and hierarchies, which facilitate cooperation, resource distribution, and survival strategies. In terrestrial animal groups, these structures significantly shape individual behaviors and group dynamics. Primates, for instance, display highly organized social hierarchies exemplified by chimpanzee communities where dominant hierarchies prevail with alpha males maintaining their positions through physical strength and strategic alliances and influencing access to food, mates, as well as conflict resolution (Kappeler et al., 2022). Similarly, elephants form matriarchal societies led by the oldest female who plays a vital role in decision-making, particularly regarding migration routes and water sources, and their social bonds are strengthened by cooperative behaviors like allo-mothering, where individuals other than the mother help care for calves (Wittemyer et al., 2022) (Figure 8.1).

Wolf packs have a clear hierarchical structure, typically led by an alpha pair responsible for decision-making and pack cohesion while subordinate wolves support the pack by hunting and caring for the alpha pair's offspring and also ensuring the group's success (Cordoni & Palagi, 2019). Meerkats on the other hand exhibit a cooperative breeding system where a dominant pair monopolizes breeding, and subordinates assist in rearing offspring, enhancing group survival through division of labor and contribution to vigilance and foraging activities (Clutton-Brock, 2002). These

FIGURE 8.1 Meerkats (photo credit: ipic).

examples demonstrate that roles and hierarchies in wild animal groups are crucial for resource allocation, protection, and the maintenance of social order, ultimately enhancing the survival and reproductive success of group members. Understanding these social structures provides insights into evolutionary strategies and social dynamics, contributing to animal success.

8.4.3 Communication and Interaction

Social biology and behavior in wild animals, especially terrestrial species, center on complex communication and interaction mechanisms crucial for survival and reproduction. These behaviors encompass vocalizations, visual signals, chemical cues, and tactile interactions, each serving specific purposes within animal communities (Figure 8.2).

Vocal communication is widespread among many mammals and birds. For example, meerkats utilize distinct calls to alert group members of predators, showcasing sophisticated social structures and cooperative behaviors (Mausbach et al., 2022). Similarly, birds like chickadees use alarm calls that convey information about predator size and threat level, demonstrating complex vocal learning and adaptation (Suzuki et al., 2023). Visual signals are equally essential, particularly in species like primates. Facial expressions and body postures in chimpanzees communicate emotional states and intentions, facilitating social bonding and conflict resolution (Waller et al., 2021). The use of bright colors in reptiles, such as chameleons, serves not only in mate attraction but also for deterring rivals and predators (Bergman et al., 2020). Chemical communication is another essential aspect, especially in mammals. Scent marking in animals like wolves and lions serves to establish territory boundaries and convey individual identity and reproductive status (Jordan et al., 2021). This form of communication often overlaps with tactile interactions, such as primates grooming, which reinforces social bonds and preserves group unity (Dunbar, 2021).

Understanding these communication methods highlights the complexity and adaptability of animal behavior in natural habitats. Analyzing these interactions offers valuable perspectives into the evolutionary pressures that shape social structures and dynamics in wild animals, shedding light on their complex adaptation and survival strategies.

FIGURE 8.2 Wolf pack (photo credit: ipic).

8.4.4 Reproduction and Parental Care

Social biology and behavior in wild animals, particularly in the context of reproduction and parental care, exhibit a fascinating array of strategies shaped by evolutionary pressures. Terrestrial mammals and birds provide compelling examples of how social structures impact reproductive success and offspring survival (Figure 8.3).

In many mammalian species, social structures such as packs, herds, or prides facilitate complex reproductive behaviors and strategies. For example, African elephants (Loxodonta africana) live in matriarchal groups that support alloparental care, where related females assist in the rearing of young and reflect how this cooperative breeding enhances calf survival rates through the provision of additional protection and resources (Wittemyer et al., 2022). Similarly, meerkats (Suricata suricatta) exhibit cooperative breeding, where non-breeding individuals help care for the dominant pair's offspring, improving overall group fitness (Clutton-Brock, 2002). Birds exhibit a range of parental care strategies, which are connected to their mating behaviors. In species like the emperor penguin (Aptenodytes forsteri), both parents invest heavily in offspring through bi-parental care, crucial in harsh Antarctic environments (Ancel et al., 2013). On the other hand, in polygynous species like the red-winged blackbird (Agelaius phoeniceus), males defend multiple females but offer little direct parental care, relying on females to raise the young (Yasukawa et al., 2009). Additionally, social monogamy in birds like the albatross (Diomedea exulans) features long-term pair bonds with shared parental responsibilities, ensuring high offspring survival rates in demanding oceanic environments (Weimerskirch et al., 2023). These examples highlight the adaptive nature of social behaviors in boosting reproductive success and offspring survival, reflecting a dynamic interplay between environmental pressures and evolutionary tactics.

8.4.5 Survival Strategies

Social biology and behavior in wild animals are crucial for understanding survival strategies in terrestrial environments. These behaviors are shaped by the need to adapt to varying ecological pressures, ensuring survival and reproductive success.

FIGURE 8.3 Elephants (photo credit: ipic).

Group living offers significant benefits, including enhanced predator detection and defense. For instance, meerkats (Suricata suricatta) exhibit cooperative vigilance, where individuals take turns acting as sentinels, reducing the risk of predation for the group (Clutton-Brock, 2002). Similarly, African elephants (Loxodonta africana) demonstrate complex social structures that aid in protecting calves and acquiring resources (Wittemyer et al., 2022). Reproductive strategies also have significant social aspects. In many primate species, social hierarchies influence mating opportunities and success. High-ranking males in baboon troops (Papio spp.) have better access to fertile females, increasing their reproductive success (Bauer, 2015). This social structure ensures the strongest genes are passed on, enhancing the population's overall fitness.

Communication within social groups is vital for maintaining cohesion and coordinating activities. For example, vocalizations in wolves (Canis lupus) strengthen pack bonds and coordinate hunting (Smith et al., 2012). This cooperative behavior is essential for tackling large prey that single individuals couldn't manage alone. In addition to predation and reproduction, resource acquisition often drives social behaviors. Honey bees (Apis mellifera) exhibit a complex social structure where worker bees perform specific roles supporting the colony, such as foraging and brood care (Seeley, 2019). This division of labor maximizes efficiency and ensures the survival of the hive. Understanding these social behaviors in wild animals reveals intricate survival strategies evolved over millennia, highlighting the importance of social structures in navigating ecological challenges.

8.5 COMPANION ANIMALS

8.5.1 Human-Animal Interactions

The study of social biology and behavior in companion animals, focusing on human-animal interaction, highlights the complex bonds between humans and their pets, including dogs, cats, and birds. This interaction is a significant research area, revealing benefits for both humans and animals (Figure 8.4).

FIGURE 8.4 Human–animal interaction (photo credit: ipic).

Companion animals often provide emotional support, reducing stress, anxiety, and depression in their human counterparts (Grajfoner et al., 2021). These pets act as social catalysts, promoting social interaction and physical activity, contributing to improved overall health (McNicholas et al., 2005). From the animal's perspective, their social behavior and attachment styles mirror human relationships, with dogs for example displaying secure attachment behaviors towards their owners, similar to children with their parents (Hawkins et al., 2017). These bonds are crucial in understanding how animals perceive and respond to human emotions and commands. Pickersgill et al. (2023) reveal that though cats are considered independent, they show significant social behaviors and form strong attachments to their owners, expressing it differently than dogs. Recent studies have expanded to provide insight on how these interactions impact animal well-being, as seen from the works of Rault et al. (2020), which reflects how positive human-animal interactions enhance the well-being of companion animals, reducing stress and promoting positive behavioral outcomes. This research field highlights the mutual benefits of companionship, advocating for better integration of pets in therapeutic settings to harness advantages for human health and animal welfare.

8.5.2 Communication and Behavior

The social biology and behavior of companion animals encompass interactions with humans and each other through intricate communication methods and are deeply influenced by their complex social structures, as observed in dogs, cats, and other terrestrial pets, which shape and define the nature of these interactions (Albuquerque & Resende, 2023). For example, dogs utilize a sophisticated array of vocalizations, body language, and facial expressions to communicate with both humans and other dogs, and they are exceptionally skilled at interpreting human emotions and responding appropriately to various human cues, a capability that significantly strengthens and enhances the human-animal bond through mutual understanding and emotional attunement (Siniscalchi, 2018) (Figure 8.5).

FIGURE 8.5 Companion cat (photo credit: ipic).

Bradshaw (2017) is of the opinion that despite their more solitary nature, cats demonstrate a remarkable array of intricate communication behaviors, employing a sophisticated combination of vocalizations such as meows and purrs, alongside a diverse repertoire of body postures, including arched backs, tail positions, and ear orientations, as well as engaging in scent marking through facial rubbing, urine spraying, and glandular secretions, all meticulously orchestrated to convey their various needs, desires, and emotional states, while simultaneously establishing and maintaining their territorial boundaries and social hierarchies. According to Vitale (2019), cats form attachments to humans in a manner that is fundamentally similar to the way dogs do, yet they manifest these bonds through a series of distinct, often understated behaviors such as head-butting, which serves as a form of scent marking that conveys affection and trust, and kneading, an instinctive action derived from kittenhood that signifies comfort and contentment, thereby highlighting the nuanced and complex nature of feline social interactions and their unique methods of expressing attachment and bonding. Additionally, social learning significantly influences companion animals' behavior. From the findings of Shajid Pyari et al. (2022), through the sophisticated process of social learning, dogs and cats acquire knowledge and behaviors by meticulously observing the actions and reactions of humans and other animals in their environment; this ability to learn vicariously significantly enhances their adaptability, allowing them to adjust their behaviors in various contexts, and positively influences their training outcomes by enabling them to respond to both direct and indirect cues. Through the process of domestication, dogs, for instance have developed the remarkable ability to understand and respond to human pointing gestures, a sophisticated form of interspecies communication that underscores their capacity to interpret human intentions and directions, thus demonstrating an advanced level of social cognition and adaptability that is deeply ingrained in their behavioral repertoire (Siniscalchi, 2018).

In summary, companion animals' social biology and behavior are marked by sophisticated communication and social interactions. These behaviors are crucial for establishing and maintaining the human-animal bond, supporting their roles as companions.

8.5.3 Social Needs and Welfare

Companion animals' social biology and behavior significantly influence their welfare as their social needs vary across species, yet most companion animals, such as dogs and cats, necessitate regular social interactions, employing a combination of vocalizations, body language, and scent marking to communicate with both humans and conspecifics (Walsh, 2009). These interactions, encompassing companionship from humans or other animals, serve crucial roles in preventing behavioral issues and psychological stress in companion animals such as dogs, where a deficiency in social interaction may manifest in conditions like anxiety, aggression, and destructive behaviors, emphasizing the necessity of regular human and canine interaction to ensure the well-being of dogs, mimicking their natural pack environment and promoting their mental and emotional health (McEvoy et al., 2022).

Cordova et al. (2023) reveal that though cats are commonly perceived as solitary creatures, they actually derive significant benefits from social interactions with their human owners and occasionally with other feline companions, as engaging in a stimulating environment featuring social play and interaction not only fosters their psychological well-being but also mitigates stress-related behaviors. Social needs in companion animals extend beyond direct interactions, and they benefit from environmental enrichment like toys and exploration opportunities, which not only fulfill their social and cognitive needs but also contribute to their overall well-being and the strengthening of the human-animal bond (Doulames et al., 2014) (Figure 8.6).

Beery and Kaufer (2015) mention that through the implementation of enrichment strategies which prevent boredom as well as promote mental well-being thereby enhancing companion animals' overall welfare, understanding their social structures is crucial as exemplified by guinea pigs which are highly social rodents and require same-species companionship to mitigate severe stress and health issues. Furthermore, the authors state that rabbits by nature are highly social animals that exhibit complex social behaviors, and in their natural habitat, they live in groups known as colonies or warrens, where they establish intricate social hierarchies and engage in various forms of social interaction (Phillips and Rind, 2002). Additionally, they emphasize how their social structure is vital for survival and

FIGURE 8.6 Companion rabbits (photo credit: ipic).

overall well-being, noting that when kept as pets, their social needs remain significant. In conclusion, addressing companion animals' social needs through human interaction, companionship, and environmental enrichment is vital for their overall welfare. These measures mitigate stress, prevent behavioral problems, and enhance their quality of life.

8.5.4 Reproductive Behaviours

The social biology and behavior of companion animals, particularly regarding reproductive behaviors, is a field that explores the complex social dynamics and physiological processes underlying reproduction. Companion animals, such as dogs and cats, display diverse reproductive behaviors influenced by both biological drives and social contexts.

Reproductive behavior in dogs is notably impacted by hormonal fluctuations, where female dogs, referred to as bitches, experience estrus approximately biannually, demonstrating behaviors such as heightened urination frequency and distinct postural alterations to entice male counterparts, while male dogs exhibit escalated interest towards females in heat, frequently manifesting marking behaviors and heightened aggression as they vie for mating opportunities (Scandurra et al., 2018). Social hierarchies within dog groups also impact reproductive success, with dominant individuals often gaining priority access to mates. Cats, which are characterized by their status as induced ovulators where female cats are known as queens and necessitate mating stimulation to initiate ovulation, display a unique reproductive strategy; during estrus, queens manifest various behaviors such as vocalizations, rolling, and heightened displays of affection, prompting male cats, or toms, to reciprocate with mating behaviors such as mounting and neck biting in response to these cues (Yamane, 2010). Social interactions among cats, such as grooming and play, can also influence reproductive behaviors, although cats are generally more solitary than dogs.

Both species' reproductive behaviors are shaped by domestication, which has altered natural mating patterns through human intervention, such as selective breeding and controlled environments. Understanding these behaviors is crucial for managing breeding programs, addressing behavioral issues, and ensuring the well-being of companion animals.

8.6 AQUATIC ANIMALS

8.6.1 Group Dynamics

The social biology and behavior of aquatic animals, particularly group dynamics, have a profound impact on their survival and reproductive success. Living in groups offers numerous benefits, including enhanced foraging efficiency, predator avoidance, and improved mate selection. Recent research highlights the complex social structures and communication methods employed by aquatic animals within their groups, such as fish forming schools, which reduce individual predation risk and enhance foraging success through collective vigilance and information sharing, with schooling behavior influenced by both environmental factors and social interactions within the group (Tapia-Fonllem et al., 2020) (Figure 8.7).

Additionally, cetaceans, such as dolphins, renowned for their highly developed intelligence and social structures, exhibit a repertoire of sophisticated social behaviors, encompassing cooperative hunting strategies and intricate vocal communications, which serve not only to maintain group cohesion but also to effectively coordinate various activities within their social units, thus highlighting the intricate dynamics inherent in aquatic animal societies (Meyer et al., 2023). Furthermore, Sbragaglia et al. (2017) explain how cetaceans, exemplified by dolphins, display sophisticated social behaviors, engaging in cooperative hunting and employing intricate vocal communications to sustain group cohesion and effectively coordinate activities while social hierarchies are observable in various aquatic species, exerting influence over resource allocation and mating opportunities, exemplified by lobsters establishing dominance hierarchies that dictate access to crucial resources and

FIGURE 8.7 Fish species (photo credit: ipic).

mates, wherein dominant individuals enjoy enhanced privileges and opportunities within the group structure. Similarly, within coral reef fish communities, exemplified by species like clownfish, intricate social hierarchies play a pivotal role in dictating breeding privileges, whereby the dominant pair exclusively engage in reproduction activities, whereas subordinate individuals await their turn, highlighting the structured dynamics governing reproductive success within these aquatic ecosystems (Gainsford et al., 2015). Moreover, group dynamics have evolutionary implications, influencing the trajectory of species. The evolution of cooperation and altruistic behavior in aquatic animals, such as reciprocal grooming in fish, suggests long-term fitness benefits and potential evolutionary advantages (Bergmuller et al., 2010)

Understanding the group dynamics of aquatic animals is crucial for conservation efforts, as social disruptions can have far-reaching effects on populations. Studies on social structures and interactions are vital for developing strategies to preserve these complex and dynamic ecosystems.

8.6.2 Communication Methods

Aquatic animals' social biology encompasses a broad range of behaviors, particularly in communication methods, crucial for survival, reproduction, and social organization. Recent research has illuminated a plethora of sophisticated communication techniques employed by aquatic species; for instance, dolphins and whales have been found to utilize intricate vocalizations not only for social interactions but also for navigation, with dolphins notably employing signature whistles as a means to identify individuals within their social groups and to coordinate complex hunting strategies, highlighting the intricacies of social behavior and the depth of communication methods within marine ecosystems (King et al., 2021). Similarly, like dolphins, humpback whales, renowned for their expansive and complex vocalizations, engage in intricate songs that serve multifaceted functions, including facilitating mate attraction and maintaining social cohesion within their groups, wherein the structure and usage of these vocalizations reflect the whales' social dynamics, highlighting the importance of acoustic communication in their cooperative behaviors and group interactions (Herman, 2016).

Fish also exhibit diverse communication methods. Many species utilize acoustic signals to attract mates, defend territories, or establish dominance hierarchies. For instance, the plainfin midshipman fish produces humming sounds during the breeding season to attract females to their nests (Frommen, 2000). In aquatic environments, visual cues play a crucial role in communication among aquatic species, wherein fish, for example, heavily depend on intricate body coloration and movements; for instance, the cuttlefish exhibits remarkable abilities to swiftly alter its skin color and texture, employing this adaptive trait not only for camouflage to evade predators but also for signaling purposes within its social interactions, thereby underscoring the significance of visual communication mechanisms in facilitating survival strategies and social dynamics among aquatic organisms (Mikami et al., 2023). Chemical communication is prevalent among aquatic animals as well. Pheromones play a significant role in reproductive behaviors and social interactions. Lobsters use urine signals to convey social status and readiness to mate (Shabani et al., 2009). Additionally, alarm substances released by injured fish warn conspecifics of predators, promoting group cohesion and survival (Browman et al., 2019). These communication methods highlight the intricate social structures and survival strategies of aquatic animals, driven by evolutionary pressures and environmental demands.

8.6.3 Social Hierarchies

Aquatic animals' social biology and behavior exhibit complex social hierarchies that influence interactions, reproduction, and survival strategies. These hierarchies are commonly observed in species like fish, cetaceans, and certain invertebrates, where dominance and rank determine resource and mate access. Fish, particularly those inhabiting structured environments such as coral reefs, frequently exhibit conspicuous social hierarchies, as seen in cichlid fish, where these hierarchies are established through aggressive interactions and sustained through consistent dominance behaviors, thereby affording dominant individuals priority access to vital resources including food, shelter, and breeding sites, consequently augmenting their overall fitness within the group dynamic (Majolo et al., 2012). These hierarchies are dynamic, with subordinate fish rising in rank if dominant individuals are removed, adapting to environmental changes (Figure 8.8).

Cetaceans, such as dolphins and whales which are renowned for their cognitive abilities and complex social dynamics, demonstrate intricate social structures, as evidenced by bottlenose dolphins (Tursiops truncatus) that engage in the formation of fluid social hierarchies predicated upon alliances and cooperative behaviors, wherein male dolphins form coalitions to strategically compete for access to females, defend against potential rivals, and thereby augment reproductive success, underscoring the sophisticated social strategies employed within cetacean communities (King et al., 2021). These social bonds are crucial for navigating aquatic challenges and accessing limited resources. Invertebrates, including shrimp and octopuses, also show social stratification. Cleaner shrimp establish hierarchies determining cleaning station positions, impacting food intake from symbiotic partners (Becker, 2005). Octopuses, though typically solitary, engage in complex social interactions where dominance influences mating success and territory control (Edsinger et al., 2020).

Social hierarchies in aquatic animals are essential for maintaining order and facilitating cooperation within groups. These structures mitigate conflicts, optimize resource allocation, and enhance individual fitness within a community. Studying these hierarchies sheds light on aquatic species' behavioral ecology and offers insights into evolutionary mechanisms driving social organization in diverse environments.

8.6.4 Breeding and Parental Care

Aquatic animals, spanning a wide array of species, demonstrate a rich spectrum of social behaviors, particularly evident in the realms of breeding and parental care, behaviors which are intricately influenced by environmental pressures and the imperative to augment reproductive success, with

FIGURE 8.8 Cetacean—dolphin (photo credit: ipic).

numerous fish species exhibiting sophisticated breeding behaviors, such as cichlids, known for their elaborate courtship rituals and meticulous nest-building activities, where males meticulously construct and safeguard nests in a bid to allure females, who subsequently deposit their eggs within these prepared shelters, exemplifying the multifaceted dynamics underlying reproductive strategies in aquatic ecosystems (Slagsvold & Wiebe, 2021). After the spawning process, which involves the release of eggs and sperm into the water, both parental figures often partake in brood care, a crucial phase wherein they meticulously attend to the offspring by engaging in behaviors such as gently fanning the eggs to ensure adequate oxygenation and shielding the vulnerable young from potential predators lurking in the aquatic environment, thereby fostering their survival and promoting the continuation of the species (Balshine, 2005). Parental care strategies in aquatic animals exhibit notable diversity, with certain species, such as the clownfish (Amphiprioninae), showcasing biparental care, a cooperative behavior involving both male and female parents in guarding and tending to eggs until hatching, thereby substantially bolstering offspring survival rates through diminished predation risks and heightened efficiency in egg maintenance (Ridley, 2007). In contrast, many marine invertebrates, such as certain species of octopus, display a more solitary approach to parental care. Female octopuses lay thousands of eggs and protect them until they hatch, often at the cost of their own health and survival. This intense maternal investment ensures that a significant number of offspring survive despite high predation rates (Mikami et al., 2023) (Figure 8.9).

Seahorses (genus Hippocampus) present a unique case of male pregnancy, where males carry the fertilized eggs in a specialized brood pouch until they hatch. This role reversal not only ensures the protection of the developing embryos but also allows females to invest energy in producing more eggs, thereby increasing reproductive output (Kloc, 2023). Environmental factors, such as habitat complexity and predation pressure, play crucial roles in shaping these behaviors. For example, in coral reefs, high predation risk and competition for space drive the evolution of cooperative breeding and extensive parental care to enhance offspring survival (Hing et al., 2017). Overall, the breeding and parental care behaviors of aquatic animals are highly adaptive, enhancing reproductive success and offspring survival in various ecological.

FIGURE 8.9 Seahorse (photo credit: ipic).

8.7 LIVESTOCK ANIMALS

8.7.1 Social Structures in Herds and Flocks

The social dynamics of livestock animals, including cattle, sheep, and poultry, are characterized by intricate hierarchies within their herds and flocks. Recognizing and understanding these complex social structures is essential for promoting their overall well-being, health, and productivity.

8.7.1.1 Hierarchies in Cattle, Sheep, and Poultry

In cattle societies, social hierarchies commonly manifest as linear structures where dominant individuals assert control over subordinates using a combination of physical actions and visual signals, thereby establishing and maintaining order within the group (Val-laillet et al., 2009; Llonch et al., 2018). Dominance within cattle social structures is established through a variety of interactions, including head butting, pushing, and posturing, with higher-ranking individuals typically enjoying priority access to resources such as food and space, which consequently exerts a significant influence on their overall health and productivity levels (Deniz et al., 2021). Sheep, in addition to cattle, engage in the formation of social hierarchies characterized by a fluidity and reduced aggression compared to bovine counterparts, where subtle behaviors such as nudging and displacing others from feeding spots serve as mechanisms for establishing dominance within the sheep social structure (Fraser, 2008) (Figure 8.10).

Poultry, particularly chickens, are known to display a distinct social hierarchy known as a pecking order, whereby individuals establish and maintain dominance through aggressive interactions, with higher-ranked birds asserting control over lower-ranked counterparts, a behavioral phenomenon crucial for resource access and social stability within the flock (Strauss et al., 2022). High-ranking birds within the social hierarchy tend to have better health and productivity due to their

FIGURE 8.10 Livestock cattle (photo credit: ipic).

superior access to essential resources like food, water, and nesting sites, as determined by this hierarchy (Emery, 2006).

8.7.1.2 Impact of Social Structures on Health and Productivity

Livestock social structures, which are pivotal for their health and productivity, particularly exemplified in cattle where established hierarchies serve to diminish occurrences of aggressive encounters, mitigate stress levels, and bolster immune function, collectively contribute to the overall well-being and efficiency of the herd (Takahashi et al., 2018). Due to their dominant status within the herd hierarchy, higher ranked cows exhibit higher milk yields attributable to superior access to food resources and consequent reduced stress levels, as dominance ensures priority access to optimal grazing areas and less competition for nutritional resources, thereby fostering physiological conditions conducive to increased milk production (Valente et al., 2022). On the other hand, Tufekci and Sejian (2023) highlights how stable social structures in sheep which are characterized by cohesive group dynamics and strong social bonds play a vital role in minimizing stress and fostering overall well-being, thereby promoting optimal immune responses and wool production; conversely, social instability within sheep populations, marked by disruptions in group cohesion and social hierarchies, results in heightened stress levels, which in turn lead to suppressed immune function and reduced wool productivity.

Effective management practices, which take into account the intricate social dynamics of livestock animals, not only promote group cohesion and social bonding but also facilitate efficient communication through both vocalizations and non-vocal cues, thereby significantly enhancing productivity levels and contributing to improved health outcomes within the herd or flock (Eke & Obafemi, 2020). Stable pecking orders in poultry, characterized by established social hierarchies, effectively diminish aggressive pecking, minimize injuries, and contribute to improved health and egg production, whereas disturbances in the social order, such as frequent regrouping, elevate stress levels, diminish egg production, and heighten susceptibility to diseases (Salak-Johnson, 2017). Managing flock composition and resources, understanding and managing social hierarchies in livestock to optimize health and productivity, and implementing strategies supporting stable social structures yield significant welfare and economic benefits, mitigating negative impacts.

8.7.2 Communication and Interaction

Improving livestock welfare, productivity, and management practices requires a thorough understanding of their social biology and behavior. Cattle, sheep, pigs, and poultry, among other livestock animals, depend on intricate communication and social structures to interact with each other and their surroundings. This discussion explores both vocal and non-vocal communication, as well as social bonding and group cohesion, in equal measure.

8.7.2.1 Vocal and Non-Vocal Communication

Livestock animals, such as pigs and cattle, employ a combination of vocal and non-vocal means to communicate needs, emotions, and social status, with vocal communication being especially common and relying on sound cues; for instance, pigs utilize grunts, squeals, and screams to express emotions ranging from contentment to distress (Briefer et al., 2019). Similarly, cattle utilize vocalizations, such as mooing, not only to establish and maintain contact with their calves but also as a means to communicate distress signals, thereby reflecting their intricate social communication system and emotional responsiveness to various environmental stimuli (Watts & Stookey, 2000).

Non-vocal communication comprises physical gestures, body language, and facial expressions. Sheep, for instance, practice allogrooming—a behavior where one animal grooms another—to strengthen social bonds and reduce tension within the flock (Mattiello et al., 2019). Pigs, highly attuned to their environment, utilize tail position and ear orientation as essential non-vocal indicators, effectively communicating their mood and intentions to conspecifics and humans alike, thereby demonstrating their acute social awareness and facilitating smoother interactions within their social groups (Marcet et al., 2018). These forms of communication are vital for maintaining harmony and preventing conflicts within groups (Figure 8.11).

8.7.2.2 Social Bonding and Group Cohesion

Social bonding and group cohesion are essential aspects of livestock behavior, impacting overall well-being and productivity. Stable social bonds and group structures reduce stress and aggression,

FIGURE 8.11 Livestock pigs (photo credit: ipic).

promoting better health and performance (Silk et al., 2003). Social bonds, especially those between mother and calf, play a vital role in the development and survival of cattle, and any disruption to these bonds results in notable stress and behavioral problems (McNeal et al. (2014). Group cohesion is crucial in species like sheep and poultry, where flocking behavior ensures safety and efficient resource use. In sheep, maintaining a cohesive group protects individuals from predators and facilitates efficient grazing (Buitenhuis et al., 2009). In poultry, social hierarchies and pecking orders minimize conflicts and ensure access to food and nesting sites (Boucherie et al., 2022). Disrupting these social structures, such as through frequent regrouping or overstocking, increases aggression and stress, negatively impacting welfare and productivity.

In conclusion, effective communication and strong social bonds are fundamental to the social biology and behavior of livestock animals. Understanding these dynamics helps design better management practices that enhance animal welfare and productivity. Incorporating knowledge of vocal and non-vocal communication, as well as fostering social bonding and group cohesion, improves livestock handling and care.

8.7.3 Behavioral Management

Behavioral management in livestock is essential for ensuring animal welfare and productivity. Understanding the social biology and behavior of these animals is crucial for implementing effective management strategies. Two critical aspects in this field are the influence of farming practices on behavior and strategies to improve welfare, particularly through housing and handling.

8.7.3.1 Influence of Farming Practices on Behavior

Farming practices have a significant impact on the behavior of livestock animals. Intensive farming, marked by high stocking densities and limited space, often leads to stress and abnormal behaviors. For example, pigs in confined environments may exhibit tail-biting and aggressive behaviors due to frustration and lack of stimulation Mkwanazi et al. (2019) .Similarly, dairy cows in high-density settings are prone to social stress, which can manifest in increased aggression and reduced lying times, affecting their overall health and milk production (Proudfoot & Habing, 2015). Additionally, early weaning of piglets disrupts their natural social development, leading to increased stress and long-term behavioral issues (Van Kerschaver et al., 2023). According to the EFSA Panel on Animal Health and Welfare et al. (2022), sheep and goats, when subjected to intensive farming conditions, often display stereotypic behaviors such as repetitive movements or vocalizations, indicating poor welfare.

8.7.3.2 Strategies to Improve Welfare: Housing and Handling

Housing and handling: Improving livestock welfare through better housing and handling practices is crucial. Enriched housing environments that provide adequate space, bedding, and environmental enrichments can significantly enhance animal well-being. For instance, providing straw bedding and access to outdoor areas for pigs has been shown to reduce stress and promote natural behaviors (Mkwanazi et al., 2019). Dairy cows benefit from housing systems that allow for sufficient lying time and comfortable resting areas, essential for their health and productivity (Leso et al., 2020).

Handling practices also play a vital role in livestock welfare. Gentle and consistent handling reduces fear and stress in animals, leading to improved behavior and productivity. Training handlers in low-stress handling techniques is essential; for example, using calm and consistent movements can minimize stress in cattle during handling and transport (EFSA Panel on Animal Health and Welfare et al. 2022) Moreover, positive human-animal interactions, such as regular gentle touching and talking, can enhance the animals' trust and reduce fear responses (Rault et al. 2020).

8.7.4 Reproductive Behavior

Social biology and behavior in livestock animals, particularly regarding reproductive behavior, encompasses various critical aspects, including breeding systems, management, maternal behaviors, offspring rearing, and overall animal well-being.

8.7.4.1 Breeding Systems and Their Management

Breeding systems in livestock, including selective breeding, controlled mating, and artificial insemination, are essential for optimizing reproductive efficiency and genetic improvement. Selective breeding involves selecting superior animals based on desirable traits, which enhances productivity and disease resistance (Hu et al.,2020). Controlled mating ensures that only selected males and females reproduce, managing genetic diversity and preventing inbreeding (Plate et al., 2019). Artificial insemination, widely used in dairy cattle, utilizes semen from genetically superior sires, improving herd genetics and reducing the risk of sexually transmitted diseases (Mohammed, 2018) (Figure 8.12).

Effective management practices are crucial in breeding systems to ensure high fertility rates and healthy offspring. Optimal nutrition, health monitoring, and environmental management significantly impact reproductive success (Paul et al., 2011). Hormonal synchronization techniques, such as estrus synchronization, control ovulation timing, facilitating planned breeding and increasing the likelihood of successful conception (Macmillan, 2010). The application of reproductive technologies like embryo transfer and genomic selection enhances breeding efficiency and accelerates genetic progress (Crowe et al., 2021).

8.7.4.2 Maternal Behaviors and Offspring Rearing

Livestock offspring rely heavily on maternal behaviors for their survival and growth, encompassing crucial aspects such as nurturing, protecting, and feeding young, particularly vital during the vulnerable early life stages (Mens et al., 2020), with maternal bonding initiating soon after birth facilitated by

FIGURE 8.12 Livestock poultry (photo credit: ipic).

hormonal changes that promote caregiving instincts, a phenomenon observed in various species such as sheep and goats, where maternal bonds are strengthened through activities like grooming and vocal communication, thereby aiding in offspring recognition and fostering a strong bond between mother and young (Nowak et al., 2000).

Livestock management involves ensuring young receive adequate nutrition and care, with colostrum intake shortly after birth being crucial, providing essential antibodies and nutrients that enhance newborn immune systems, thereby bolstering their resilience to pathogens and environmental stressors, facilitating optimal growth and development, and laying the foundation for long-term health and productivity within the herd (Weaver & Hernandez, 2016). Dairy cattle calf rearing systems may involve early separation and individual or group housing, presenting both benefits and challenges for health and social development, while proper weaning management, including a gradual transition to solid feed, minimizes stress and promotes healthy growth (Khan et al., 2011).

Understanding and managing livestock reproductive behaviors through effective breeding and supportive maternal behaviors is vital for productivity and animal welfare. Advancements in reproductive technologies and improved management practices enhance breeding efficiency and overall livestock health and well-being.

8.8 CONCLUSION AND RECOMMENDATIONS

8.8.1 Conclusion

Social biology and behavior studies across various animal groups reveal intricate social structures and interactions crucial for survival, reproduction, and well-being. Wildlife social behaviors, which often center around survival and reproduction, exemplify intricate dynamics, such as those observed in elephant herds where complex matriarchal social structures prevail, with older females taking the lead and sharing vital knowledge crucial for the success of migrations, highlighting the nuanced interplay between social organization and ecological adaptation in ensuring the perpetuation of species (Karenina et al., 2018). Similarly, within the intricate dynamics of wolf pack behavior, exemplifying the deeply ingrained principles of cooperation and mutual support, packs display sophisticated strategies of cooperative hunting and communal pup rearing, wherein the collective efforts of pack members are seamlessly orchestrated to ensure the procurement of sustenance and the nurturing of offspring, pivotal components essential for the perpetuation of pack cohesion, resilience, and ultimately, survival amidst the challenges of their natural environment (Mech & Boitani, 2019).

Companion animals, especially dogs and cats, showcase a myriad of intricate social behaviors deeply influenced by their interactions with humans. Having evolved alongside humans for thousands of years, dogs exhibit a remarkable array of traits ranging from unwavering loyalty to intricate pack mentality behaviors, all of which have proven instrumental not only in their domestication but also in their seamless integration into the fabric of human families, where they often assume roles akin to beloved companions and steadfast protectors (Hare & Woods, 2013). Cats as domestic companions demonstrate a remarkable capacity for engaging in social interactions not only with humans but also with other animals, a phenomenon that underscores their adaptive ability to navigate a delicate equilibrium between independence and companionship, exhibiting behaviors ranging from seeking affection and forming bonds with their human caregivers to engaging in playful interactions and establishing hierarchies within multi-cat households, despite their inherent solitary nature (Turner, 2021).

Aquatic animals, including fish and marine mammals, display diverse social behaviors that facilitate survival in challenging environments, exemplified by dolphins, which are known for their sophisticated social networks and cooperative behaviors, such as coordinated hunting and alloparental care (Erbe et al., 2016). Schooling in fish, such as herring and sardines, exemplifies the intricate and highly coordinated group dynamics that these species have evolved to optimize, whereby individual fish benefit from the increased likelihood of predator avoidance due to the confusion effect and dilution effect that large, cohesive groups create, as well as improved foraging efficiency through the collective behavior that enables them to locate and exploit food sources more effectively than they could if operating alone (Pitcher, 2012).

Social behavior significantly impacts livestock management and productivity exemplified by cattle which exhibit complex social hierarchies that influence feeding, resting, as well as reproductive behaviors, and as such understanding these social structures can improve welfare and productivity by reducing stress and aggression through proper herd management (Grandin, 2020). Similarly, pigs display a range of social behaviors, including intricate hierarchy formation, complex social bonding, and nuanced interactions, all of which are critical for their well-being, overall health, and social harmony, and these behaviors are significantly influenced by the conditions of their housing, such as space availability, group size, and environmental enrichment (Fraser, 2008).

Comparative studies across these groups highlight the adaptability and evolutionary significance of social behavior; for instance, the evolution of cooperation and altruism in animals, including kin selection and reciprocal altruism, underscores the adaptive value of social behavior in diverse ecological contexts (Nowak & Highfield, 2021). Moreover, understanding these animals' social biology can inform conservation efforts, improve animal welfare, and enhance human-animal interactions. In conclusion, wildlife, companion, aquatic, and livestock animals' social biology and behavior reveal adaptive strategies enhancing survival and reproduction. The intricate social structures observed across these groups emphasize the importance of social interactions in animal welfare and management. Continued research in this field is crucial for advancing the understanding of animal behavior and improving conservation, domestication, and animal husbandry practices.

8.8.2 Recommendations

Understanding the social biology and behavior of various animal groups is vital for their well-being, conservation, and effective management. However, current research has significant gaps that must be addressed to improve our understanding and practices across these animal categories.

8.8.2.1 Wildlife Animals: Current Gaps

Research on the social structures and behaviors of many wildlife species remains limited. The lack of long-term field studies and behavioral data hinders our understanding of their social dynamics. Numerous species, particularly those in remote or inaccessible habitats, are understudied (Silk et al., 2009).

8.8.2.2 Recommendations for Improvement

Increased field studies: There should be an emphasis on long-term observational studies in natural habitats that can provide comprehensive data on social interactions.

Technological integration: Leveraging technological advancements like GPS tracking, remote sensing, and drones can facilitate data collection in hard-to-reach areas (Wilson et al., 2016).

Citizen science: Citizen science programs can engage the public in data collection, augmenting research efforts for widespread species.

8.8.2.3 Companion Animals: Current Gaps

Companion animal research often focuses on domesticated dogs and cats, neglecting other pets like birds, reptiles, and small mammals. Moreover, most research is confined to controlled environments, failing to reflect real-world conditions (Bradshaw, 2017).

8.8.2.4 Recommendations for Improvement

Diversified research: Expanding studies to include a broader range of companion animals can yield a more comprehensive understanding of pet behavior.

Naturalistic studies: Conducting research in naturalistic, home-like settings can provide insights applicable to everyday scenarios.

Behavioral enrichment: Further research is necessary to understand the effectiveness of enrichment strategies for improving companion animal welfare and mental health (Serpell, 2018).

8.8.2.5 Aquatic Animals: Current Gaps

The social structures and behaviors of many aquatic species remain poorly understood due to the challenges of underwater research. Most research has focused on a few species, such as dolphins and certain fish (Browman et aI., 2019).

8.8.2.6 Recommendations for Improvement

Enhanced monitoring techniques: Adopting advanced underwater technologies, like hydrophones and robotic submersibles, can enhance the study of aquatic behaviors.

Interdisciplinary research: Interdisciplinary collaboration between marine biologists, ecologists, and engineers can lead to innovative research methods.

Conservation-oriented studies: Conservation efforts can benefit from research focusing on how social behaviors impact survival and reproduction (Krause et al., 2015).

8.8.2.7 Livestock Animals: Current Gaps

Research on livestock behavior often prioritizes production efficiency rather than welfare. There is a lack of studies on the social needs and mental health of livestock animals (Fraser, 2008).

8.8.2.8 Recommendations for Improvement

Welfare-focused research: Shifting the focus to understanding and improving livestock social welfare can enhance their overall well-being.

Innovative farming practices: Alternative farming practices like free-range and mixed-species farming can promote social interactions and improve livestock welfare

Policy and education: Developing guidelines and educational programs for farmers can highlight the importance of social behaviors in livestock welfare (Grandin, 2020).

In conclusion, bridging the gaps in our understanding of social biology and behavior across these diverse animal groups requires innovative technologies, expanding research scope and welfare considerations into study designs. Addressing these gaps can improve animal welfare, contributing to conservation efforts and sustainable management practices.

REFERENCES

Albuquerque, N., & Resende, B. (2023). Dogs functionally respond to and use emotional information from human expressions. *Evolutionary Human Sciences*, 5(e2). https://doi.org/10.1017/ehs.2022.57

Ancel, A., Beaulieu, M., & Gilbert, C. (2013). The different breeding strategies of penguins: A review. *Comptes Rendus Biologies*, 336(1), 1–12.

Balshine, S. (2005). The benefits of uniparental versus biparental mouth brooding in Galilee St. Peter's fish. *Journal of Fish Biology*, 50, 371–381. https://doi.org/10.1111/j.1095-8649.1997.tb01365.x

Bauer, C. (2015). The baboon (Papio sp.) as a model for female reproduction studies. *Contraception*, 92(2), 120–123. https://doi.org/10.1016/j.contraception.2015.06.007

Becker, J. H. A., Curtis, L. M., & Grutter, A. S. (2005). Cleaner shrimp use a rocking dance to advertise cleaning service to clients. *Current Biology: CB*, 15(8), 760–764. https://doi.org/10.1016/j.cub.2005.02.067

Beery, A. K., & Kaufer, D. (2015). Stress, social behavior, and resilience: Insights from rodents. *Neurobiology of Stress*, 1, 116–127. https://doi.org/10.1016/j.ynstr.2014.10.004

Bergman, T. J., Kitchen, D. M., & Cheney, D. L. (2020). Examining social interactions in non-human primates: Implications for understanding the evolution of sociality. *Current Zoology*, 66(3), 265–278.

Bergmuller, R., Schürch, R., & Hamilton, I. M. (2010). Evolutionary causes and consequences of consistent individual variation in cooperative behaviour. *Philosophical Transactions of the Royal Society of London. Series B, Biological Sciences*, 365(1553), 2751–2764. https://doi.org/10.1098/rstb.2010.0124

Boucherie, P. H., Gallego-Abenza, M., Massen, J. J. M., & Bugnyar, T. (2022). Dominance in a socially dynamic setting: hierarchical structure and conflict dynamics in ravens' foraging groups. *Philosophical Transactions of the Royal Society of London. Series B, Biological Sciences*, 377(1845). https://doi.org/10.1098/rstb.2020.0446

Bradshaw, J. (2017). Behaviour of cats. In *The Ethology of Domestic Animals – An Introductory Text.* Glasgow: CABI, 241–254.

Briefer, E., Vizier, E., Gygax, L., & Hillmann, E. (2019). Expression of emotional valence in pig closed-mouth grunts: Involvement of both source- and filter-related parameters. *The Journal of the Acoustical Society of America*, 145, 2895–2908. https://doi.org/10.1121/1.5100612

Browman, H. I., Cooke, S. J., Cowx, I. G., Derbyshire, S. W., Kasumyan, A., Key, B., ... & Arlinghaus, R. (2019). Welfare of aquatic animals: Where things are, where they are going, and what it means for research, aquaculture, recreational angling, and commercial fishing. *ICES Journal of Marine Science*, 76(1), 82–92.

Buitenhuis, B., Hedegaard, J., Janss, L., & Sørensen, P. (2009). Differentially expressed genes for aggressive pecking behaviour in laying hens. *BMC Genomics*, 10(1), 544. https://doi.org/10.1186/1471–2164–10–544

Carter, G. G., & Wilkinson, G. S. (2017). Social benefits of non-kin food sharing by female vampire bats. *Proceedings of the Royal Society B: Biological Sciences*, 284(1852), 20170284.

Clutton-Brock, T. (2002). *Breeding Together: Kin Selection and Mutualism in Cooperative Vertebrates.* Princeton University Press.

Cordoni, G., & Palagi, E. (2019). Back to the future: A glance over wolf social behavior to understand dog–human relationship. *Animals: An Open Access Journal from MDPI*, 9(11), 991. https://doi.org/10.3390/ani9110991

Cordova, A., Caballero-García, A., Drobnic, F., Roche, E., & Noriega, D. C. (2023). Influence of stress and emotions in the learning process: The example of COVID-19 on university students: A narrative review. *Healthcare (Basel, Switzerland), 11*(12), 1787. https://doi.org/10.3390/healthcare11121787

Cornwallis, C. K., West, S. A., & Griffin, A. S. (2022). The evolution of cooperative breeding. *Nature*, 601(7893), 350–354.

Crowe, A. D., Lonergan, P., & Butler, S. T. (2021). Invited review: Use of assisted reproduction techniques to accelerate genetic gain and increase value of beef production in dairy herds. *Journal of Dairy Science*, 104(12), 12189–12206. https://doi.org/10.3168/jds.2021-20281

Deniz, M., de Sousa, K. T., do Vale, M. M., & Dittrich, J. R. (2021). Age and body mass are more important than horns to determine the social position of dairy cows. *Journal of Ethology*, 39(1), 19–27. https://doi.org/10.1007/s10164-020-00667-x

Doulames, V., Lee, S., & Shea, T. B. (2014). Environmental enrichment and social interaction improve cognitive function and decrease reactive oxidative species in normal adult mice. *The International Journal of Neuroscience*, 124(5), 369–376. https://doi.org/10.3109/00207454.2013.848441

Dunbar, R. I. M. (2021). Grooming and the origins of mutualistic relationships. *Biological Reviews*, 96(3), 789–810.

Dunston, E. J., Abell, J., Doyle, R. E., Kirk, J., Hilley, V. B., Forsyth, A., Jenkins, E., & Freire, R. (2016). An assessment of African lionPanthera leosociality via social network analysis: Prerelease monitoring for an ex situ reintroduction program. *Current Zoology*, zow012. https://doi.org/10.1093/cz/zow012

Erbe, C., Reichmuth, C., Cunningham, K., Lucke, K., & Dooling, R. (2016). Communication masking in marine mammals: A review and research strategy. *Marine Pollution Bulletin*, 103(1–2), 15–38.

Edsinger, E., Pnini, R., Ono, N., Yanagisawa, R., Dever, K., & Miller, J. (2020). Social tolerance in Octopus laqueus—A maximum entropy model. *PloS One,* 15(6), e0233834. https://doi.org/10.1371/journal.pone.0233834

EFSA Panel on Animal Health and Welfare (AHAW), Nielsen, S. S., Alvarez, J., Bicout, D. J., Calistri, P., Canali, E., Drewe, J. A., Garin-Bastuji, B., Gonzales Rojas, J. L., Gortázar Schmidt, C., Michel, V., Miranda Chueca, M. Á., Padalino, B., Pasquali, P., Roberts, H. C., Spoolder, H., Stahl, K., Velarde, A., Viltrop, A., … Herskin, M. (2022). Welfare of cattle during transport. *EFSA Journal*, 20(9). https://doi.org/10.2903/j.efsa.2022.7442

Eke, F., & Obafemi, F. (2020). *Health Outcomes and Labour Productivity in Nigeria.* In Nigeria's Socio-Economic Challenges and Development Options: Essays in Honour of Professor (Mrs) Christiana Ehimen Edenomen Okojie (pp. 134–148), Mindex Publishers.

Emery, N. J. (2006). Cognitive ornithology: the evolution of avian intelligence. *Philosophical Transactions of the Royal Society B: Biological Sciences*, 361(1465), 23–43. https://doi.org/10.1098/rstb.2005.1736. PMID: 16553307; PMCID: PMC1626540.

Fraser, D. (2008). Understanding animal welfare. *Acta Veterinaria Scandinavica*, 50(Suppl 1), S1.

Frommen, J. G. (2020). Aggressive communication in aquatic environments. *Functional Ecology*, 34(2), 364–380.

Gainsford, A., Herwerden, L., & Jones, G. (2015). Hierarchical behaviour, habitat use and species size differences shape evolutionary outcomes of hybridization in a coral reef fish. *Journal of Evolutionary Biology*, 28, 205–222. https://doi.org/10.1111/jeb.12557.

Gill, R. J., Ramos-Rodriguez, O., & Raine, N. E. (2016). Combined pesticide exposure severely affects individual-and colony-level traits in bees. *Nature*, 491(7422), 105–108.

Goodall, J. (1986). *The Chimpanzees of Gombe: Patterns of Behavior.* Harvard University Press.

Grandin, T. (Ed.). (2020). *Improving Animal Welfare: A Practical Approach.* Cabi.

Grajfoner, D., Ke, G. N., & Wong, R. M. M. (2021). The effect of pets on human mental health and wellbeing during COVID-19 lockdown in Malaysia. *Animals: An Open Access Journal from MDPI*, 11(9), 2689. https://doi.org/10.3390/ani11092689

Hare, B., & Woods, V. (2013). *The Genius of Dogs: Discovering the Unique Intelligence of Man's Best Friend.* Simon & Schuster.

Hawkins, R., Williams, J., & Scottish Society for the Prevention of Cruelty to Animals (Scottish SPCA). (2017). Childhood attachment to pets: Associations between pet attachment, attitudes to animals, compassion, and humane behaviour. *International Journal of Environmental Research and Public Health*, 14(5), 490. https://doi.org/10.3390/ijerph14050490

Herman, L. (2016). The multiple functions of male song within the humpback whale (Megaptera novaeangliae) mating system: Review, evaluation, and synthesis: Humpback whale male song. *Biological Reviews*, 92. https://doi.org/10.1111/brv.12309.

Hing, M., Klanten, O., Dowton, M., & Wong, M. (2017). The right tools for the job: Cooperative breeding theory and an evaluation of the methodological approaches to understanding the evolution and maintenance of sociality. *Frontiers in Ecology and Evolution*, 5. https://doi.org/10.3389/fevo.2017.00100.

Hu, G., Do, D. N., Gray, J., & Miar, Y. (2020). Selection for favorable health traits: A potential approach to cope with diseases in farm animals. *Animals (Basel)*, 10(9), 1717. https://doi.org/10.3390/ani10091717. PMID: 32971980; PMCID: PMC7552752.

Hunt, G. R. (2020). *Crows: Cognition and Behavior.* Cambridge University Press.

Jordan, N. R., Golabek, K. A., & Apps, P. J. (2021). Chemical communication in territorial mammals. *Journal of Mammalogy*, 102(4), 993–1002.

Kappeler, P. M., Huchard, E., Baniel, A., Canteloup, C., Charpentier, M. J., Cheng, L., & Girard-Buttoz, C. (2022). Sex and dominance: How to assess and interpret intersexual dominance relationships in mammalian societies. *Frontiers in Ecology and Evolution*, 10, 918773.

Karenina, K., Giljov, A., de Silva, S., & Malashichev, Y. (2018). Social lateralization in wild Asian elephants: Visual preferences of mothers and offspring. *Behavioral Ecology and Sociobiology*, 72, 1–11.

Khan, M. A., Weary, D. M., & Von Keyserlingk, M. A. G. (2011). Invited review: Effects of milk ration on solid feed intake, weaning, and performance in dairy heifers. *Journal of Dairy Science*, 94(3), 1071–1081.

King, S. L., Guarino, E., Donegan, K., McMullen, C., & Jaakkola, K. (2021). Evidence that bottlenose dolphins can communicate with vocal signals to solve a cooperative task. *Royal Society Open Science*, 8(3). https://doi.org/10.1098/rsos.202073

Kloc, M. (2023). Seahorse male pregnancy as a model system to study pregnancy, immune adaptations, and environmental effects. *International Journal of Molecular Sciences*, 24(11), 9712. https://doi.org/10.3390/ijms24119712. PMID: 37298663; PMCID: PMC10253279.

Krause, J., James, R., Franks, D. W., & Croft, D. P. (Eds.). (2015). *Animal Social Networks.* Oxford University Press.

Krützen, M., Mann, J., Heithaus, M. R., Connor, R. C., Bejder, L., & Sherwin, W. B. (2005). Cultural transmission of tool use in bottlenose dolphins. *Proceedings of the National Academy of Sciences*, 102(25), 8939–8943.

Leso, L., Barbari, M., Lopes, M. A., Damasceno, F. A., Galama, P., Taraba, J. L., & Kuipers, A. (2020). Invited review: Compost-bedded pack barns for dairy cows. *Journal of Dairy Science*, 103(2), 1072–1099. https://doi.org/10.3168/jds.2019-16864

Llonch, P., Mainau, E., Ipharraguerre, I. R., Bargo, F., Tedó, G., Blanch, M., & Manteca, X. (2018). Chicken or the egg: The reciprocal association between feeding behavior and animal welfare and their impact on productivity in dairy cows. *Frontiers in Veterinary Science*, 5. https://doi.org/10.3389/fvets.2018.00305

Macmillan, K. L. (2010). Recent advances in the synchronization of estrus and ovulation in dairy cows. *The Journal of Reproduction and Development,* 56(Suppl), S42–S47. https://doi.org/10.1262/jrd.1056s42. PMID: 20629216.

Majolo, B., Lehmann, J., Vizioli, A., & Schino, G. (2012). Fitness-related benefits of dominance in primates. *American Journal of Physical Anthropology*, 147, 652–660. https://doi.org/10.1002/ajpa.22031.

Marcet, R., Míriam Pageat, P., Bienboire-Frosini, C., Teruel, E., Monneret, P., Leclercq, J., Lecuelle, C., & Cozzi, A. (2018). Tail and ear movements as possible indicators of emotions in pigs. *Applied Animal Behaviour Science*, 205. https://doi.org/10.1016/j.applanim.2018.05.012.

Mattiello, S., Battini, M., De Rosa, G., Napolitano, F., & Dwyer, C. (2019). How can we assess positive welfare in ruminants? *Animals*, 9(10), 758. https://doi.org/10.3390/ani9100758

Mausbach, J., Manser, M. B., & Graw, B. (2022). Alarm calls in meerkats: Vocal signals of predation risk. *Animal Behaviour*, 184, 53–62.

McComb, K., Moss, C., Durant, S. M., Baker, L., & Sayialel, S. (2011). Leadership in elephants: The adaptive value of age. *Proceedings of the Royal Society B: Biological Sciences*, 278(1722), 3270–3276.

McEvoy, V., Espinosa, U., Crump, A., & Arnott, G. (2022). Canine socialisation: A narrative systematic review. *Animals: An Open Access Journal from MDPI*, 12(21), 2895. https://doi.org/10.3390/ani12212895

McNeal, N., Scotti, M.-A. L., Wardwell, J., Chandler, D. L., Bates, S. L., LaRocca, M., Trahanas, D. M., & Grippo, A. J. (2014). Disruption of social bonds induces behavioral and physiological dysregulation in male and female prairie voles. *Autonomic Neuroscience: Basic & Clinical*, 180, 9–16. https://doi.org/10.1016/j.autneu.2013.10.001

McNicholas, J., Gilbey, A., Rennie, A., Ahmedzai, S., Dono, J.-A., & Ormerod, E. (2005). Pet ownership and human health: A brief review of evidence and issues. *BMJ*, 331(7527), 1252–1254. https://doi.org/10.1136/bmj.331.7527.1252

Mech, L. D., & Boitani, L. (Eds.). (2019). *Wolves: Behavior, Ecology, and Conservation*. The University of Chicago Press.

Mens, A., Van Krimpen, M., & Kwakkel, R. (2020). Nutritional approaches to reduce or prevent feather pecking in laying hens: Any potential to intervene during rearing? *World's Poultry Science Journal*, 76, 1–20. https://doi.org/10.1080/00439339.2020.1772024.

Meyer Sørensen, P., Haddock, A., Guarino, E., Jaakkola, K., McMullen, C., Jensen, F., Tyack, P., & King, S. (2023). Anthropogenic noise impairs cooperation in bottlenose dolphins. *Current Biology*, 33. https://doi.org/10.1016/j.cub.2022.12.063.

Mikami, K., Watanabe, N., Tochio, T., Kimoto, K., Akama, F., & Yamamoto, K. (2023). Impact of gut Microbiota on host aggression: Potential applications for therapeutic interventions early in development. *Microorganisms*, 11(4), 1008. https://doi.org/10.3390/microorganisms11041008

Mkwanazi, M. V., Ncobela, C. N., Kanengoni, A. T., & Chimonyo, M. (2019). Effects of environmental enrichment on behaviour, physiology and performance of pigs – A review. *Asian-Australasian Journal of Animal Sciences*, 32(1), 1–13. https://doi.org/10.5713/ajas.17.0138. Epub 2018 Dec 21. PMID: 28728387; PMCID: PMC6325398.

Mohammed, A. (2018). *Artificial Insemination and its Economical Significance in Dairy Cattle: Review.* https://doi.org/10.20431/2454-9428.0401005.

Nowak, M. A., & Highfield, R. (2021). *SuperCooperators: Altruism, Evolution, and Why We Need Each Other to Succeed.* Free Press.

Nowak, R., Porter, R. H., Lévy, F., Orgeur, P., & Schaal, B. (2000). Role of mother-young interactions in the survival of offspring in domestic mammals. *Reviews of Reproduction*, 5(3), 153–163.

Pacheco, X. P., & Madden, J. R. (2021). Does the social network structure of wild animal populations differ from that of animals in captivity? *Behavioural Processes*, 190, 104446. https://doi.org/10.1016/j.beproc.2021.104446.

Paul, V. K., Sachdev, H. S., Mavalankar, D., Ramachandran, P., Sankar, M. J., Bhandari, N., ... & Kirkwood, B. (2011). Reproductive health, and child health and nutrition in India: Meeting the challenge. *The Lancet*, 377(9762), 332–349.

Plate, M., Bernstein, R., Hoppe, A., & Bienefeld, K. (2019). The importance of controlled mating in honeybee breeding. *Genetics, Selection, Evolution: GSE*, 51(1). https://doi.org/10.1186/s12711-019-0518-y

Peterson, R. O., Vucetich, J. A., & Fenton, M. B. (2021). The ecology and social dynamics of wild canids. *Ecological Monographs*, 91(4), e01485.

Phillips, C., & Rind, M. (2002). The effects of social dominance on the production and behavior of grazing dairy cows offered forage supplements. *Journal of Dairy Science,* 85, 51–59. https://doi.org/10.3168/jds.S0022-0302(02)74052-6.

Pickersgill, O., Mills, D. S., & Guo, K. (2023). Owners' beliefs regarding the emotional capabilities of their dogs and cats. *Animals (Basel),* 13(5), 820. https://doi.org/10.3390/ani13050820. PMID: 36899676; PMCID: PMC10000035.

Pitcher, T. J. (2012). *The Behaviour of Teleost Fishes.* Springer Science & Business Media.

Proudfoot, K., & Habing, G. (2015). Social stress as a cause of diseases in farm animals: Current knowledge and future directions. *The Veterinary Journal*, 206. https://doi.org/10.1016/j.tvjl.2015.05.024.

Rault, J., Waiblinger, S., Boivin, X., & Hemsworth, P. (2020). The power of a positive human–Animal relationship for animal welfare. *Frontiers in Veterinary Science*, 7, 590867. https://doi.org/10.3389/fvets.2020.590867.

Ridley, A. (2007). Factors affecting offspring survival and development in a cooperative bird: Social, maternal and environmental effects. *The Journal of Animal Ecology*, 76, 750–760. https://doi.org/10.1111/j.1365-2656.2007.01248.x.

Salak-Johnson, J. (2017). Social status and housing factors affect reproductive performance of pregnant sows in groups. *Molecular Reproduction and Development*, 84. https://doi.org/10.1002/mrd.22846.

Sapolsky, R. M. (2018). *Behave: The Biology of Humans at Our Best and Worst.* Penguin Books.

Sasaki, T., & Biro, D. (2017). Cumulative culture can emerge from collective intelligence in animal groups. *Nature Communications*, 8(1), 15049.

Sbragaglia, V., Leiva, D., Arias, A., García del Arco, J., Aguzzi, J., & Breithaupt, T. (2017). Fighting over burrows: The emergence of dominance hierarchies in the Norway lobster (Nephrops norvegicus). *The Journal of Experimental Biology*, 220, jeb.165969. https://doi.org/10.1242/jeb.165969.

Scandurra, A., Alterisio, A., Di Cosmo, A., & D'Aniello, B. (2018). Behavioral and perceptual differences between sexes in dogs: An overview. *Animals (Basel)*, 8(9), 151. https://doi.org/10.3390/ani8090151. PMID: 30142932; PMCID: PMC6162565

Seeley, T. D. (2019). *The Lives of Bees: The Untold Story of the Honey Bee in the Wild.* Princeton University Press.

Serpell, J. A. (2018). Companion animals. In *Anthrozoology: Human-animal Interactions in Domesticated and Wild Animals*, G. Hosey, V. Melfi (eds), pp. 17–31. U.K.: Oxford University Press.

Shabani, S., Kamio, M., & Derby, C. D. (2009). Spiny lobsters use urine-borne olfactory signaling and physical aggressive behaviors to influence social status of conspecifics. *The Journal of Experimental Biology,* 212(Pt. 15), 2464–2474. https://doi.org/10.1242/jeb.026492. PMID: 19617440.

Shajid Pyari, M., Vékony, K., Uccheddu, S., & Pongrácz, P. (2022). Companion cats show no effect of trial-and-error learning compared to dogs in a transparent-obstacle detour task. *Animals,* 13(1), 32. https://doi.org/10.3390/ani13010032. PMID: 36611642; PMCID: PMC9817999.

Silk, J. B. (2007). The adaptive value of sociality in mammalian groups. *Philosophical Transactions of the Royal Society B: Biological Sciences*, 362(1480), 539–559.

Silk, J. B., Alberts, S. C., & Altmann, J. (2003). Social bonds of female baboons enhance infant survival. *Science*, 302(5648), 1231–1234.

Silk, J. B., Beehner, J. C., Bergman, T. J., Crockford, C., Engh, A. L., Moscovice, L. R., ... & Cheney, D. L. (2009). The benefits of social capital: close social bonds among female baboons enhance offspring survival. *Proceedings of the Royal Society B: Biological Sciences,* 276(1670), 3099–3104.

Siniscalchi, M., d'Ingeo, S., Minunno, M., & Quaranta, A. (2018). Communication in dogs. *Animals*, 8(8), 131. https://doi.org/10.3390/ani8080131. PMID: 30065156; PMCID: PMC6116041.

Slagsvold, T., & Wiebe, K. L. (2021). Egg covering in cavity nesting birds may prevent nest usurpation by other species. *Behavioral Ecology and Sociobiology*, 75(8), 116. https://doi.org/10.1007/s00265-021-03045-w. Epub 2021 Jul 31. PMID: 34759442; PMCID: PMC8570351.

Smith, J. E., & Holekamp, K. E. (2023). Hunting success in the spotted hyena: Morphological adaptations and behavioral strategies. In *Social Strategies of Carnivorous Mammalian Predators: Hunting and Surviving as Families* (pp. 139–175). Cham: Springer International Publishing.

Smith, J. E., Swanson, E. M., Reed, D., & Holekamp, K. E. (2012). Evolution of cooperation among mammalian carnivores and its relevance to hominin evolution. *Current Anthropology*, 53(S6), S436–S452.

Strauss, E. D., Curley, J. P., Shizuka, D., & Hobson, E. A. (2022). The centennial of the pecking order: current state and future prospects for the study of dominance hierarchies. *Philosophical Transactions of the Royal Society B: Biological Sciences*, 377(1845), 20200432. https://doi.org/10.1098/rstb.2020.0432. Epub 2022 Jan 10. PMID: 35000437; PMCID: PMC8743894.

Suzuki, T. N., Wheatcroft, D., & Griesser, M. (2023). Complexity of alarm calls and responses in birds. *Nature Communications*, 14, 1125.

Takahashi, A., Flanigan, M. E., McEwen, B. S., & Russo, S. J. (2018). Aggression, social stress, and the immune system in humans and animal models. *Frontiers in Behavioral Neuroscience*, 12. https://doi.org/10.3389/fnbeh.2018.00056

Tapia-Fonllem, C., Fraijo-Sing, B., Corral-Verdugo, V., Garza-Terán, G., & Moreno-Barahona, M. (2020). School environments and elementary school children's well-being in northwestern Mexico. *Frontiers in Psychology*, 11. https://doi.org/10.3389/fpsyg.2020.00510

Tufekci, H., & Sejian, V. (2023). Stress factors and their effects on productivity in sheep. *Animals: An Open Access Journal from MDPI*, 13(17), 2769. https://doi.org/10.3390/ani13172769.

Turner, D. C. (2021). The mechanics of social interactions between cats and their owners. *Frontiers in Veterinary Science*, 8, 650143.

Valente, J. P. S., Deniz, M., de Sousa, K.T., Mercadante, M. E. Z., Dias, L. T. (2022). socialh: An R package for determining the social hierarchy of animals using data from individual electronic bins. *PLoS One*, 3(17), e0271337.

Val-Laillet, D., Guesdon, V., von Keyserlingk, M. A. G., de Passillé, A. M., & Rushen, J. (2009). Allogrooming in cattle: Relationships between social preferences, feeding displacements and social dominance. *Applied Animal Behaviour Science*, 116(2–4), 141–149. https://doi.org/10.1016/j.applanim.2008.08.005

Van Kerschaver, C., Turpin, D., Michiels, J., & Pluske, J. (2023). Reducing weaning stress in piglets by pre-weaning socialization and gradual separation from the sow: A review. *Animals: An Open Access Journal from MDPI*, 13(10), 1644. https://doi.org/10.3390/ani13101644

Vitale, K., Behnke, A., & Udell, M. (2019). Attachment bonds between domestic cats and humans. *Current Biology*, 29, R864–R865. https://doi.org/10.1016/j.cub.2019.08.036.

Waller, B. M., Cray, J. J., & Burrows, A. M. (2021). Primate facial expressions and social communication. *Current Anthropology*, 62(2), 159–176.

Walsh, F. (2009). Human-animal bonds I: The relational significance of companion animals. *Family Process*, 48, 462–480. https://doi.org/10.1111/j.1545-5300.2009.01296.x.

Watts, J., & Stookey, J. (2000). Vocal behaviour in cattle: The animal's commentary on its biological processes and welfare. *Applied Animal Behaviour Science*, 67, 15–33. https://doi.org/10.1016/S0168-1591(99)00108-2.

Weaver, S. R., & Hernandez, L. L. (2016). Autocrine-paracrine regulation of the mammary gland. *Journal of Dairy Science*, 99(1), 842-853.

Weimerskirch, H., Corbeau, A., Pajot, A., Patrick, S. C., & Collet, J. (2023). Albatrosses develop attraction to fishing vessels during immaturity but avoid them at old age. *Proceedings of the Royal Society B, 290*(1990), 20222252.

Wilson, E. O. (1975). *Sociobiology: The New Synthesis*. Harvard University Press.

Wilson, E. O., & Hölldobler, B. (2005). Eusociality: Origin and consequences. *Proceedings of the National Academy of Sciences*, 102(38), 13367–13371.

Wilson, R. P., Holton, M. D., Walker, J. S., Shepard, E. L., Scantlebury, D. M., Wilson, V. L., ... & Jones, M. W. (2016). A spherical-plot solution to linking acceleration metrics with animal performance, state, behaviour and lifestyle. *Movement Ecology*, 4, 1–11.

Wittemyer, G., et al. (2022). Elephant matriarchal leadership. *Biology Letters*, 18(3), 20220008.

Yamane, A. (2010). Male homosexual mounting in the group-living feral cat (Felis catus). *Ethology Ecology & Evolution*, 399–406. https://doi.org/10.1080/08927014.1999.9522822.

Yasukawa, K., Enstrom, D. A., Parker, P. G., & Jones, T. C. (2009). Epaulet color and sexual selection in the Red-winged Blackbird: A field experiment. *The Condor*, 111(4), 740–751.

9 Self-Defensive Behavior

Vipin Maurya, Manish Pandey, K. Tamilarasan and Tanmoy Rana

9.1 INTRODUCTION

The responses and patterns of response in animals to fear or to help them deal with threats are known as defensive behaviors; fear or threats can be from predators, animals of the same species, and the environment (Blanchard *et. al.*, 2010). The fear response is essential for animals' protection and survival (Bakshi *et. al.*, 2000). Defensive behavior decreases the chances of animals being injured, killed, or affected but influences normal behaviors (Elliott *et al.*, 2016), which in turn can affect body condition and influence their reproduction and survival (Lind and Cresswell, 2005). Understanding self-defensive behavior is crucial for maintaining animal welfare, preventing injuries (Cafe *et al.*, 2011), and promoting effective farm management.

9.2 TYPES OF SELF-DEFENSIVE BEHAVIOR IN FARM ANIMALS

Common self-defensive behaviors exhibited by farm animals to protect themselves from predators, stress, and harm are flight, fight response, freezing, and defensive postures or defensive threats (Roelofs, 2017). When animals sense danger, their first instinct is to flee, which is known as the "flight response." Every animal has a personal space, or "flight zone," and will run away when this space is invaded. However, animals that are far from a safe place or engaged in important activities like eating or defending their territory may be less likely to flee immediately (Graziano and Cooke, 2006). Research has consistently shown that larger animals and those in good physical condition tend to have a greater distance between themselves and a potential threat before they start to flee, compared to smaller animals or those in poor condition. The factors influencing flight response are grouped into four main categories: the predator's characteristics, such as size and speed; the organism's physical condition, including health, size, and overall well-being; environmental factors like terrain, weather, and vegetation; and the effects of experience and learning, which shape the animal's response to potential threats based on its past encounters and acquired knowledge (Stankowich and Blumstein, 2005) (Figure 9.1).

In various animal species, conflicts with members of the same species have been observed, often stemming from competition for scarce resources such as mating opportunities, food, or shelter (Rico-Guevara & Hurme, 2019). The "fight or flight" response, as described by Cannon (1935), is the initial and active response to stressors, marked by territorial defense and aggressive behavior. In contrast, the conservation-withdrawal response, characterized by Forkman et al. (2007), is a passive reaction to stress, typified by immobility, urination, defecation, and subdued aggression.

Freezing is the tense body with a complete absence of movement except for body vitals viz. heartbeat and respiration that result due to increased muscle tone in a defensive state (Fanselow, 1994; Kalin and Shelton, 1989). This response is also associated with bradycardia (Vianna and Carrive, 2005), reduced vocalizations, and changes in body temperature (Hagenaars et al., 2014). The distance between the animal and predator and the escape routes play a key role in determining species freezing behavior. A longer freezing response is observed during distal threat, while short freezing occurs when escape routes are available (Blanchard et al., 2011). This freezing act is effective in preventing detection and thereby an attack from predators (Palmer, 1909; Curio, 1976).

DOI: 10.1201/9781003710271-9

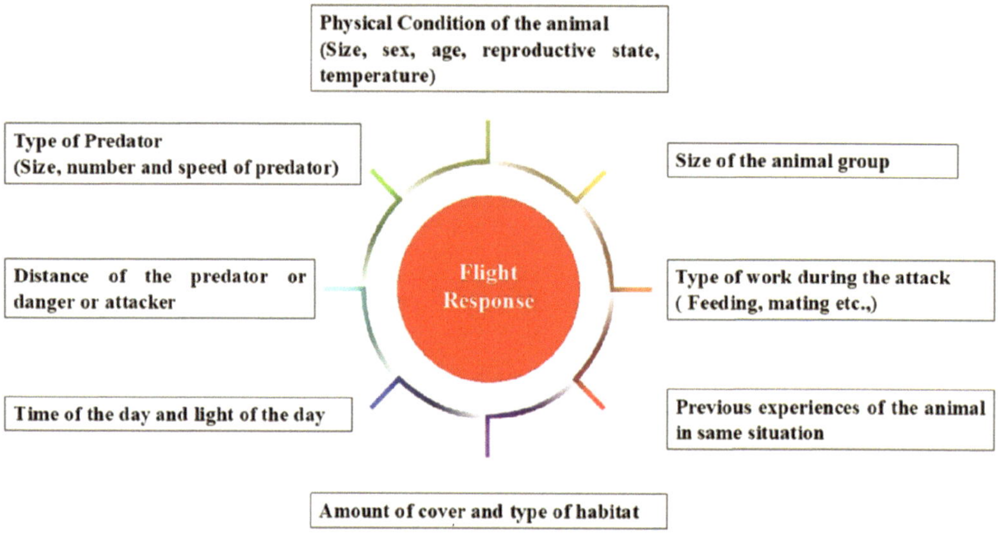

FIGURE 9.1 Factors affecting flight response and decision of the animal.

When approached by a potential predator or a car, rabbits often freeze, making them less noticeable. During hunting, deer may freeze in place, relying on their camouflage to avoid detection. During any threat condition, farm animals sometimes adopt a posture called defective posture/threat. The function of defensive posture is to communicate an attacking disposition to the opponent, which is often successful in deferring the attack. Furry animals often raise their body hair, looking bigger and more frightening to the assailing animal (Sih, 1998). The stressed farm animal usually shows tail swishing, kicking, pawing at bedding, and loud vocalization (Ekesbo and Gunnarsson, 2018)

9.3 CAUSES OF SELF-DEFENSIVE BEHAVIOR

Self-defensive behavior in farm animals is a multifaceted phenomenon, driven by a combination of factors. Fear and stress are the primary triggers, with an animal's temperament, previous experiences, and inherited traits influencing their stress response (Grandin and Shivley, 2015). Temperament can be shaped by various factors, including breed, sex, age, experience, production systems, handling routines, and human-animal interactions (Estévez-Moreno *et al.*, 2021). Certain breeds, such as more reactive cattle and sheep, may exhibit higher cortisol levels and more agitated behavior when handled, transported, or restrained (Hall *et al.*, 1998). For instance, *B. indicus* cattle tend to be more excitable than *Bos taurus* cattle during handling (León-Llanos *et al.*, 2022). Traumatic experiences, such as abuse or stressful situations, can lead to long-term behavioral changes, making animals more prone to defensive behaviors. Research has shown that animals with negative handling experiences develop a stronger defensive response, as seen in pigs that display increased aggression and fearfulness after rough handling (Hemsworth *et al.*, 2000).

Other factors such as loud noises, unfamiliar environments, and predators or perceived threats can also trigger self-defensive behaviors. It is found that cattle exposed to unpredictable environments showed heightened fear responses (Boissy and Le Neindre 1997). Social hierarchy and group interactions can significantly impact defensive behaviors. When confronted by dominant individuals, animals that are lower in the social hierarchy may display more defensive behaviors. Social stress and competition for resources can result in heightened aggression and increased self-defensive behaviors (Estevez et al., 2007). The maternal instincts in animals like cows and pigs can significantly influence their self-defensive behaviors, especially when protecting their offspring (Weary *et al.*, 2008).

9.4 SPECIES-SPECIFIC SELF-DEFENSIVE BEHAVIOR

Defensive responses and the pattern of reactions to defend themselves in animals vary from species to species (Bolles, 1970). In vertebrates, defensive behavior is mainly evolved targeting their survival from predators. As these defensive behaviors are shown immediately after the recognition of a nearby threat or attack from the predator, it affects the normal feeding and reproductive behavior that the organism exhibits (Ficken and Witkin, 1977; Magurran and Girling, 1986; Morse, 1993). In young rodent species (rat pups), they produce ultrasonic vocalization to direct maternal attention to them; this might be associated with their immature sensory and motor system and require maternal care (Noirot, 1972; Hofer, 1996). Likewise, in non-human primates, the constellation of responses that include vocalizations, freezing, fleeing, or defensive hostility and aggression are common defensive behaviors. The characteristic responses or patterns exhibited by the different species depend upon the nature of the perceived threat (Kalin and Shelton, 1989; Kalin, 1993).

Farm animals may exhibit aggressive behaviors such as biting, kicking, or using their body weight to gore, trample, knock down, or crush others, including other animals or humans. They may also use their strength to scratch, drag, or ram into objects or individuals, potentially causing harm (Langley and Morrow 2010). In cattle, horns may have evolved through natural selection to offer adaptive benefits, including enhanced defense against predators, improved competitiveness, and establishing dominance for mates and resources among members of the same species (Knierim, 2015). Additionally, they also use their horns for self-grooming of hard-to-reach parts. Cattle employ kicking as a highly effective defense mechanism, utilizing powerful hind leg thrusts to target vulnerable areas like the head, neck, or torso of perceived threats, making it a crucial survival strategy due to its formidable strength and effectiveness (Grandin and Shivley, 2015). Charging is also a threatening behavior exhibited by cattle to deter predators or intruders. Cattle will often lower their heads, paw the ground, and charge towards the perceived threat, using their size and momentum to intimidate. Pigs exhibit several self-defensive behaviors to protect themselves from perceived threats and predators. They use loud squeals and grunts as alarm calls to warn other pigs of danger and startle predators (Marchant et al., 2001). Pigs can bite attackers using their strong jaws and sharp teeth, particularly sows protecting their piglets, and wild pigs can use their teeth for fighting. They also use their snouts to root and dig, throwing up dirt and creating barriers between themselves and the predator (Meijaard et al., 2011).

9.5 ENVIRONMENTAL FACTORS INFLUENCING SELF-DEFENSIVE BEHAVIOR

Environmental factors such as weather extremes, fire, flood, and other natural disasters threaten farm animals (Blanchard *et. al.*, 2010). In addition, environmental factors, viz. space, overcrowding, noise, and animal-human interactions, influence animals' self-defensive behaviors. Overcrowding and less space increase aggression and behavioral problems and reduce performance in most farm animals (Fregonesi & Leaver, 2002). Overcrowded pens increase the frequency of aggressive interactions and displacements from the feed bunk (DeVries *et al.*, 2004). Subordinate animals are most affected by overcrowding, as a dominant animal would often displace them, and these submissive animals wait a long time to access feed bunks, ultimately consuming a poorer quality diet; lying time and feeding time will also be affected (Munksgaard *et al.*, 2005). Adverse housing facilities affect the overall health, production, and welfare in farm animals (Pandey *et al.*, 2021).

Noise is also one of the important factors that influence self-defensive behavior in animals. The animals often get frightened due to sudden, loud, intermittent sounds (Waynert *et al.*, 1999). The response noted to noise in heifers is increased heart rate and anxiety (Arnold *et al.*, 2007), while sudden sound with an intensity above 110 dB can cause animals to jump, reduce activity, and stay huddled. Animals may freeze into a motionless stance; sometimes, animals may become aggressive, increase defecation, and reduce sniffing, grooming, crawling, and social activities (Waynert *et al.*, 1999).

Furthermore, human-animal interaction is another important factor that affects defensive behavior. The behavior of animal attendants determines the quality of human-animal interaction. Farm animals that were handled roughly were highly fearful of humans and felt stressed; consequently, their welfare and productivity were compromised (Zulkifli, 2013). Regular positive contact with animals at farms alters their physiology, behavior, and performance (Hemsworth *et al.*, 2000; Hemsworth, 2003). While caring for animals, a relationship develops between an animal attendant and an animal responding to tactile, visual, olfactory, gustatory, and auditory stimuli, during which mutual individual recognition takes place. Animals may respond to tactile, visual, olfactory, gustatory, and auditory stimuli from humans (Zulkifli, 2013).

9.6 RECOGNIZING AND MANAGING SELF-DEFENSIVE BEHAVIOR

Knowledge of and understanding animals' body language and behavioral cues enhance interaction with animals and make working with them easier, safer, and enjoyable. Animals' body language indicates the state of animals or affective state and in what way they must be handled. A cow and calf will hang their tails loosely, away from their bodies while in fear or anxiety, a cow's tail tucked up next to her body. In severely stressful situations, they wring the tail, curl it, and flip it (Wagner, 2006). Several dangerous situations in animals can result in fear and anxiety, which are closely related emotions. Fear is a reaction to the perception of actual danger, whereas anxiety is the reaction to a potential danger that threatens the integrity of the individual (Boissy, 1998). Fear and anxiety change the natural animal behavior. The behavioral change depends upon the related type of threat encountered. The reaction to the threat can be contradictory, such as active defense (attack, threat) or active avoidance (flight, hiding, escape), and passive avoidance (immobility) can be viewed as expressing fear (Wagner, 2006). Animals also express stress through frequent urination and defecation; under high stress and agitation, animals display signs such as bellowing, butting, kicking, and fence-walking (Wagner, 2006). One should be very careful around cows in a high level of stress, as they may become aggressive. A bull performs a broadside threat prior to attack; a person can make a bull back off by responding with a frontal stance or just going back slowly away from the bull. It is always recommended to never run away and not turn your back on a bull.

Stress reduction and enrichment strategies: Stress in farm animals can be minimized by not separating animals from other animals and avoiding loud noise or yelling. Animals generally become agitated, stressed, or aggressive when around strangers when they feel threatened (Wagner, 2006). Animals with their offspring become aggressive in their protection. Animals showing aggression with signs like vocalization, pawing, or head-shaking, entering their flight zone, may cause them to attack. It is always better to avoid agitated or defensive animals. Intense pain due to several diseases, viz. displaced abomasum, colic, etc., can affect the temperament of docile animals; they can behave in an agitated or even aggressive way, and the animals should be handled with caution. Moreover, kind human-animal relations are very important. Pats, gentle touch, good food provisions, and talking to the animals show positive effects (Ceballos *et al.*, 2018, Mota-Rojas *et al.*, 2020). Handling large animals to perform routine activities like moving, feeding, and clinical examination often involves force. Such conditions can result in agitation, fear, or anxiety in animals, risking the safety of the handlers. To avoid such situations, animals can be given positive reinforcement training (PRT), which alters the behavior of animals (Skinner, 1938) and minimizes the negative impact of such procedures on animals (Jønholt *et al.*, 2021) and animal handlers involved with handling aggressive animals (Minier *et al.*, 2011), thereby providing better animal welfare and improved control (Bassett and Buchanan-Smith, 2007). PRT involves rewarding animals showing desired behavior and no punishment when animals do something unwanted. The punishment is only taking away/withholding the reward (Sørensen *et al.*, 2011). PRT lowers cortisol levels, improves the rate of learning, creates a more positive emotional state while handling (Heinsius *et al.*, 2024), shows calm behaviors (O'Brien *et al.*, 2008), and reduces stress (Lambeth *et al.*, 2006). Other important techniques to control stress or aggression can be desensitization and habituation. Desensitization means associating the process with a positive stimulus and a negative event, which results in the loss of ability of the negative event to adversely influence behavior (Chance, 2003). On, the other hand, habituation is simple exposure to a particular stimulus that, over some time, results in decreased response to that stimulus (Webster, 1994).

9.7 IMPLICATIONS FOR ANIMAL WELFARE AND FARM MANAGEMENT

The negative complications of self-defensive behavior shown towards threats or fear can be reduced by avoiding exposure of animals to the sources of threats. This can be achieved through proper care, better housing, and other facilities that contribute to better animal welfare. Not treating animals in a proper way or being unethical creates problems in animals that generally behave normally (Orihuela and Solano, 1994). Improper housing increases the risk that animals will suffer from sickness and injuries or reduce opportunities to express their natural behavior (Rushen, 2017). In intensive production systems, threats from animals of the same species and from the environment are extremely significant, and providing proper housing and space becomes essential, as it prevents animals from breaching the individual space of another, which results in agonistic interactions. Animals can be managed well if they are provided with sufficient space and stocking density. Overcrowding results in a significant reduction in feeding activity, resting, and decreased ruminal activity (Grant and Albright, 2001). There are reports (Menke *et al.*, 1999; Kondo et al., 1989) that mention a negative correlation between general space allowances per cow in the barn and agonistic behaviors. Recognizing and prioritizing the well-being of animals within the livestock production system contributes to a holistic approach to sustainability that extends beyond mere production metrics.

9.8 NEUROPHYSIOLOGICAL BASIS OF SELF-DEFENSIVE BEHAVIOR

Environmental stressors trigger physiological responses, which can be categorized into two primary types for simplicity. Firstly, the "fight or flight" response is a swift and automatic reaction, mediated by the sympathetic-adrenal-medullary axis (Chen *et al.*, 2015). When faced with acute challenges, the adrenal medulla rapidly releases epinephrine and norepinephrine, triggering a cascade of physiological changes. These changes include vasoconstriction, increased heart rate and blood pressure, rapid breathing, and elevated blood glucose levels, all of which prepare the body to either confront the challenge or flee from it. This rapid response is crucial for increasing energy availability, oxygen supply, and cardiovascular function, thereby enhancing the body's ability to cope with emergency situations (Jurkovich *et al.*, 2024).

Secondly, the rapid response is followed by a slower reaction, characterized by the release of adrenocorticotropic hormone from the anterior pituitary gland (Lay *et al.*, 1996), leading to the production of corticosteroids from the adrenal cortex. These hormones amplify and prolong the metabolic effects of catecholamines. These hormones influence behaviors including the self-defensive behavior of animals during short-term stress (Jurkovich *et al.*, 2024) (Figure 9.2).

9.9 BREEDING AND GENETICS IN SELF-DEFENSIVE BEHAVIOR

Domestication alters the intensity of behavioral responses in farm animals, reducing fear and stress, and increasing adaptability to various conditions. For instance, domesticated Norway rats exhibit less caution than their wild counterparts in diverse situations (Price, 1999). Individual variation in domestic animals is well-documented and serves as the basis for selection and further domestication through breeding (Koolhaas and Van Reenen, 2016). Factors influencing individual coping

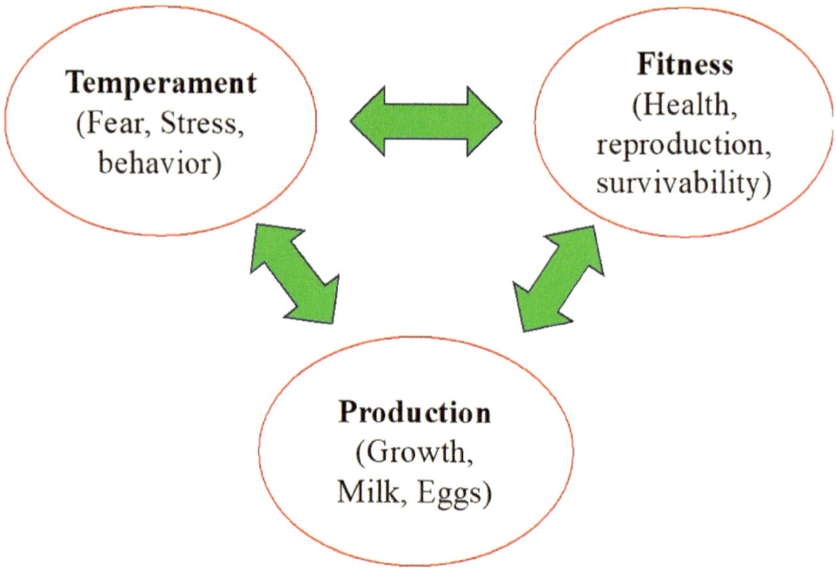

FIGURE 9.2 Interrelationship between temperament, fitness, and production.

capacity include genotype, ontogeny, adult experience, age, social support, and more (Carere and Maestripieri, 2013).

The complex relationship between an animal's temperament, productivity, and overall fitness is influenced by multiple factors. Key personality traits like fearfulness and stress tolerance impact well-being and resilience through interactions between the nervous, endocrine, and immune systems. Selective breeding for desirable production traits, such as milk yield or growth rate, can enhance performance but may inadvertently compromise fitness by altering resource allocation. Moreover, selection's impact on temperament can have a ripple effect, further influencing overall fitness and adaptability. To address this, a holistic breeding approach is necessary, considering biologically relevant temperamental traits to cultivate more resilient and adaptable domestic farm animals. This integrated approach can lead to improved overall welfare and performance (Koolhaas and Van Reenen, 2016).

9.10 ETHOLOGICAL AND EVOLUTIONARY PERSPECTIVES

Understanding self-defensive behaviors from an ethological and evolutionary perspective sheds light on the mechanisms driving these survival strategies and their significance in natural selection. Ethological research has documented numerous self-defensive behaviors across species, providing empirical data on their occurrence, effectiveness, and variation. For example, field studies on meerkats have shown that sentinel behavior—where one individual watches for predators while others forage—enhances group survival (LeDoux, 2000). Self-defense is essential for survival, helping animals protect themselves from predators, competitors, and harsh environments (Dawkins, 1976). Animals that lack effective self-defensive behaviors are more likely to die or fail to reproduce (Endler, 1991). These strategies vary depending on the predator and environment (Sih 1998) and are often context-dependent. Self-defensive behaviors are heritable, influenced by environmental factors, and associated with specific brain regions and neurotransmitters (Flint et al., 1995; LeDoux, 2000). From an evolutionary perspective, self-defensive behavior is a crucial adaptation driven by the threat of predation and competition for resources and mates (Endler, 1991; Dawkins, 1976). These behaviors have evolved through natural selection and often involve trade-offs, such as energy expenditure or reduced foraging time (Sih, 1998). Phylogenetic and comparative analyses reveal the independent evolution and diversification of these traits across different lineages, illustrating convergent evolution and offering insights into their evolutionary history (Harvey & Pagel, 1991; Carroll, 2005).

REFERENCES

Arnold, N. A., Ng, K. T., Jongman, E. C., and Hemsworth, P. H. 2007. "The behavioural and physiological responses of Dairy Heifers to tape-recorded milking facility noise with and without a pre-treatment adaptation phase." *Applied Animal Behaviour Science* 106(1–3): 13–25.

Bakshi, V. P., S. E. Shelton, and N. H. Kalin. 2000. "Neurobiological correlates of defensive behaviors." *Progress in Brain Research* 122: 105–115.

Bassett, L., and Buchanan-Smith, H. M. 2007. "Effects of predictability on the welfare of captive animals." *Applied Animal Behaviour Science* 102: 223–245. https://doi.org/10.1016/j.applanim.2006.05.029.

Blanchard, D. C., Defensor, E. B., and Blanchard, R. J. 2010. "Fear, anxiety, and defensive behaviors in animals." In *Encyclopedia of behavioral neuroscience*, edited by G. F. Koob, M. Le Moal, and R. F. Thompson, 532–537. Academic Press. https://doi.org/10.1016/B978-0-08-045396-5.00090-7.

Blanchard, D. C., Griebel, G., Pobbe, R., and Blanchard, R. J. 2011. "Risk assessment as an evolved threat detection and analysis process." *Neuroscience & Biobehavioral Reviews* 35: 991–998. https://doi.org/10.1016/j.neubiorev.2010.10.016.

Boissy, A. 1998. "Fear and fearfulness in determining behavior." In *Genetics and the behaviour of domestic animals*, edited by T. Grandin, 67–111. San Diego: Academic Press.

Boissy, A., and Le Neindre, P. 1997. "Behavioral, cardiac and cortisol responses to brief peer separation and reunion in cattle." *Physiology & Behavior* 61(5): 693–699.

Bolles, R. C. 1970. "Species-specific defense reactions and avoidance learning." *Psychological Review* 77(1): 32.

Cafe, L. M., Robinson, D. L., Ferguson, D. M., Geesink, G. H., and Greenwood, P. L. 2011. "Temperament and hypothalamic-pituitary-adrenal axis function are related and combine to affect growth, efficiency, carcass, and meat quality traits in Brahman steers." *Domestic Animal Endocrinology* 40(4): 230–240.

Cannon, W. B. 1935. "Stresses and strains of homeostasis." *American Journal of Medical Science* 189: 1–14.

Carere, C., and Maestripieri, D. (Eds.). 2013. *Animal Personalities: Behavior, Physiology, and Evolution.* Chicago and London: University of Chicago Press.

Carroll, S. B. 2005. *Endless Forms Most Beautiful: The New Science of Evo Devo.* New York: Norton & Company.

Ceballos, M. C., Sant'Anna, A. C., Boivin, X., Oliveira Costa, F., Carvalhal, M. V. L., and Paranhos da Costa, M. J. R. 2018. "Impact of good practices of handling training on beef cattle welfare and stock-people attitudes and behaviors." *Livestock Science* 216: 24–33.

Chance, P. 2003. *Learning and Behavior.* Belmont, CA: Wadsworth Cengage Learning.

Chen, Y., Arsenault, R., Napper, S., and Griebel, P. 2015. "Models and methods to investigate acute stress responses in cattle." *Animals* 5: 1268–1295.

Curio, E. 1976. *The Ethology of Predation.* New York: Springer-Verlag.

Dawkins, R. 1976. *The Selfish Gene.* Oxford: Oxford University Press.

DeVries, T. J., Von Keyserlingk, M. A. G., and Weary, D. M. 2004. "Effect of feeding space on the inter-cow distance, aggression, and feeding behavior of free-stall housed lactating dairy cows." *Journal of Dairy Science* 87(5): 1432–1438.

Eibl-Eibesfeldt, I. 1970. *The Biology of Behaviour.* New York: Holt, Reinhart & Winston.

Ekesbo, I., and Gunnarsson, S. 2018. *Farm Animal Behaviour: Characteristics for Assessment of Health and Welfare.* CABI.

Elliott, K. H., Betini, G. S., Dworkin, I., and Norris, D. R. 2016. "Experimental evidence for within-and cross-seasonal effects of fear on survival and reproduction." *Journal of Animal Ecology* 85(2): 507–515.

Endler, J. A. 1991. "Interactions between predators and prey." In *Behavioral Ecology*, edited by J. R. Krebs and N. B. Davies, 169–196. Oxford: Blackwell Scientific Publications.

Erhard, H. W., Mendl, M., Christhiansen, S. B. 1999. "Individual differences in tonic immobility may reflect behavioural strategies." *Applied Animal Behaviour Science* 64: 31–46.

Estevez, I., Andersen, I. L., and Nævdal, E. 2007. "Group size, density and social dynamics in farm animals." *Applied Animal Behaviour Science* 103(3–4): 185–204.

Estévez-Moreno, L. X., Miranda-de la Lama, G. C., Villarroel, M., García, L., Abecia, J. A., Santolaria, P., and María, G. A. 2021. "Revisiting cattle temperament in beef cow-calf systems: Insights from farmers' perceptions about an autochthonous breed." *Animals* 11(1): 82.

Fanselow, M. S. 1994. "Neural organization of the defensive behavior system responsible for fear." *Psychonomic Bulletin & Review* 1: 429–438. https://doi.org/10.3758/BF03210947.

Ficken, M. S., and Witkin, S. R. 1977. "Responses of Black-capped chickadees to predators." *The Auk* 94: 156–157.

Flint, J., et al. 1995. "The genetic basis of anxiety-related behaviors in mice." *Behavioral Genetics* 25(5): 407–420.

Forkman, B., Boissy, A., Meunier-Salaün, M. C., Canali, E., and Jones, R. B. 2007. "A critical review of fear tests used on cattle, pigs, sheep, poultry, and horses." *Physiology & Behavior* 92: 340–374.

Fregonesi, J. A., and Leaver, J. D. 2002. "Influence of space allowance and milk yield level on behaviour, performance, and health of dairy cows." *Applied Animal Behaviour Science* 78: 35–50.

Grandin, T., and Shivley, C. 2015. "How farm animals react and perceive stressful situations such as handling, restraint, and transport." *Animals* 5(4): 1233–1251.

Grant, R. J., and Albright, J. L. 2001. "Effect of animal grouping on feeding behavior and intake of dairy cattle." *Journal of Dairy Science* 84: E156–E163.

Graziano, M. S., and Cooke, D. F. 2006. "Parieto-frontal interactions, personal space, and defensive behavior." *Neuropsychologia* 44(6): 845–859.

Hagenaars, M. A., Oitzl, M., and Roelofs, K. 2014. "Updating freeze: aligning animal and human research." *Neuroscience & Biobehavioral Reviews* 47: 165–176. https://doi.org/10.1016/j.neubiorev.2014.07.021.

Hall, S. J. G., Broom, D. M., and Kiddy, G. N. S. 1998. "Effect of transportation on the plasma cortisol and packed cell volume in different genotypes of sheep." *Small Ruminant Research* 29: 233–237.

Harvey, P. H., and Pagel, M. D. 1991. *The Comparative Method in Evolutionary Biology*. Oxford University Press.

Heinsius, J. L., Lomb, J., Lee, J. H. W., Von Keyserlingk, M. A. G., and Weary, D. M. 2024. "Training dairy heifers with positive reinforcement: Effects on anticipatory behavior." *Journal of Dairy Science* 107(2): 1143–1150.

Hemsworth, P. H. 2003. "Human–animal interactions in livestock production." *Applied Animal Behaviour Science* 85: 185–198.

Hemsworth, P. H., Coleman, G. J., Barnett, J. L., and Borg, S. 2000. "Relationships between human-animal interactions and productivity of commercial dairy cows." *Journal of Animal Science* 78(11), 2821–2831.

Hofer, M. A. 1996. "Multiple regulators of ultrasonic vocalizations in the infant rat." *Psychoneuroendocrinology* 21: 203–217.

Jønholt, L., Bundgaard, C. J., Carlsen, M., and Sørensen, D. B. 2021. "A case study on the behavioural effect of positive reinforcement training in a novel task participation test in göttingen mini pigs." *Animals (Basel)* 11: 1610. https://doi.org/10.3390/ani11061610.

Jurkovich, V., Hejel and Kovács. 2024. A Review of the Effects of Stress on Dairy Cattle Behaviour. Animals, 14(14), 2038.

Koolhaas, J. M., and Van Reenen, C. G. 2016. "Animal behavior and well-being symposium: Interaction between coping style/personality, stress, and welfare: Relevance for domestic farm animals." *Journal of Animal Science* 94(6): 2284–2296.

Kalin, N. H. 1993. "The neurobiology of fear." *Scientific American* 268: 94–101.

Kalin,N. H., and Shelton, S. E. 1989. "Defensive behaviors in infant rhesus monkeys: environmental cues and neurochem- ical regulation." *Science* 243: 1718–1721. https://doi.org/10.1126/science.2564702.

Knierim, U., Irrgang, N., and Roth, B. A. 2015. "To be or not to be horned—consequences in cattle." *Livestock Science* 179: 29–37.

Kondo, Seiji, Junjiro Sekine, Masahiko Okubo, and Yasushi Asahida. 1989. "The effect of group size and space allowance on the agonistic and spacing behavior of cattle." *Applied Animal Behaviour Science* 24(2): 127–135. https://doi.org/10.1016/0168-1591(89)90040-3.

Lambeth, S. P., Hau, J., Perlman, J. E., Martino, M., Schapiro, S. J., and Keeling, M. E. 2006. "Positive reinforcement training affects hematologic and serum chemistry values in captive chimpanzees (Pan troglodytes)." *American Journal of Primatology* 68: 245–256. https://doi.org/10.1002/ajp.20148.

Langley, R. L., and Morrow, W. M. 2010. "Livestock handling—Minimizing worker injuries." *Journal of Agromedicine* 15(3): 226–235.

Lay, D. C., Friend, T. H., Randel, R. D., Jenkins, O. C., Neuendorff, D. A., Kapp, G. M., and Bushong, D. M. 1996. "Adrenocorticotropic hormone dose response and some physiological effects of transportation on pregnant Brahman cattle." *Journal of Animal Science* 74: 1806.

LeDoux, J. E. 2000. "Emotion circuits in the brain." *Annual Review of Neuroscience* 23: 155–184.

León-Llanos, L. M., Flórez-Díaz, H., Duque-Muñoz, L. G., Villarroel, M., and Miranda-de la Lama, G. C. 2022. "Influence of temperament on performance and carcass quality of commercial Brahman steers in a Colombian tropical grazing system." *Meat Science* 191: 108867.

Lind, J., and Cresswell, W. 2005. "Determining the fitness consequences of antipredation behavior." *Behavioral Ecology* 16(5): 945–956.

Magurran, A. F., and Girling, S. L. 1986. "Predator model recognition and response habituation in shoaling minnows." *Animal Behaviour* 34: 510–518.

Marchant, J. N., Whittaker, X., and Broom, D. M. 2001. Vocalisations of the adult female domestic pig during a standard human approach test and their relationships with behavioural and heart rate measures. *Applied Animal Behaviour Science* 72(1): 23–39.

Meijaard, E., d'Huart, J., and Oliver, W. 2011. "Family suidae (pigs)." In *Handbook of the Mammals of the World, Vol. 2. Hoofed Mammals*, edited by D. E. Wilson and R. A. Mittermeier, 248–291.

Menke, C., Waiblinger, S., Fölsch, D. W., and Wiepkema, P. R. 1999. "Social behaviour and injuries of horned cows in loose housing systems." *Animal Welfare* 8(3): 243–258. https://doi.org/10.1017/S0962728600021734.

Minier, D. E., Tatum, L., Gottlieb, D. H., Cameron, A., Snarr, J., Elliot, R., Cook, A., Elliot, K., Banta, K., Heagerty and McCowan, B. 2011. "Human-directed contra-aggression training using positive reinforcement with single and multiple trainers for indoor-housed rhesus macaques." *Applied Animal Behaviour Science* 132: 178–186. https://doi.org/10.1016/j.applanim.2011.04.009.

Morse, D. H. 1993. "Interactions between tit flocks and sparrowhawks (*Accipiter nisus*)." *Ibis* 115: 591–593.

Mota-Rojas, D., Broom, D. M., Orihuela, A., Velarde, A., Napolitano, F., and Alonso-Spilsbury, M. 2020. "Effects of human-animal relationship on animal productivity and welfare." *Journal of Animal Behaviour and Biometeorology* 8(3): 196–205.

Munksgaard, L., Jensen, M. B., Pedersen, L. J., Hansen, S. W., and Matthews, L. 2005. "Quantifying behavioural priorities—Effects of time constraints on behaviour of dairy cows, Bos taurus." *Applied Animal Behaviour Science* 92(1–2): 3–14.

Noirot, E. 1972. "Ultrasounds and maternal behavior in small rodents." *Developmental Psychobiology* 5: 371–387.

O'Brien, J. K., Heffernan, S., Thomson, P. C., and Mcgreevy, P. D. 2008. "Effect of positive reinforcement training on physiological and behavioural stress responses in the hamadryas baboon (Papio hamadryas)." *Animal Welfare* 17: 125–138. https://doi.org/10.1017/S0962728600027639.

Orihuela, J., and Solano, J. J.. 1994. "Relationship between order of entry."

Palmer, W. 1909. "Instinctive Stillness in Birds." *The Auk* 26(1): 23–36.

Pandey, M., Patel, J. V., Madhavatar, M., and Shah, S. V. 2021. "Influence of housing and housing facilities on bovine health, production, and welfare." In *Recent Research Trends in Veterinary Sciences and Animal Husbandry*, edited by A. N. Publications, 71–92. India: AkiNik Publications. https://doi.org/10.22271/ed.book.1186.

Price, E. O. 1999. "Behavioral development in animals undergoing domestication.' *Applied Animal Behaviour Science* 65: 245–271.

Rico-Guevara, A., and Hurme, K. J. 2019. "Intrasexually selected weapons." *Biological Reviews* 94: 60–101.

Roelofs, K. 2017. "Freeze for action: neurobiological mechanisms in animal and human freezing." *Philosophical Transactions of the Royal Society B: Biological Sciences* 372(1718): 20160206.

Rushen, J. 2017. "Housing and the welfare of dairy cattle." Retrieved from https://www.researchgate.net/profile/Jeffrey_Rushen/publication/319168965_Housing_and_the_welfare_of_dairy_cattle/links/5a4c2d86458515a6bc6bfb1a/Housing-and-the-welfare-of-dairy-cattle.pdf. https://doi.org/10.19103/AS.2016.0006.03.

Sih, A. 1998. "Behavioral syndromes: An ecological and evolutionary overview." *Trends in Ecology & Evolution* 19(7): 372–378.

Skinner, B. F. 1938. *The Behavior of Organisms: An Experimental Analysis*. New York: D. Appleton-Century Company Inc.

Sørensen, D. B., Dragsted, N., and Glerup, P. 2011. "Positive reinforcement training in large experimental animals." *Altex* 28: 250.

Stankowich, T., and Blumstein, D. T. 2005. “Fear in animals: A meta-analysis and review of risk assessment.” *Proceedings of the Royal Society B: Biological Sciences* 272(1581): 2627–2634.

Vianna, D. M. L., and Carrive, P. 2005. “Changes in cutaneous and body temperature during and after conditioned fear to context in the rat.” *European Journal of Neuroscience* 21: 2505–2512. https://doi.org/10.1111/j.1460-9568.2005.04073.x.

Wagner, S. (2006, September). “Understanding bovine behavior for safe handling.” In *American Association of Bovine Practitioners Conference Proceedings* Veterinary Technician Session (pp. 227–228). https://doi.org/10.21423/aabppro20074555.

Waynert, D. F., Stookey, J. M., Schwartzkopf-Genswein, K. S., Watts, J. M., and Waltz, C. S. 1999. “The response of beef cattle to noise during handling.” *Applied Animal Behaviour Science* 62: 27–42.

Weary, D. M., Jasper, J., and Hötzel, M. J. 2008. “Understanding weaning distress.” *Applied Animal Behaviour Science* 110(1–2): 24–41.

Webster, J. 1994. *Animal Welfare: A Cool Eye towards Eden*. Oxford: Blackwell Science.

Zulkifli, I. 2013. “Review of human-animal interactions and their impact on animal productivity and welfare.” *Journal of Animal Science and Biotechnology* 4: 1–7.

10 Reproductive and Maternal Behavior

Ranjini Manuel

10.1 INTRODUCTION

The evolutionary prosperity of animals is directly linked to their reproductive behaviors. Ecological pressures, natural selection, and sexual conflicts highly influence evolutionary changes. Successful reproduction incorporates a range of actions, from seeking an appropriate partner, charming the partner, and mating to a fruitful creation. Over the millennia, these actions and the genetic traits related to reproduction face significant selection pressures and may undergo gradual evolutionary changes. The conflicting goals of the male and female partners across the species usually steer new reproductive adaptations (Anholt et al., 2020).

The study of reproductive and maternal behaviors in the animal kingdom characterizes a fascinating tapestry of adaptations and ecological dynamics that shaped the continuity of species. The elaborate courtship behaviors, reproduction, and the nurturing given by the mothers have paved the way for interesting evolutionary adaptations through natural selection. A rich assortment of reproductive strategies are finely weaved to suit different taxa. This chapter aims to dissect these reproductive strategies and various adaptations embraced by animals. It critically examines the mating systems, courtship rituals, and parental care employed to raise a successful future generation. The diverse paths adopted by various species to ensure fruitful reproduction and better survival of the future generation are reflected by the multilayered array of maternal behaviors in harmony with the environmental variables. The key advantages of such adaptations encompass offspring survival, reproductive success, environmental synchronization, allocation of resources, social cohesion, and transfer of knowledge and skills.

The chapter highlights the mating schemes, intricate courtship rituals, and environmental factors that play a pivotal role in breeding, and, most importantly, the parental investment across diverse species of mammals, birds, reptiles, and amphibians. Understanding the earlier factors will better clarify how such behaviors influence evolutionary changes.

10.2 MAMMALS

Mammals are distinct vertebrates with characteristic features like mammary glands, hair/fur, homeostasis, advanced brains, specialized teeth, muscular diaphragm, and the ability to give birth. Mammals demonstrate expansive adaptations that help them thrive in diverse ecological niches. These adaptations include locomotion, reproductive strategies, various social behaviors, and feeding patterns. The diversity of species in mammals is so elaborate that they range from tiny rodents to humongous marine creatures demonstrating extraordinary modifications, which allowed them to survive over centuries (Kielan-Jaworowska et al., 2004).

DOI: 10.1201/9781003710271-10

10.2.1 Reproductive Strategies in Mammals

Mammals engage in various reproductive tactics, which vary substantially among species based on ecological, social, and environmental factors (Hamlett 2011). The primary reproductive strategies include

10.2.1.1 Mating Systems

The different species in the class Mammalia have different social and reproductive structures. Understanding mating systems in mammals sheds light on the evolutionary pressures they had to undergo for survival. The selection of mating schemes can impact a population's social structures, reproductive success, and genetic diversity. Mating schemes refer to the behavioral orientation regarding reproduction and breeding. Standard mating systems include monogamy, polygamy, and polygynandry.

10.2.1.1.1 Monogamy

Monogamy refers to a system where one male and female share a long-term bond and commitment towards each other. They share responsibilities in terms of parenting and protection of the territory. This system is less common in mammals compared to other systems. However, several species exhibit monogamy. Gibbons are known for their monogamy. They form unbreakable social bonds between a male and a female. They equally share the burden of territorial defense and parenting, which helps to rule over areas with abundant food resources. Research suggests they engage in exquisite duet singing and courtship (Brockelman et al., 1998) (Figure 10.1).

California mice also exhibit monogamy. They share a long-term bond, and both parents participate in nurturing the offspring (Ribble and Salvioni, 1990) (Figure 10.2).

10.2.1.1.2 Polygamy

In polygamy, individuals may consider having multiple partners during their reproductive lives or a particular breeding season. It generally showcases two forms: polygyny and polyandry.

FIGURE 10.1 Gibbons pair bonding in natural environment.

FIGURE 10.2 California mouse pair bonding in natural environment.

Polygyny refers to one male considering multiple females, and polyandry refers to one female considering multiple males in the reproductive course. In a polygynous system, males often have to compete with other males within the population to access females and the associated resources. Lions are one of the classic examples of polygynous mating systems. In a lion pride, the dominant males copulate with multiple females, defending the pride's territory from other prides and strangers. (Packer and Pusey, 1987) (Figure 10.3).

Elephant herds are all about matriarchy, and the herd consists of mainly females and their young ones. At the same time, males prefer solitary lives and form relaxed associations.

Nevertheless, they join the herd during the breeding season, and males compete for dominance. Proven dominant males earn exclusive rights to mate with multiple females (Lee and Moss, 2012) (Figure 10.4).

Polyandry is the system where one female considers mating with multiple males. It is rarely seen in mammals, but the spotted hyena is one animal in which it occurs. In hyena societies, females lead in defending the territory; they are particularly dominant and chivalrous. Due to their superiority, they tend to have multiple mates in one breeding season, which makes unclear the paternity line of their young (East et al., 2003).

10.2.1.1.3 Polygynandry

Polygynandry, also known as the promiscuous mating system, refers to both males and females indulging in multiple copulations with multiple partners in a breeding season. They do not form long-lasting bonds. Notable examples are bonobos, who are known for their complicated social structures and promiscuous mating schemes (De Waal and Lanting, 2023). Bottlenose dolphins also exhibit polygynandrous mating systems but are primarily within a social group called pods (Connor et al., 2000).

FIGURE 10.3 Lion pride guarding the territory.

FIGURE 10.4 Elephant bulls fighting for dominance on a savannah.

10.2.1.2 R Selection vs. K Selection

R' and K' selections are reproductive strategies. R' selection strategy refers to the intrinsic growth rate, denoted by the letter R, in population ecology models. The critical goal in an R's strategy is to prioritize quantity over quality. It emphasizes a high reproductive rate, often at the expense of the survivability of the young. These animals concentrate on maximizing the number of offspring in a reproductive event, boosting the chances of at least some offspring surviving to reproductive maturity. These mammals often easily attain sexual maturity, allowing them to reproduce early. However, they have shorter lifespans because of higher mortality rates. These offspring may be born in big batches, but the parental investment in nurturing is limited. Species adopting the R selection system capitalize on swift reproduction when there are abundant resources and favorable environmental conditions. Rodent species are suitable examples of quantity over quality missions. They produce large numbers of litters quickly and do not invest in nurturing and protecting the young. The offspring often face challenging living conditions on their own, which leads to a high mortality rate among them (Pianka, 1970).

K' selection strategy refers to the carrying capacity, denoted by the letter K, in population ecology models. The fundamental goal of a K' strategy is to prioritize quality over quantity. They aim at the quality and survival of the young ones rather than increasing the reproductive output. Their litter size is often small, which aids the mother in ensuring better survival of their babies. These species have long gestation periods and delayed attainment of reproductive maturity. The parents, especially the mother, significantly invest time and energy in nursing, protecting, and guiding the offspring, contributing to their increased survival rates. Most species that adopt K selection are large mammals such as elephants, whales, certain ungulates, and primates. They have a much more extended gestation period with a limited number of offspring compared to R selection. Most of the time, they successfully seek optimal resources for raising their little ones, leading to low mortality rates (Figueredo et al., 2005).

10.2.1.3 Seasonal Breeding

Seasonality in reproduction suggests that the mating and birthing cycles happen at a specific period in a year. Environmental conditions like photoperiod, temperature, and resource availability highly prompt this phenomenon. Considering that different seasons bring different environmental changes, only some are suitable for reproduction. They rely on the changing seasonal cues to prepare for breeding. Once the environment is set, they actively seek mates and engage in courtship rituals. After successful copulation, females undergo gestation with the predictability of births occurring in favorable seasons and specific times. The adaptation that helps in coping with the changing seasons is hormone levels that directly influence reproductive processes. Wolves display seasonal breeding. Breeding typically occurs in late winter or spring, which ensures suitable availability of resources when they are born (Mech and Boitani, 2019). White-tailed deer in North America prefer breeding in a season called rut. Rut falls between October and December. In this season, the bucks compete with other bucks for mating rights. They enter estrus, and births occur seven months later (Pierce et al., 2022).

10.2.1.4 Delayed Implantation

Delayed implantation, otherwise known as embryonic diapause, is another reproductive strategy in certain mammals. It describes the temporary suspension of the development of fertilized embryos within the animal's uterus. These embryos remain viable until favorable conditions are met. Rather than immediate implantation of the fertilized embryo into the uterus, after fertilization and blastocyst formation, further embryo development is paused. Hormonal levels, especially progesterone, play a pivotal role in the process of delayed implantation. The blastocyst remains dormant, preserved in the uterus with insignificant activity until the hormones trigger further development. The duration of the dormancy may vary across the species depending upon their breeding adaptations. It may range from days to months. Another advantage of delayed implantation is that it aids in the synchronization of the birth timings with suitable environmental conditions. Therefore, the offspring are born when food and shelter are plentiful. Large carnivores like bears, weasels, minks, and marsupials exhibit delayed implantation.

10.2.2 Parental Behavior and Investment in Mammals

Parental investment describes the resources, time, energy, and work parents assign to care for and foster their young ones. This investment plays a potential role in the offspring's survival, development, and breeding success. Depending on the environmental niches and social configurations, parental care can vary. It refers to an array of behaviors a mother demonstrates to guarantee her young one's endurance, well-being, and growth. Parental care includes gestation, lactation, protection, nurturing, teaching, and socialization.

10.2.2.1 Gestation

Maternal care starts from gestation. It is the process where the mother protects the growing embryo inside her uterus. In this period, the mother provides the fetus with the nutrients and oxygen required for growth and development. After fertilization, the embryo is transported from the reproductive tract and gets implanted in the uterus. The period the embryo gets nourished within the uterus may vary depending on the species. The time taken from fertilization to the delivery of the fetus is called the gestation period. The gestation period of mice is 19–21 days; for dogs, it is 58–68 days; for humans, it is 38–42 weeks. The role of the placenta in gestation is irreplaceable. It facilitates the exchange of nutrients, gases, and the removal of waste products between a mother and a fetus. The development of the fetus takes place stage by stage, from the fusion of egg and sperm to a multi-organ system.

More extensive and complex mammals may take longer to complete fetal development. During this time, the mother undergoes different bodily changes to support the growth of the fetus. The changes are mainly physiological and hormonal. Hormonal changes are responsible for increased blood volume, alterations in the function of specific organs, and other changes needed to accommodate the growing fetus. Labor begins at the end of gestation. It involves extensive uterine contractions, which will result in the expulsion of the baby along with the placenta. It is vital to note that gestation periods are highly influenced by factors like environment, maternal health, and fetal development (Martin and MacLarnon, 1985).

10.2.2.2 Lactation

One of the characteristic features of a mammal is its ability to nourish its offspring with milk. Lactation is a strong pillar in maternal care. The milk nourishes the offspring with nutrients, antibodies, and other immune factors required for growth and immunity development. It is enabled by the specific organ called mammary glands. It is the organ synthesizing and secreting milk from a mother's body. Hormones play a crucial role in the secretion of the milk. During the gestation period, hormones like prolactin, estrogen, and progesterone aid in the development and enlargement of the mammary glands. The initial milk produced by the mammary glands is called colostrum. It is a nutrient-rich yellowish milk that acts as the source of antibodies required to protect the offspring. The mature milk that secretes after colostrum combines fats, proteins, carbohydrates, vitamins, minerals, and water. It forms a complete balanced diet for the newborn. The composition of the milk changes from time to time depending on the needs and developmental stage of the newborn.

The unbreakable bond, or imprinting, mostly happens when the offspring suckles on the mother's nipple. It stimulates the nerve endings, initiating the release of oxytocin. *Oxytocin* is the love hormone that binds the mother to the offspring. Oxytocin causes the muscles around the glands to contract, releasing the milk into the mammary ducts. Newborns feed very frequently, as their energy requirements for development are higher. As they grow, the frequency of nursing reduces. *Weaning* is a period when the newborn switches from the mother's milk to solid or semi-solid food. It is also the period when the offspring learns to be independent (Oftedal, 2000).

TABLE 10.1
Gestation Period in Mammals

Animal	*Average Gestation Period (Days)*
Cattle	280
Cat	65
Dog	63
Donkey	365
Goat	150
Horse	335–342
Llama, Alpaca	335–365
Pig	114
Rabbit	31
Sheep	150
Bear	210
Bison	280
Camel	365–400
Chimpanzee	236
Coyote	63
Deer	200
Elephant	660
Giraffe	425
Gorilla	260
Lion	108
Monkey	180
Panther	90
Tiger	103
Whale	450

10.2.2.3 Protection

Mothers are known for their instincts when it comes to protecting their young ones. They build nests/shelters, camouflage their offspring, and demonstrate defensive behaviors when they spot a threat. Nests or shelters are the foremost factor that any mother with maternal instincts looks for before parturition (the act of giving birth). This could be a burrow, nest, sheltered area, or den. Finding a shelter aims to protect the litter from predators, adverse climatic conditions, and other potential dangers. On exposure to danger, the mother takes aggressive postures, vocal warnings, and shows a tendency to attack. Another tactic to deter predators is to hide the defenseless young ones. This includes camouflaging them in thick vegetation or rocks. Some mammals use their mouth and paws to relocate them occasionally to keep them safe and away from predators. When the offspring are defenseless, mothers hold high vigilance to detect danger. In specific herds, the young ones' protection is vested by the whole pack or group itself. The group takes care and protects the young population. This type of behavior is typically found in herds with strong social structures and ecological niches (Cameron and Garcia, 2013).

10.2.2.4 Nurturing and Grooming

Grooming, cuddling, and providing warmth and comfort are different forms of expressing love and care among animals. Maternal care is no different. These gestures help the mother deepen the bond with the offspring and help them regulate their body temperature. The mother aids in eliminating moisture and redistributing natural oils across the body. In addition, nurturing behaviors contribute substantially to newborns' well-being, growth, and adaptation to their surroundings. The grooming behavior by the mother involves licking and cleaning the offspring's body. Grooming behavior improves the young one's hygiene, eliminates parasites, and reinforces emotional bonds. Other advantages of grooming behavior are that it relieves stress and establishes a sense of security, particularly in dangerous scenarios. The tactile sensations during grooming assist in developing senses and improving interaction with the environment. Maternal nurturing behaviors work as a tool that helps young ones recognize their mother. These meaningful interactions pave the way for understanding social structure, ecological dynamics, and acceptance (Broad et al., 2006).

10.2.2.5 Teaching and Socialization

A mother familiarizes the social skills to their offspring through various behaviors and communication. These skills prepare them to face real-world problems like survival, adaptations, and social group interactions. Mothers impart essential knowledge regarding foraging, hunting, seeking shelter, and defense. Most of the skills are taught by demonstration techniques, correcting while young ones engage in activities, presenting opportunities for scrutiny and imitation. When the offspring are young, mothers help them locate food, teach them nutritional wisdom, and how to handle different food situations. A mother and offspring interact through different body language, vocalizations, and species-specific communication cues. Littermates also play an essential role in teaching and socialization. Littermates mingle with each other and help to determine social relationships; they help each other recognize differences in dominant and submissive behaviors. Mothers encourage play behaviors to develop motor skills and problem-solving tactics. After weaning, mothers slowly withdraw nurturing behaviors and encourage independence. The transition from being dependent on the mother to being self-sufficient is gradual (Rosenbaum and Gettler, 2018).

10.2.2.6 Extended Care

The prolonged support and care that the mothers give for extended periods is called extended maternal care. Usually, maternal care is given only until the weaning age or only in the initial stages of life of the young animals. Certain species of mammals require prolonged care. The period of maternal care may vary across the species depending upon the contributing factors. The contributing factors include the developmental necessities of that species, social structures, and other environmental factors. In such extended care situations, the mother provides either partly predigested food or helps them locate food. They continue to protect the young animals from danger, predators, and unknown situations. Due to the extended time that young animals get, mothers teach elaborate hunting and survival skills. They also learn navigation and ways to cope with stressful environmental conditions. The mother continues altering unacceptable behaviors and reinforces desirable ones. In the school of extended training, mothers provide them the opportunity to make decisions, explore, and learn from their mistakes. The transition from maternal dependence to independence is gradual in certain species. Such young animals are better survivors and leaders (Van Noordwijk et al., 2012). Prominent examples are the majestic elephants. They display extended maternal care among mammals. Young elephants are cared for by their mothers and other females in the herd. Mothers nurse their young ones for over five years, and the herd provides protection to all the young animals present in the herd. In the herd, they get the opportunity to learn navigation, obstacle training, and, most importantly, survival. The females and young animals in the herd share a very close and strong bond. Even after separating from the herd, they can recognize each other and do not hesitate to help when needed (Lynch et al., 2019).

The humpback whale is one other example of extended maternal care. The mother continues to support their growth and nourish them with milk for extended periods. In addition, mothers share knowledge about migrating routes, which might have been passed down through generations. The young whales remain with their mothers and depend upon them until they become self-sufficient (Szabo and Duffus, 2008) (Figure 10.5).

FIGURE 10.5 Mother humpback whale guiding the calf.

In primates, orangutans have one of the most extended maternal care periods. Mother orangutan gives intensive care and guidance to the offspring. Young animals remain with their mothers and the group until they are eight years or older. Mothers train young animals with all the essential skills, including seeking shelter and curating tools for specialized purposes (Van Noordwijk et al., 2018).

10.3 BIRDS

Birds are vertebrate creatures categorized under the class Aves. The essential characteristics of birds are their pneumatic skeletal system, beaks, presence of air sacs, and ability to lay eggs. Most birds are adapted to fly. However, not all birds can fly. Such birds have unique adaptations like swimming and diving or might be more terrestrial (Campbell et al., 2005).

10.3.1 Reproductive Strategies in Birds

10.3.1.1 Mating Systems

The standard mating systems found in birds are monogamy and polygamy.

10.3.1.1.1 Monogamy

As mentioned earlier in the section on mammals, monogamy in birds is a reproductive strategy in which one male bird mates with one female bird. They share a longer-lasting bond that is sealed with commitment and cooperation. Most of such monogamous pairs last until they die. The bond encourages them to cooperate in nest building, incubation of eggs, and raising their offspring. Monogamous reproductive systems can be swayed by resource availability, ecosystem, and parental care for fruitful procreation. Bald eagles share long-term commitment with their partner. The male and female together build their nests, incubate their eggs, and share the burden of raising their chicks. This mating system positively impacts the species' survival (Dawe, 2016). Swans are also known for their monogamy. They often share an everlasting solid bond with their partner. Swans also collaborate in various aspects of breeding, territorial defense, egg incubation, and upbringing of the cygnets (Bush, 2018).

10.3.1.1.2 Polygamy

Polygamy is another mating system in birds. It involves an individual bird copulating with multiple mates. There are two main types of polygamy, namely polygyny and polyandry.

In a breeding season, if a male bird copulates with multiple female birds, it is known as a polygyny mating system. Just like mammals, it happens when the male birds compete with other male birds to mate with multiple females, and the male takes up the duty to defend the territory. Red-winged blackbirds follow the polygynous mating system. The male red-winged blackbirds are responsible for defending a specific territory containing numerous nests. The dominant male bird has exclusive rights to mate with multiple females and is allowed to nest in his territory under his protection (Westneat, 1993).

In a breeding season, if a female bird copulates and establishes a bond with multiple male birds, it is called a polyandry mating system. It is less typical among birds compared to the polygyny system. The polyandry system is adopted only when females invest intensively in parental care or when males contribute substantially to the upbringing of the chicks. The spotted sandpiper is polyandrous; the female spotted sandpiper works relentlessly to establish territory, initiate courtship rituals, and copulate with multiple males. Once the eggs are laid, females are not responsible for raising the chicks. The males take over the brooding process (Oring et al., 1994).

10.3.1.2 Courtship Behaviors

Courtship behaviors are an assortment of behaviors demonstrated for seeking a potential mate. These rituals are species-specific and play a vital role in enticing a potential mate for successful breeding.

Lekking: A particular courtship behavior display area is called leks. Males gather in leks to allure females for courtship and breeding. Males demonstrate elaborate and driven dancing and vocalizations to gain the attention of the females. Such behavior is called lekking. Females are the spectators in leks. They visit to grade and choose potential mates for the breeding season. The permutations and combinations of talent demonstrated by the males can influence the choice of the females. The greater sage-grouse is one of the incredibly talented birds that showcase lekking behaviors. They gather in communal grounds and display chest puffing, tail fanning, and vocalizations. Females choose the best performers (Schroeder et al., 2004) (Figure 10.6).

The Indian peafowl also displays intricate dance moves, flaunts their flamboyant tail feathers, and calls the females (Loyau et al., 2005).

Gift giving: In certain species, males try to bribe females with gifts to be the chosen one. The gifts are usually food items or nesting objects. The gift-giving behavior showcases the ability of the males to provide. The male European bee-eaters give insects as part of courtship behavior. These behaviors are only displayed during mating season (Urban et al., 2020).

Nest building: Nest building is considered an essential winning factor in courtship behavior in birds. It can be a collaborative or individual courtship pursuit for birds. Nesting skills are often noticed and credited by females. The females often choose the male birds with great nesting skills. They might even share a long-term bond with such mates. Great crested grebes construct sophisticated floating nests as part of the courtship ritual. Such a grand gesture provides a site for laying eggs and the possibility of raising chicks with the potential partner. It serves as the epitome of long-term commitment (Harrison & Cherry, 1997) (Figure 10.7).

Feeding displays: Male birds often engage in sharing food with females. Food sharing is also a part of courtship behavior. The type of food shared depends upon the species. Feeding display indicates the readiness of the male to share resources and willingness to share the burden of raising the chicks. The male Eurasian wryneck traps a favorite insect and presents it to the female bird. It demonstrates the ability to share commitments and care for each other (Cramp et al., 1983).

FIGURE 10.6 Greater sage-grouse lekking behavior.

FIGURE 10.7 Male crested grebe building a nest and protecting the chicks.

Chasing pursuits: Chasing pursuits are another courtship behavior used to attract a female partner. For the female, this behavior symbolizes agility, speed, and persistence of the partner. The northern lapwing's chasing pursuits consist of aerial displays and chasing of the females. This behavior presents the male's strength and agility (Harrison and Cherry, 1997).

10.3.1.3 Cooperative Breeding

Cooperative breeding describes a social arrangement where the group members, especially the non-breeding members, share the burden of raising and protecting the chicks. This cooperation is to increase the reproductive success of the breeding pair. The social structure of the acorn woodpecker is similar. The assisting members are the offspring from previous breeding seasons. They protect the territory, locate food for chicks, and guard the nests in the absence of breeding pairs (Stacey and Ligon, 1987).

10.3.1.4 Brood Parasitism

Brood parasitism is a famous reproductive strategy adopted by certain species of birds. When a female belonging to a species lays eggs in another female's nest of a different species and depends on the host species to incubate and raise the chicks, this phenomenon is called brood parasitism. The common cuckoo lays eggs in a crow's nest and hopes that the crow will incubate and raise the chicks. Crows get easily fooled because of the close resemblance of their egg's appearance with that of the cuckoo. Often, the female crow raises the chicks of the cuckoo along with her own or raises the cuckoo chicks alone. This confusion mainly occurs due to the short incubation period of the cuckoo's egg in comparison to the crow's egg. This reproductive strategy is advantageous for cuckoos, as it helps to avoid the tedious task of nest building and nurturing the chicks (Davies, 2011).

10.3.2 Parental Behavior and Investment in Birds

10.3.2.1 Incubation/Brooding

Incubation of eggs refers to the process by which eggs are kept at a specific temperature for the development of the embryo. The time taken from the laying of eggs to hatching is called the incubation period. Brooding is a process in which the parental birds sit on the eggs to attain a specific temperature for the development of the embryo. The heat from the body of the birds is transferred to the eggs, thereby maintaining a particular temperature over the eggs. A stable and constant temperature and relative humidity are crucial for the hatching of the eggs. The temperature required for eggs varies across the species. The warmer the temperature, the shorter the incubation period will be. Brooding parents adjust the feather alignment in such a way that it serves as a protective barrier for the eggs or for the chicks. While brooding, the eggs are turned by the parents to ensure even distribution of the heat and to avoid membranes sticking to the eggshells. In most species, both parents take turns to brood over the eggs. It is commonly referred to as biparental incubation. When one parent broods over the eggs, the other parent takes a short break to feed. Mourning doves follow biparental incubation; they take turns to incubate the eggs. The incubation period can range from a few days to weeks, depending upon the species (Table 10.2).

10.3.2.2 Nest Building

The construction of a structure to lay eggs and raise chicks is known as a nest. Upon the arrival of the breeding season, all the females start building their nests. According to the needs of the species and ecological structure, nests can vary in size, location, materials used, etc. Male American robins provide the materials for nesting, and females assemble a cup-shaped nest with grass and twigs. An American robin's nest is often seen in the crotch of the trees. They might have more than one brood in a season; therefore, they may be involved in assembling more than one nest (Wauer, 1999). Nest building can be categorized based on the nest type, materials used, location and habitat, and nesting cycles (Tables 10.3, 10.4, 10.5, and 10.6).

10.3.2.2 Feeding

Parental feeding is a common nurturing behavior exhibited by parents. It involves locating and providing food to their young ones. Depending on the species and nature of their ecology, the feeding techniques and type of food may vary. Some birds may feed their nestlings by regurgitating partially digested food. Some may place the fresh catch on the beak of the chick.

TABLE 10.2
Incubation Period of Eggs in Birds

Birds	*Average Incubation Period (Days)*
Swan	35
Chicken	21
Quail	16–18
Pigeon	16–19
Duck	28
Pheasant	24–26
Ostrich	42
Peafowl	28–30
Crested Grebe	27–29
Eagle	35
Red winged blackbird	13
Great sage grouse	25–29
European bee-eater	20
Wryneck	15
Northern lapwing	28–30
Woodpecker	14
Cuckoo	13
Mourning dove	14
American robin	14
Eastern blue bird	19
Killdeer	22–28
Common terns	22–27

TABLE 10.3
Classification of Nests Based on Nesting Cycles

Type	Nesting Cycles
Seasonal	Nesting activity is often seasonal, with birds choosing specific times to build nests, lay eggs, and raise chicks
Multiple broods	Some species may build multiple nests for successive broods in a single breeding season

TABLE 10.4
Classification of Nests Based on Location and Habitat of the Nest

Type	Location and Habitat
Tree	Build nests in trees, utilizing branches and foliage for protection
Ground	Create shallow depressions in the soil and may line the nest with softer materials
Cliff	Birds adapted to cliff environments

TABLE 10.5
Classification of Nests Based on Materials Used to Build the Nest

Type	Materials Used
Natural	Twigs, leaves, grass, feathers, and mud are commonly used
Synthetic	May incorporate human-made materials like paper, plastic, or string into their nests.

TABLE 10.6
Classification of Nests Based on Nest Type

Type	Description
Cup	Common and are often built in trees or shrubs
Platform	Flat and can be found on ledges or platforms, such as cliffs or buildings
Ground	build nests directly on the ground

The duration of parental feeding is determined by the level of nutrition required for the chick and the stage. As with any other young ones, the frequency of feeding reduces as the chick grows. Eastern bluebirds follow biparental feeding. They catch live feed and give it to the chicks. The chicks are often encouraged to pick up live insects as a part of training to be independent (Glotfelty et al., 2017). Feeding by parents contributes directly to the upbringing and survival of the chicks. These adaptations have significantly contributed to the evolution of species.

10.3.2.2 Protection

Parental protection refers to the behaviors and schemes employed by the parent birds to protect the nestlings. The parent birds spend a lot of time guarding the nest from potential threats and dangers. They actively defend by displaying resistance, emitting high-pitched vocal calls, provoking the predator, or distracting the predators away from the nests. Distraction methods are commonly used by birds where they lead predators away from the nests or they can behave as though injured and vulnerable. During danger, they can be very vocal to alert the nestlings about the danger. Some birds aggressively defend their territory; they attack by flapping wings at high speed and pecking, accompanied by aggressive calls. Mobbing is also commonly found in various bird species. The dominant males gather and attack the predators like an army marching forward at the war front. Birds take caution in selecting the nest area. They try to build the nest where they can easily hide their offspring and are unreachable to predators; it is a form of passive protection. They use natural nest materials like grass, leaves, mud, and twigs to blend in with nature and surroundings, thereby not grabbing any unwanted attention from predators. Birds like killdeers build their nests on ground; they use distraction methods to divert predators. They often demonstrate broken wing gestures or behave as though injured, thereby misleading the predators and driving them away from the nesting site (Mahr et al., 2015).

10.3.2.2 Teaching and Guidance

Survival is only for the fittest. In a world full of predators, learning to fight is the only way to sustain life. Just like mammalian parents, bird parents also focus on teaching and guiding the young generation about survival skills. Learning starts soon after hatching; every day they learn something new. When they open their eyes, the first thing they learn is to recognize their mother, her call, and her scent. When the mother feeds the nestlings, they start to understand food and how it's important to maintain life. Mother gradually teaches them to locate food, recognize potential danger, and navigate the surroundings. Once they develop body feathers, the mothers teach them how to fly and land. Nestlings often learn better by the trial-and-error method. In migratory birds, when chicks are ready to fly, they accompany the flock, where they learn migrating routes, how to feed during migration, when to be alert, and when to expect danger. Chicks learn by observing and imitating their mother and the flock. Understanding ecological structure comes from social interaction and play behavior within the flock. These understandings help them to be self-sufficient. Birds that have both terrestrial and aquatic lifestyles teach the young chicks to swim and to catch fish underwater. Parents gradually encourage the young to catch, handle, and eat the prey on their own. They also provide real-life situations to escape and defend themselves from predators. Female common terns teach and guide their chicks by presenting live, half-dead prey to them; this teaches them how to handle the prey (Whittam and Leonard, 2000).

10.4 REPTILES AND AMPHIBIANS

Reptiles and amphibians are cold-blooded vertebrates belonging to two different groups. Reptiles possess scaly and dry skin whereas amphibians possess moist skin. Reptile eggs are laid on land, and they contain amniotic fluid. Amphibian eggs are laid in water or moist areas. Amphibians undergo more obvious metamorphosis than reptiles. Reptiles prefer a terrestrial lifestyle, while amphibians need both water and a terrestrial lifestyle, as it is important for their respiration. Amphibians have the ability to respire through their skin and their lungs, while reptiles can respire only through their lungs.

10.4.1 Reproductive Strategies in Reptiles and Amphibians

10.4.1.1 Mating Systems

The reproductive strategy in reptiles and amphibians is polygamy, including polygyny and polyandry.

10.4.1.1.1 Polygyny

The polygynous system is one of the common types of mating systems found in reptiles and amphibians. Male iguanas defend their territory and compete to mate with multiple females (Pratt et al., 1992). The common garter snake is another example of a polygynous mating system.

10.4.1.1.2 Polyandry

Research suggests that Alpine newt species have a polyandrous mating system. Females lead a dominant life; they mate with multiple males within the group. In the study, upon investigation researchers found that a considerable number of young ones were sired by more than one male. However, polyandry is rarely seen in reptiles and amphibians (Bergo, 2004).

10.4.1.1.3 Mating Aggregation

This is a communal congregation where the members of the same species gather around the same time of the season for breeding. Other names for these gatherings are mating balls and mating congresses. Male and female groups take part in mating aggregation. It acts like a common ground to select the mate for the breeding season. These gatherings encourage and facilitate fruitful reproduction by connecting all the potential mates of the season. In this aggregation, males often take the opportunity to display courtship behaviors, and females get to choose a mate. The place and time of the congregation are decided based on the seasonal cues, chemical substances released by the specific species during mating season, and communication within the community. In spring, thousands of red-sided garter snakes gather in limestone caves in Manitoba, Canada, looking for opportunities to mate (Shine et al., 2000).

10.4.1.1.4 Explosive Breeding

Synonyms for explosive breeding are explosive reproduction and explosive breeding assemblages. These are reproductive strategies where the individuals of the same species gather in a particular spot for a short period for synchronized reproduction. The breeding event is flagged off by synchronized release of gametes by both males and females into the desired area or surroundings. The climatic changes are often the cues for explosive breeding. Sudden seasonal changes like rainfall or changes in temperature lead to explosive reproduction. The advantages of this strategy are that it maximizes breeding success and minimizes the chances of predation and unsuccessful mating. The European toad displays mass migration soon after rainfall, when a large number of eggs and sperm are released into water bodies for fruitful procreation. This increases the chances of survival for the next generations (Gramapurohit and Radder, 2012). Explosive reproduction may have mating aggregations as a part of it, but the corollary is not true.

10.4.1.2 Courtship Behavior

10.4.1.2.1 Visual Display and Ritualized Movements

Visual displays are major forms of gestures for courtship. Many reptiles and amphibians engage in visual displays like rhythmic body movements, head bobbing, contraction and relaxation of dewlap, and changes in skin color. In certain snakes, they play an intricate engaging dance-like movement to show their interest in breeding. Amphibians exhibit extension of vocal sacs, flaunt bright colors, and ritualized body movements to attract the opposite sex. Male anole lizards demonstrate head bobbing to attract females to mate.

10.4.1.2.2 Vocalization

Lizards and snakes creatively use their voices to attract females. These voices include hissing and growling sounds. They are capable of curating complex sounds in a pattern for attention.

Male frogs and toads give out composite calls on the arrival of the breeding season, especially after a rainfall. Female frogs can locate and communicate with the males based on the calls. The calls serve as proof of male's reproductive fitness.

10.4.1.2.3 Nuptial Gifts

The males of the reptile and amphibian family offer gifts to females of interest. Nuptial gifts are usually food items. The gifts signify the ability of the male to provide resources.

10.4.1.2.4 Amplexus

Amplexus is a significant courtship gesture present in amphibians. It is mostly seen in anurans; the males cling to the female to form a particular mating position. Males may cling to axillary, inguinal, or dorsal regions based on their preference and species. It helps to ensure effective conception.

10.4.2 Parental Behavior and Investment in Reptiles and Amphibians

Parental care varies among the species of reptiles and amphibians. However, reptiles and amphibians do not possess great maternal or paternal instincts.

10.4.2.1 Nesting and Guarding

Crocodiles and lizards build structures resembling nests or find shelter to lay eggs. Crocodiles construct nests at the bank of a waterbody where they have ready access to water and food. They guard their eggs and protect the young ones from predators. Frogs and toads also have parental instincts to guard their eggs. The female frog hides their egg mass in submerged vegetation, or they carry it on their back to protect them.

10.4.2.2 Egg Brooding

Female pythons and boas brood over the eggs; they coil around to incubate them. It helps with the development of the growing embryo inside the eggs. Amphibian eggs are delicate and are easily prone to desiccation or drying. Hence, Darwin's frog has adapted a behavior in which the male parent carries the eggs in his pouch until they hatch. It helps to protect and preserve the next generation. Australian gastric brooding frogs show extensive maternal care in nurturing the offspring. The female swallows the eggs and keeps them inside the stomach. The growing tadpoles derive their nutrition from the stomach lining of the mother. Once they are fully developed into a froglet, the female regurgitates them carefully.

10.4.2.3 Feeding

Parental feeding behaviors in reptiles are relatively uncommon compared to other groups of animals, such as birds or mammals. Reptiles generally tend to exhibit minimal parental care, and many species do not provide direct assistance or food to their offspring. Parents may actively care for the young by providing them with food in a few amphibian species, such as certain salamanders.

10.5 CONCLUSION

Advantages and disadvantages regarding reproductive and parental behavior are co-dependent across species and their environmental conditions. Nevertheless, they play irreplaceable roles in evolution, genetic diversity, and species continuation. The downside includes resource depletion, competition, conflict, and disease transmission. Therefore, maintaining the balance between procreation, births, survivability, and deaths is the only way to conserve life on earth.

REFERENCES

Anholt, R.R., O'Grady, P., Wolfner, M.F. and Harbison, S.T., 2020. Evolution of reproductive behavior. *Genetics*, *214*(1), pp.49–73.

Bergo, P.E., 2004. Living at high altitude: testing the effects of life history traits upon the conservation of Salamandra lanai (Amphibia, Salamandridae). *Ital. J. Zool*, *71*(1), pp.35–43.

Broad, K.D., Curley, J.P. and Keverne, E.B., 2006. Mother–infant bonding and the evolution of mammalian social relationships. *Philosophical Transactions of the Royal Society B: Biological Sciences*, *361*(1476), pp.2199–2214.

Brockelman, W.Y., Reichard, U., Treesucon, U. and Raemaekers, J.J., 1998. Dispersal, pair formation and social structure in gibbons (Hylobates lar). *Behavioral Ecology and Sociobiology*, *42*, pp.329–339.

Bush, D.A., 2018. *Incubation behavior and genetic attributes of Trumpeter Swans restored to Grays Lake National Wildlife Refuge* (Doctoral dissertation, Idaho State University). Y2018.B717,https://etd.iri.isu.edu/ViewSpecimen.aspx?ID=717

Cameron, N.M. and Garcia, J.R., 2013. Maternal effect and offspring development. In *Evolution's empress: Darwinian perspectives on the nature of women*, pp.133–150.

Campbell, N.A., Reece, J.B. and Mitchell, L.G., 2005. From gene to protein. In *Biology*, 7th ed. San Francisco, CA: Pearson Education, pp. 1–52.

Connor, R.C., Wells, R.S., Mann J.A.N.E.T. and Read, A.J., 2000. The bottlenose dolphin. In *Cetacean societies*, Environmental Science, Sociology, Biology pp.91–125.

Cramp, S., Simmons, K.L.E., Brooks, D.C., Collar, N.J., Dunn, E., Gillmor, R., Hollom, P.A.D., Hudson, R., Nicholson, E.M., Ogilvie, M.A. and Olney, P.J.S., 1983. *Handbook of the birds of Europe, the Middle East, and North Africa. The birds of the Western Palearctic: 3. Waders to gulls.*

Davies, N.B., 2011. Cuckoo adaptations: trickery and tuning. *Journal of Zoology*, *284*(1), pp.1–14. Oxford University Press.

Dawe, N.K., 2016. Cooperative breeding by the Bald Eagle (Haliaeetus leucocephalus) on Vancouver Island, British Columbia. *British Columbia Birds*, *26*, pp.35–40.

De Waal, F.B. and Lanting, F., 2023. *Bonobo: The forgotten ape*. University of California Press.

East, M.L., Burke, T., Wilhelm, K., Greig, C. and Hofer, H., 2003. Sexual conflicts in spotted hyenas: male and female mating tactics and their reproductive outcome concerning age, social status, and tenure. *Proceedings of the Royal Society of London. Series B: Biological Sciences*, *270*(1521), pp.1247–1254.

Figueredo, A.J., Vásquez, G., Brumbach, B.H., Sefcek, J.A., Kirsner, B.R. and Jacobs, W.J., 2005. The K-factor: Individual differences in life history strategy. *Personality and individual differences*, *39*(8), pp.1349–1360.

Glotfelty, E., Pandit, M., Perryman, D. and Grindstaff, J., 2017. The effects of supplemental feeding on eastern bluebird (Sialia sialis) parental care. *Research Reports from Life Science Freshmen Research Scholars, 3*(1).

Gramapurohit, N.P. and Radder, R.S., 2012. Mating pattern, spawning behavior, and sexual size dimorphism in the tropical toad Bufo melanostictus (Schn.). *Journal of Herpetology, 46*(3), pp.412–416.

Hamlett, W.C. ed., 2011. *Reproductive biology and phylogeny of chondrichthyes: sharks, batoids, and chimeras, volume 3* (Vol. 3), pp.435–453. CRC Press.

Harrison, J.A. and Cherry, M., 1997. The atlas of southern African birds. *Nature, 389*(6646), pp.32–32.

Kielan-Jaworowska, Z., Cifelli, R.L. and Luo, Z.X., 2004. *Mammals from the age of dinosaurs: origins, evolution, and structure.* Columbia University Press.

Lee, P.C. and Moss, C.J., 2012. Wild female African elephants (Loxodonta africana) exhibit personality traits of leadership and social integration. *Journal of Comparative Psychology, 126*(3), p.224.

Loyau, A., Jalme, M.S. and Sorci, G., 2005. Intra-and intersexual selection for multiple traits in the peacock (Pavo cristatus). *Ethology, 111*(9), pp.810–820.

Lynch, E.C., Lummaa, V., Htut, W. and Lahdenperä, M., 2019. Evolutionary significance of maternal kinship in a long-lived mammal. *Philosophical Transactions of the Royal Society B, 374*(1780), p.20180067.

Mahr, K., Riegler, G. and Hoi, H., 2015. Parental risk management to offspring defense: bad news for kids. *Proceedings of the Royal Society B: Biological Sciences, 282*(1798), p.20141670.

Martin, R.D. and MacLarnon, A.M., 1985. Gestation period, neonatal size and maternal investment in placental mammals. *Nature, 313*(5999), pp.220–223.

Mech, L.D. and Boitani, L. eds., 2019. *Wolves: behavior, ecology, and conservation.* The University of Chicago Press.

Oftedal, O.T., 2000. Use of maternal reserves as a lactation strategy in large mammals. *Proceedings of the Nutrition Society, 59*(1), pp.99–106.

Oring, L.W., Reed, J.M. and Maxson, S.J., 1994. Copulation patterns and mate guarding in the sex-role reversed polyandrous Spotted Sandpiper, Actitis macularia. *Animal Behaviour, 47*(5), pp.1065–1072.

Packer, C. and Pusey, A.E., 1987. Intrasexual cooperation and the sex ratio in African lions. *The American Naturalist, 130*(4), pp.636–642.

Pianka, E.R., 1970. On r-and K-selection. *The American naturalist, 104*(940), pp.592–597.

Pierce, R.A., Sumners, J. and Flinn, E., 2022. Ecology and management of white-tailed deer in Missouri. University of Missouri Extension, Natural Resources, http://extension.missouri.edu/p/g9479, pp. 1–8.

Pratt, N.C., Alberts, A.C., Fulton-Medler, K.G. and Phillips, J.A., 1992. Behavioral, physiological, and morphological components of dominance and mate attraction in male green iguanas. *Zoo Biology, 11*(3), pp.153–163.

Ribble, D.O. and Salvioni, M., 1990. Social organization and nest co-occupancy in Peromyscus californicus, a monogamous rodent. *Behavioral Ecology and Sociobiology, 26*, pp.9–15.

Rosenbaum, S. and Gettler, L.T., 2018. With a little help from her friends (and family) part I: the ecology and evolution of non-maternal care in mammals. *Physiology & Behavior, 193*, pp.1–11.

Schroeder, M.A., Aldridge, C.L., Apa, A.D., Bohne, J.R., Braun, C.E., Bunnell, S.D., Connelly, J.W., Deibert, P.A., Gardner, S.C., Hilliard, M.A. and Kobriger, G.D., 2004. Distribution of sage-grouse in North America. *The Condor, 106*(2), pp.363–376.

Shine, R., Harlow, P., LeMaster, M.P., Moore, I.T. and Mason, R.T., 2000. The transvestite serpent: why do male garter snakes court (some) other males?. *Animal Behaviour, 59*(2), pp.349–359.

Stacey, P.B. and Ligon, J.D., 1987. Territory quality and dispersal options in the acorn woodpecker, and a challenge to the habitat-saturation model of cooperative breeding. *The American Naturalist, 130*(5), pp.654–676.

Szabo, A. and Duffus, D., 2008. Mother–offspring association in the humpback whale, Megaptera novaeangliae: following behavior in an aquatic mammal. *Animal Behaviour, 75*(3), pp.1085–1092.

Urban, E.K., Fry, C.H. and Keith, S., 2020. *The birds of Africa: Volume V.* Bloomsbury Publishing.

van Noordwijk, M.A., 2012. From maternal investment to lifetime maternal care. In *The evolution of primate societies*, pp.321–342.

van Noordwijk, M.A., Atmoko, S.S.U., Knott, C.D., Kuze, N., Morrogh-Bernard, H.C., Oram, F., Schuppli, C., van Schaik, C.P. and Willems, E.P., 2018. The slow ape: High infant survival and long interbirth intervals in wild orangutans. *Journal of Human Evolution, 125*, pp.38–49.

Wauer, R.H., 1999. *The American Robin.* University of Texas Press.

Westneat, D.F., 1993. Polygyny and extrapair fertilizations in eastern red-winged blackbirds (Agelaius phoeniceus). *Behavioral Ecology, 4*(1), pp.49–60.

Whittam, R.M. and Leonard, M.L., 2000. Characteristics of predators and offspring influence nest defense by Arctic and Common Terns. *The Condor, 102*(2), pp.301–306.

11 Conservation and Behavior

Manoj Kumar Ahirwar, Pramod Kumar, Satyendra Kumar Maurya and Rajesh Kumar

11.1 INTRODUCTION

In the past few decades there has been a growing understanding of the role animal behavior research can play in improving the effectiveness and success of conservation management programs. Animal behavior can help us understand and predict the impacts of anthropogenic disturbance on wildlife populations, can be used as a tool in conservation interventions, and can serve as a powerful indicator of conservation problems (Berger, 2011). Overall, the emergent field of conservation behavior (applying animal behavior research to conservation and management) has already contributed to many successful conservation outcomes from devising individual-specific diets to manage sex ratios in the critically endangered Kakapo (Robertson, 2006) to promoting life skills that enhance survival after reintroduction of species into the wild (Vargas, 1999; Beck, 2002; Shier, 2006; Whiteside, 2015). Nevertheless, there is tremendous room for improvement; for example, olfactory deterrents can fail because they do not adequately recognize or manipulate context in the meaning of animal signals (Parsons, 2018). Meanwhile, traps designed in the laboratory to attract and control invasive species can prove ineffective under field conditions (Vargas, 1999). In many such cases, we simply do not understand the underlying causes of failures, which prevent us from offering sound and cost-effective guidance on conservation management.

11.2 ROLE OF BEHAVIOR IN CONSERVATION

As a component of a biotic community, different species of animals need to behave and interact with one another in a wide range of natural conditions, which helps in the understanding of the function of and variation in animal behaviors. However, human activities, such as poaching, wildlife trading, habitat destruction and fragmentation, and the introduction of invasive species are commonly observed to aggravate the rates of vulnerability and extinction of many wild animal species and their natural behaviors. To combat these multifaceted problems, conservation interventions depend on the knowledge of species biology, behavior, and ecology. Understanding how human activities can modify behaviors and subsequently, cause negative impacts on the biodiversity of an area is crucial. From a management perspective, searching for the fitness consequences of behavioral responses may provide insights into the impacts of certain human activities on wild animals. For example, a strategy to better manage the negative impacts of tourism on wild animals is to understand the proximate mechanisms underlying the response of wild animals to humans. Under such conditions, management decisions should be focused on manipulating behavior because it can be used as a leading indicator of anthropogenic disturbances or population health conditions. Moreover, unlike population density, animal behavior can instantaneously change with changes in the environment. The leading question is "How does understanding the behaviors of an animal species contribute to its conservation?" The answer might vary depending on the focal species, but it has profound consequences for our efforts toward conserving endangered and rare species. By doing so, the conceptual framework can provide novel management questions and can help to structure a scientific study of an endangered species. For example, behavioral knowledge implicitly guides conservation and monitoring strategies because ranging patterns, breeding times, and locations affect spatial

DOI: 10.1201/9781003710271-11

and temporal aspects of population censuses. Previous theoretical and empirical studies suggested that behavioral responses of organisms to environmental changes are quick and can represent true leading indicators which are useful to conservation biology. However, the linkage between animal behavior and conservation biology is very weak due to historical and cultural setbacks and scientific roadblocks. For example, differences in scales, themes, professional biases, and approaches have hindered such important progress. Recently, a growing interest in integrating an understanding of animal behavior into conservation biology has clearly been present. By virtue of this reality, recent books and journals imply a new movement toward integrating animal behavior and conservation biology. For example, a number of behavioral biologists have written several reviews and book chapters on the role and importance of animal behavior in conservation, arguing for a fundamental understanding of behavioral processes for contributing to conservation biology. Without a clear conceptual framework for such integration, conservation biologists may have difficulties recognizing how behavioral knowledge can help them to solve real-world conservation problems. For example, if behavioral research becomes more central to conservation, it needs more professionals working at the interface, and its nature should also be changed. Conservation biologists have already begun to apply general principles of animal behavior to solve conservation problems. In serving to examine both the sources of conservation principles and the expanded concept of biodiversity conservation, it is clear that animal behavior should be considered an explicit part of conservation biology. Animal behavioral studies should change their focus from individuals to populations of organisms, as conservation biology has successfully emerged from population ecology, population genetics, and population sociobiology. Leading ethologists should mention the ways in which animal behavior and conservation interact, investigate the conservation impacts of people, and identify novel behavioral threats to wild populations of animals. Similarly, animal behaviorists may also act to conserve biodiversity by applying their professional skills to conservation problems, or they may act on a personal level by supporting conservation groups and acting locally to preserve animal habitats. Consequently, animal behaviorists should develop theoretical and empirical frameworks which can help to identify conservation priorities and behavioral approaches used for conservation applications. In recent years, a holistic approach has been sought as another conservation alternative for better success. The holistic approach deals with higher levels of ecological organization, i.e., community, ecosystem, and landscape. In the holistic approach, any effort to conserve the behaviors of one animal species may be based on the account of some other species. For example, by studying several species simultaneously, one species can gain a much better understanding of how different species respond to the same ecological problems (e.g., disturbance, climate change, disease prevalence, etc.). The basic concept of behavior includes two dimensions that can be depicted on a 2D coordinate map (Figure 11.1).

However, whether different animal species respond to stimuli in the same or different ways can be used as good information for answering both conservation and behavioral questions. Each approach has its own advantages and limitations in comparison to the other; however, the optimal approach should be chosen based on the existing realities.

11.2.1 Conservation Behavior

Conservation behavior is a relatively new interdisciplinary field aimed at investigating how proximate and ultimate aspects of animal behavior can be of value in preventing the loss of biodiversity. The usefulness of this new discipline in promoting practical conservation matters is subject to debate, with some scientists arguing that the importance of behavior in conservation practice is overemphasized. Here, we propose a conceptual model that identifies the key linkages between animal behavior and conservation biology. The model is a simply structured, hierarchical, and parsimonious framework that will help bridge the gap between the two disciplines and establish a common ground on which the field of conservation behavior can evolve and from which paradigms can be developed.

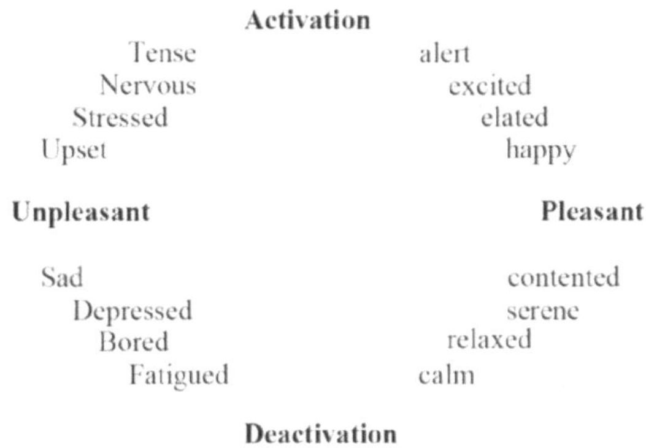

FIGURE 11.1 Two dimensions of emotion.

Separately, the disciplines of animal behavior and conservation biology are well-established, thriving fields of scientific inquiry, each with its own history and approach. Animal behavior is a chiefly theoretical discipline, while conservation biology is more of an applied science, established in response to the ongoing biodiversity crisis. Recent attempts have been made to apply principles of animal behavior to conservation problems to create a formally integrated discipline, commonly called conservation behavior. There are several reasons why the union of these fields seems intuitive and profitable. First, behavioral ecologists focus on understanding how the state of an individual influences behavior. State-dependent behaviors include those that are plastic or flexible rather than fixed, whose expression can depend on environmental influences, such as temperature, resources, social cues, or maternal condition. Because the environment can influence the experience or condition of an individual, which can consequently influence its behavior, these state-dependent behaviors are of particular conservation relevance; in effect, they have the potential to be impacted by anthropogenic factors that alter the environment, such as disturbance and global climate change. There is growing evidence that anthropogenic disturbances are disruptive to behavioral processes. Disturbance may increase stress levels, thereby affecting behavior and physiology, and may also disrupt normal foraging, movement, communication, and mating patterns, among other behaviors.

Second, behavior is expected to have conservation relevance because it can affect demographic processes, such as survival and reproduction, and hence fitness. Quantification of fitness consequences is a major research area in behavioral ecology and is essential to conservation behavior because it is these fitness links that allow the development of individual-based models to better understand anthropogenic impacts. In fact, a shortcoming of most studies that investigate the effects of disturbance on behavior is that they fail to formalize the link to demography and fitness.

Third, many forms of behavioral information have been identified as useful conservation tools. For example, captive breeding and reintroduction programs have clearly demonstrated the importance of managing natural behavior (e.g., mating behavior, social behavior, and experience with predators and prey) to increase captive breeding and reintroduction success.

Fourth, behavioral diversity is a potential currency to consider, along with species and genetic diversity, when setting future conservation priorities. Behavior plays a key role in animal survival and evolution, and its variation may allow species to respond to mounting changes in the environment. Conservation plans that incorporate the diversity of interesting behavioral traits also have the added benefit of harnessing public interest and support.

Several barriers have impeded the full integration of these disciplines, and these hurdles are being crossed only gradually. In this chapter, the history and current status of the field of conservation behavior are reviewed, barriers to the integration of behavior and conservation as well as tools to overcome them.

11.3 CONSERVATION BEHAVIOR FRAMEWORK

Our framework is composed of three basic themes by which conservation and behavior are linked (Figure 11.2) i.e. direct and indirect anthropogenic impacts on animal behavior that, in turn, impact biodiversity; behavior-based management, representing the use and consideration of behavior in conservation practice; and behavioral indicators to other processes that are of conservation concern. All three require knowledge of animal behavior. The key element of behavioral ecology is the adaptive nature of behavior. Behavioral strategies in a population are the outcome of evolutionary processes that depend on the fitness of particular strategies under prevailing environmental conditions (Krebs and Davies, 1997; Norris, 2004). Behaviors should evolve to maximize the fitness of the individuals showing those behaviors (Krebs and Davies, 1981; Owens, 2006). Within the field of behavioral ecology, we recognize three key behavior domains that are central to the attainment of high fitness in individuals of all species and are therefore of key concern in conservation:

1) Movement and space-use patterns.
2) Foraging and predator–prey related behaviors.
3) Social behavior and reproduction.

All the different behaviors in the three domains affect survival and reproduction (hence recruitment), thus providing invaluable information on population and community dynamics. Within each of the three basic themes, we identified two focal pathways on which behavior-oriented conservation studies should focus:

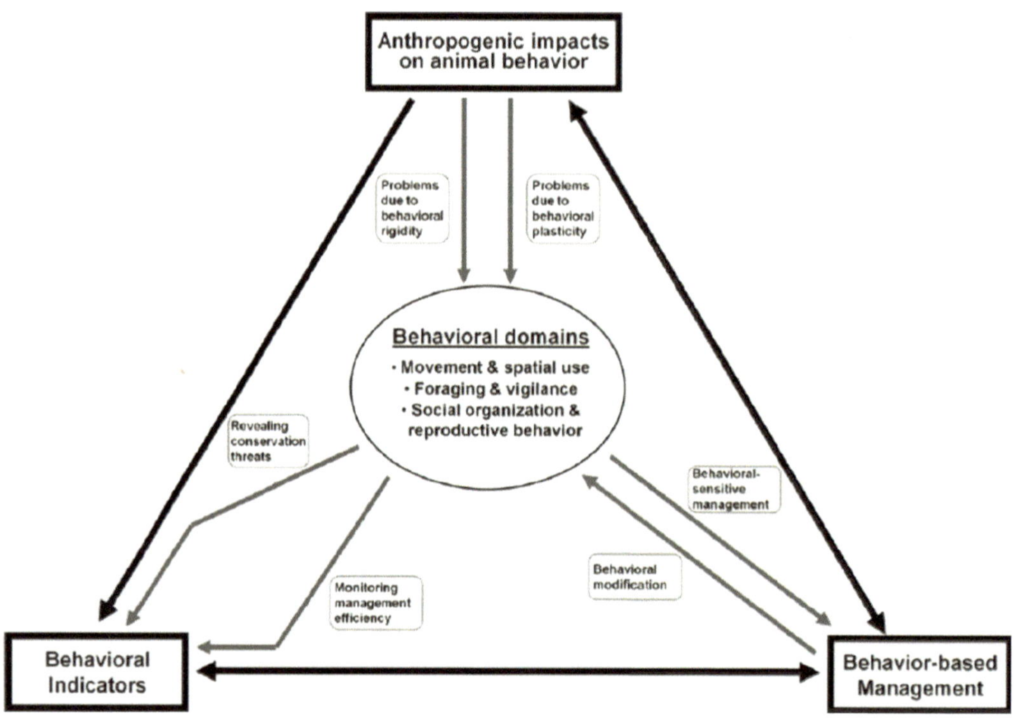

FIGURE 11.2 The conservation behavior framework is composed of three basic interrelated conservation themes: 1) anthropogenic impacts on animal behavior; 2) behavior-based management; and 3) behavioral indicators. The black arrows represent interactions between the conservation themes. Gray arrows represent the pathways that connect each theme to the behavioral domains.

Theme 1: Anthropogenic impacts on animal behavior. Anthropogenic impacts on animal behavior come about by direct human disturbances, such as overharvesting, fragmentation, and nuisance disturbances, and by indirect disturbances, such as the introduction of alien species or the creation of ecological traps (Robertson and Hutto 2006). These disturbances can affect behavior-dependent animal fitness through two distinct pathways (Figure 11.2):

First, when humans alter the environment, the fitness value of existing behavioral strategies changes. If fitness is drastically reduced and the strategies are either not sufficiently "plastic" to respond to the environmental change or an evolutionary response to the altered environment is slow relative to the rate of environmental change, the population will decline (Norris, 2004). Second, and in contrast to the first, if behavior is plastic it may be altered by anthropogenic changes. Although this may be adaptive in the short term, the behavioral response may change other fitness-related behaviors, such as social structure or mating success, thus altering the evolutionary trajectory of the species or the ecosystems in question, which may facilitate the closure of evolutionary options (Ehrlich, 2001), creating conservation concerns on a longer evolutionary timescale (Manor and Saltz, 2003). Furthermore, a change in the behavior of one species may alter the dynamics of an entire community or ecosystem (Wright et al., 2010). In the cases where anthropogenic impacts on animal behavior lead to conservation concerns, the best solutions are in most cases behavioral-based management schemes that lead us to the next conservation behavior theme.

Theme 2: Behavior-based management. Here too, we recognize two pathways incorporating animal behavior into active management for conservation (Figure 11.2). In the first, the species' behavior is considered in conservation decision-making and protocols. We term this pathway "behavior-sensitive management". Behavioral considerations may play a crucial role in reserve design and corridor planning (Schultz, 1998; Pe'er et al., 2004; Afonso et al., 2008), wildlife epidemiology (Craft et al., 2009), and planning of reintroductions and translocations (Saltz et al., 2000; Bar-David et al., 2005; Shier, 2006; Zidon et al., 2009). The proximate goals of behavior-based management will usually have a strong demographic nature whether they aim to stabilize or increase the numbers of small or declining populations or to control populations of invasive or pest species. However, in cases where the change of the animal's behavior is the cause for conservation concern, the proximate goal of the management efforts may be changing the behavior of the target population. Thus, in contrast to the first pathway in this theme where management decisions are made based on the species' behavior, in the second pathway, the manager seeks to change or preserve the behavior itself. This approach is commonly applied in training captive-bred individuals designated for reintroduction to become predator-savvy, etc. (e.g., McLean et al., 1996; Griffin et al., 2000; Alberts, 2007). This pathway is termed as "behavioral modification".

Theme 3: Behavioral indicators. The various adaptive behaviors of organisms give us a great deal of information about the evolutionary forces shaping these behaviors, the environments which the organisms inhabit, and any recent changes to either the selection forces or the environment. Thus, we can use the behavior itself as an indicator of the organism's state as well as of the state of its environment (Kotler et al., 2007). Such indicators include foraging and patch use behaviors (e.g., Whelan and Jedlicka, 2007), diving behaviors (e.g., Mori et al., 2007), habitat selection (e.g., Heithaus et al., 2007), and home range use (Owen-Smith and Cain, 2007). The two pathways in which behavioral indicators have been used in conservation are as follows:

1. Behavioral indicators that provide an early warning of population decline or habitat degradation before numerical responses are evident (e.g., Searle et al., 2007; van Gils et al. 2009).
2. Behavioral indicators used to monitor the effectiveness of management programs, or evaluate the success of a management program at its early stages, before population or ecosystem-level responses are evident (e.g., Ikuta and Blumstein, 2003; Lindell, 2008).

The three behavioral conservation themes are strongly linked (Figure 11.2). For example, anthropogenic impact on animal behavior may be detected using behavioral indicators and can suggest the need for behavior-sensitive management (Ikuta and Blumstein, 2003; Zidon et al., 2009). Alternatively, behavior-sensitive active management can be evaluated using behavioral indicators, and this knowledge may change the management plan accordingly. However, in many situations, the behavioral aspect of one theme may dictate a non-behavioral component of another theme (e.g., behavioral indicators may often be indicative of disturbances impacting non-behavioral elements, such as dynamics). One can view the three themes as entry points for behavioral ecologists aiming to use their knowledge and expertise of animal behavior in conservation. In each of the themes, new research should be based on previous knowledge of animal behavior as well as on pressing conservation concerns. In this way, conservation behavior can serve as a much-needed link between the ever-expanding knowledge in behavioral ecology and the more practical needs of conservation biologists. Although behavioral ecologists may, in many cases, address only one of the proposed three themes in any given research, the role of the conservation biologist facing a conservation issue is to consider and integrate all three themes into one adaptive management scheme.

11.4 RECENT ADVANCES IN CONSERVATION METHODS

Although the formal integration of behavior and conservation is a recent effort, wildlife biologists have been using basic behavioral data to inform management decisions since the inception of the discipline of wildlife management. For the last century, wildlife biologists have been documenting the basic natural history of animals, including descriptions of fundamental behavior such as movement patterns, habitat selection, sociality, mating systems, and foraging. Indeed, the early history of wildlife biology was often dominated by naturalistic observations of the behavior of animals in the wild, employing traditional tools such as visual observations, binoculars, spotting scopes, and trapping. Ironically, the advent of new technologies, such as radio telemetry, trail cameras and remote-sensed satellite and GIS (Geographic Information System) data, has allowed wildlife researchers to spend less time in the field directly observing wildlife, including their behavior. These new approaches, combined with an increasing focus on population-level processes such as estimating sizes and dynamics of populations, may have contributed to the decline of the importance of animal behavior in the field of wildlife biology in recent decades. For example, to become a certified wildlife biologist through The Wildlife Society, a course in animal behavior is not required but rather an elective; such is also the case for most undergraduate wildlife biology programs. Conservation practitioners should work together with behavioral ecologists to create easily accessible and thorough systematic reviews and guidelines (Dicks et al., 2014) that will allow for the creation of evidence-based directories of potential and actual behavioral solutions to conservation and management problems. Being aware of the importance of behavior in conservation does not necessarily translate into efficient use of behavior in conservation. There are many cases where the potential of behavioral knowledge for conservation has been reported but no concrete and useful advice was provided. Managers can encounter various constraints and problems when attempting to apply behavioral principles, such as optimal foraging theory or associative learning, to real-life situations (Caro, 2007; Schakner et al., 2014). For behavioral ecology to become an effective tool, behavioral ecologists must turn their implied insights into applied principles. Our heat map highlights the research areas that have been sufficiently studied to facilitate such systematic reviews and points out the areas where more research is needed before a systematic review can be conducted.

11.5 BIODIVERSITY AND ITS CONSERVATION METHODS

Biodiversity refers to the variability of life on earth. It can be conserved in the following ways (Figure 11.3):

- In-situ conservation
- Ex-situ conservation

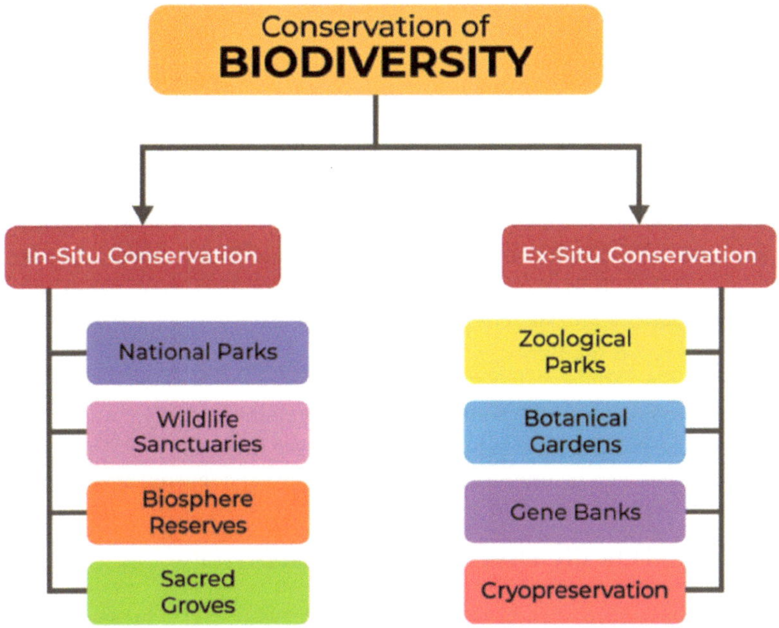

FIGURE 11.3 Conservation of biodiversity.

11.6 IN-SITU CONSERVATION

In-situ conservation of biodiversity is the conservation of species within their natural habitat. In this method, the natural ecosystem is maintained and protected.

In-situ conservation has several advantages. The following are the important advantages of in-situ conservation:

1. It is a cost-effective and convenient method of conserving biodiversity.
2. A large number of living organisms can be conserved simultaneously.
3. Since the organisms are in a natural ecosystem, they can evolve better and can easily adjust to different environmental conditions.

Certain protected areas where in-situ conservation takes place include national parks, wildlife sanctuaries, and biosphere reserves.

11.6.1 National Parks

These are small reserves maintained by the government. Their boundaries are well demarcated, and human activities such as grazing, forestry, habitat, and cultivation are prohibited, for example, Kanha National Park and Bandipur National Park.

11.6.2 Wildlife Sanctuaries

These are the regions where only wild animals are found. Human activities such as timber harvesting, cultivation, collection of wood, and other forest products are allowed here as long as they do not interfere with the conservation project. Also, tourists visit these places for recreation.

11.6.3 Biosphere Reserves

Biosphere reserves are multi-purpose protected areas where the wildlife, traditional lifestyle of the inhabitants, and domesticated plants and animals are protected. Tourist and research activities are permitted here.

11.7 EX-SITU CONSERVATION

Ex-situ conservation of biodiversity involves the breeding and maintenance of endangered species in artificial ecosystems such as zoos, nurseries, botanical gardens, gene banks, etc. There is less competition for food, water, and space among the organisms.

Ex-situ conservation has the following advantages:

1. The animals are provided with a longer time and breeding activity.
2. The species bred in captivity can be reintroduced into the wild.
3. Genetic techniques can be used for the preservation of endangered species.

11.8 STRATEGIES FOR BIODIVERSITY CONSERVATION

The following are important strategies for biodiversity conservation:

1. All the varieties of food, timber plants, livestock, microbes, and agricultural animals should be conserved.
2. All the economically important organisms should be identified and conserved.
3. Unique ecosystems should be preserved first.
4. The resources should be utilized efficiently.
5. Poaching and hunting of wild animals should be prevented.
6. The reserves and protected areas should be developed carefully.
7. The levels of pollutants should be reduced in the environment.
8. Deforestation should be strictly prohibited.
9. Environmental laws should be followed strictly.
10. The useful and endangered species of plants and animals should be conserved in their natural as well as artificial habitats.
11. Public awareness should be created regarding biodiversity conservation and its importance.

11.9 CHALLENGES OF WILDLIFE CONSERVATION AND MANAGEMENT

Conservation practitioners should have access to an extremely diverse and constantly evolving toolkit of expertise that includes population dynamics, computer modeling, genetics, remote sensing, social sciences, and law. Behavioral ecology is one such important tool. An animal's behavior lies at the center of gene-environment interactions, and, as such, it serves as a mediator between an animal's fitness and anthropogenic disturbances (Berger-Tal et al., 2011). Many wildlife managers explicitly appreciate this role and have been using their knowledge of animal behavior to manage species for decades (e.g., Geist and Walther, 1974; Harcourt, 1999). But within the scientific community, attention to the potential role of behavioral ecology in conservation has greatly increased in the past 10 to 15 years, and the use of animal behavior knowledge in conservation came to be known as conservation behavior (Blumstein, 2001). Several recent books and various publications in journals aim to raise awareness among wildlife managers to the advantages of using the existing theories of behavioral ecology to improve conservation (e.g., Clemmons and Buchholz, 1997; Festa-Bianchet and Apollonio, 2003; Blumstein and Fernández-Juricic, 2010; Greggor et al., 2014).

Although there are cases in which knowledge of animal behavior cannot inform the protection or recovery of species, either because it is not relevant to the problem or because it is not cost-effective (Blumstein and Berger-Tal, 2015), the direct link between animal behavior and fitness makes behavioral-ecology knowledge an important tool. However, in many cases, conservation practitioners do not apply knowledge from behavioral ecology in their management plans, or they consider it only after a problem has emerged, resulting in setbacks, which in many cases can lead to the failure of management efforts (Knight, 2001; Blumstein and Fernández-Juricic, 2004). In an attempt to evaluate the success of the emerging field of conservation behavior in resolving such setbacks, some authors have assessed the integration of behavioral ecology into conservation and management by searching the academic literature for combinations of derivatives of the words *behavior* and *conservation* and analyzing whether reported trends have changed in the last decade (Angeloni et al. 2008; Nelson 2014). These authors concluded that currently there is little association between the fields of behavioral ecology and conservation biology and that attention to the potential role of behavioral ecology in conservation has not had an impact on conservation and management.

11.10 FACTORS AFFECTING THE BEHAVIOR OF ANIMALS

This topic covers the variety of conservation behavior applications, aimed at better understanding mechanisms to design interventions, all of which have been flagged as research priorities for conservation behavior (Greggor, 2016). From exploring interventions that attract animals via scent lures (Price, 2019) or acoustic playbacks (Putman, 2019) to mapping interventions that tap into learning to deter animals from human-conflict scenarios (Snijders, 2019), the potential applications highlighted in this issue are diverse.

The entries in this section emphasize that behavior and conservation will intersect in important ways when behavior is state-dependent and therefore potentially influenced by environmental disturbance, when behavioral change results in fitness or demographic change, and when behavioral knowledge can be used to provide useful tools in conservation programs. For example, how anthropogenic disturbance, specifically noise, can disrupt normal animal behavior. Noise can induce stress in animals, which can affect any number of behaviors (e.g., reproductive or anti-predator behavior) that depend on being in good physiological state, but it can also interfere with animal communication by masking intentional acoustic signals as well as cues used to find prey and avoid predators. Anthropogenic noise has been shown to affect animal behavior in both terrestrial and aquatic systems for diverse groups including insects, fish, frogs, birds, bats, and whales. When it induces stress, when it masks sounds that are used to find prey and avoid predators, and when it interferes with courtship calls that are used to find mates or begging calls that are used to provision young, noise can impact individual fitness, which may translate to demographic consequences at the population level. In the end, further research is needed to formally assess the effects of anthropogenic noise on reproductive success and population viability.

Human disturbance can negatively impact signaling between animals, in particular sexual signaling by males and the ability of females to assess male ornaments. Noise is not the only form of disturbance that can interfere with signaling between potential mates. Chemical pollution and acidification of aquatic habitats also interfere with chemical cues that fish use to find appropriate mates. If these male ornaments advertise their quality to potential mates, then disruption of these signals may result in the choice of a lower-quality mate, potentially causing reduced fitness of the female or her offspring. Worse yet, anthropogenic disturbance may interfere with species recognition cues, causing individuals to hybridize across species boundaries, not only reducing fitness but also potentially influencing biodiversity. The conservation relevance of disrupted sexual signaling will depend on discovering its relative importance in determining fitness and population demography. Humans have introduced exotic species worldwide both intentionally and accidentally, which can negatively impact populations of native species. In addition to ecological interactions between native and exotic species (e.g., competition or predation), there can also be behavioral interactions that interfere with the reproduction of the native species. The signals of exotic species may mask the acoustic mating signals of the native species or may even jam their chemical pheromone receptors. Males may expend energy on courtship with females, and on rivalry with males of the other species. Mating interference can reduce the ability of native females to choose the highest quality mate, can reduce the rate of successful copulation for the native species, and can result in hybridization across species boundaries. Humans impact not only the behavior of animals in the wild but also those in captivity.

Only a small number of reintroduction programs have successfully established self-sustaining populations in the wild, in part because captive animals have not had the opportunity to develop normal foraging, anti-predator, and social behavior. High densities and limited space in captivity can cause stress and aggression and can impede the development of territoriality, seasonal migratory behavior, exploratory behavior, and social behavior. The regular provisioning of food reduces the ability to develop foraging behavior, and the lack of predators reduces the ability to develop anti-predator behavior. Reduced exposure to parents and the opposite sex and a lack of mate choice may hinder the development of appropriate reproductive behavior. Abnormal repetitive behavior sometimes develops as a result of frustration, fear, or discomfort, and it may be a sign of stress during early development. Fortunately, the negative behavioral effects of captive breeding can often be reversed once they are discovered, through greater exposure to conspecifics and predators, through environmental enrichment (including rearing in semi-natural environments) and by acclimatizing individuals to natural conditions prior to release. These methods have enhanced the captive breeding and reintroduction programs for a number of birds and mammals. A number of birds and rodents have been shown to learn about food aversions and food preferences, as well as the timing and location of food sources. Some species (e.g., some birds, mammals, and fishes) also learn to identify appropriate mates and develop mating preferences from their social experiences. Thus, detailed knowledge about the behavior of the particular species in nature, although sometimes difficult to obtain for an endangered species, can provide useful tools for captive management.

Outside of captivity, learning can also have conservation implications for wild populations that live in human-modified environments. Species that are thought to have greater capacity for learning, including larger-brained birds and mammals, seem to have greater flexibility in their behavior. Thus, knowledge about behavioral flexibility may help conservation biologists predict which species are most likely to be successful invaders, which are most likely to adjust to habitat modification and urbanization, and which to target for protection because of their vulnerability. However, while large-brained species may be more flexible in novel environments, they may also have life history characteristics such as delayed maturation and slow reproduction that make them more vulnerable to extinction and more difficult to rear in captivity. Understanding the link between animal behavior and its complicated effects on seed dispersal and plant demography will be important in predicting how habitat fragmentation, climate change, invasive species, and the loss of seed dispersers will affect plant populations.

However, behavioral ecologists must first consider the conservation problem and then select the most relevant tool to address that problem, acknowledging that the tool may not always be a behavioral tool. They must also share lessons from the application of behavioral research to make conservation practitioners more aware of behavioral tools. Additional research is required to distinguish between these two types of cells, allowing for the development of novel and useful approaches to conservation on one hand and preventing a waste of resources on inefficient tools on the other hand. Being aware of the importance of behavior in conservation does not necessarily translate into efficient use of behavior in conservation. There are many cases where the potential of behavioral knowledge for conservation has been reported but no concrete and useful advice was provided. Managers can encounter various constraints and problems when attempting to apply behavioral principles, such as optimal foraging theory or associative learning, to real-life situations for behavioral ecology to become an effective tool, behavioral ecologists must turn their implied insights into applied principles.

Studies concerning anthropogenic effects on wildlife behavior are important both in their own right and for their unique contribution to effective conservation strategies (Caro, 1999; Sutherland,

1998; Berger-Tal et al., 2011). Behavioral data from areas of low disturbance can serve as a baseline of comparison for behaviors in disturbed areas and as standards for individuals in captive breeding programs. For example, an understanding of how wild animals select habitats in undisturbed areas allows conservationists to effectively create protected areas and release individuals in captive breeding programs in preferred habitats, thus increasing the chances of species survival (Swaisgood, 2007; Indermaur and Schmidt, 2011).

Protected areas such as national parks become integral for observing the species, processes, and interactions that are otherwise altered, unobservable, or nonexistent in areas manipulated by humans (Davis, Graber and Acker, 2004). Alternatively, in areas of moderate to high disturbance, behavioral studies can help assess the degree and complexity of anthropogenic impacts as well as aid in designing management decisions. An understanding of animal ranging across large areas, for instance, informs preserve design and possible optimization of connectivity between preserves (Reed, 2002). Along these lines, protected areas with high levels of human disturbance offer unique opportunities for research to directly influence management policies within focal preserves or parks, as well as in similar ecosystems outside the protected areas (Parsons, 2004; Ramirez, 2004). For instance, the effects of recreation and tourism on wildlife have become a concern in many protected areas. Thus, researchers are now performing studies to quantify these effects, such as observing the minimum distance from humans tolerated by large ungulates (Malo, Acebes and Traba, 2011), or the maximum number of boats tolerated by marine animals (Williams and Ashe, 2007). These studies are highly informative in the creation of park regulations that protect animals from behavioral disturbances, while allowing visitors to observe the animals they have come to see (Figure 11.4).

The number of papers found in a search of the ISI Web of Science database for intersections between animal behaviors search terms (rows, 10 categories) and conservation themes search terms (columns, 16 categories divided into threats and solutions). The number of papers is represented by color; hotter colors represent a larger number of papers (color scale is logarithmic).

11.11 EMERGENCE OF CONSERVATION BEHAVIOR

Conservation behavior will not provide the solutions to all conservation problems, but we conservation behaviorists need to show how useful a behavioral perspective can be. By understanding the type of questions conservation behavior can address, one can better evaluate how and when wildlife conservation could benefit from behavioral knowledge. Addressing both proximate and

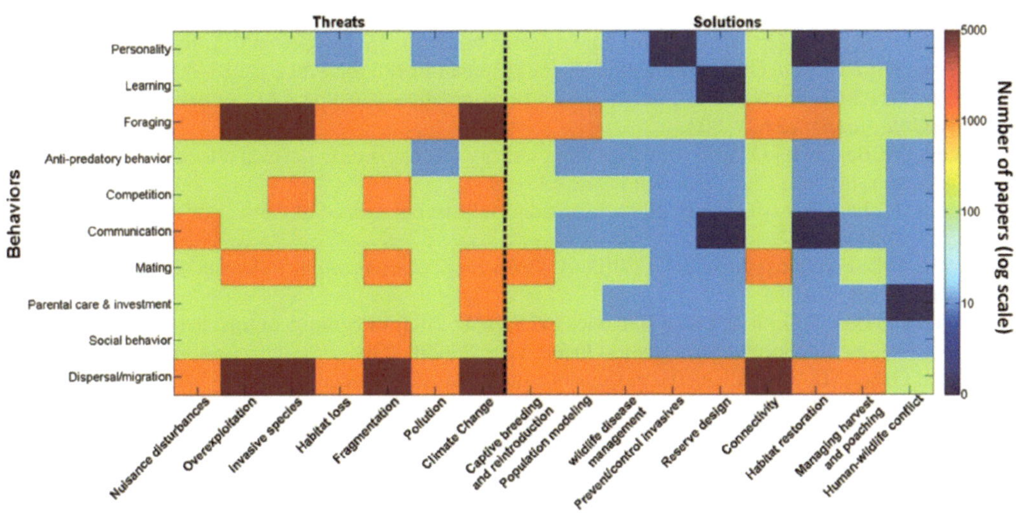

FIGURE 11.4 A systematic survey of the integration of animal behavior into conservation.

ultimate questions of conservation behavior is essential to improving our management decisions. Two research areas in particular will not only help bridge the gulf between conservationists and behaviorists but will also help research on conservation behavior to produce specific conservation outcomes. First, we must develop predictive models of behavior that are accessible to wildlife managers. Advances have been made in foraging theory and modeling habitat selection, wherein animal decisions have been modeled with increased sophistication. However, the sophistication that comes with deeper understanding may impede application by its complexity. Behavioral biologists should strive to make their models accessible to others by identifying specific behavioral mechanisms. These mechanisms should be the basis of bottom-up models that predict the behavior, movement, habitat use, and distribution of species of conservation concern. Second, we must develop an ultimate understanding of how and why species are vulnerable to human impacts. Study of conservation behavior is much more than simply documenting whether or not humans affect wildlife. The explosion in comparative biology and macro ecology in the past decade offers a framework for future advances in predictive evolutionary models. Future studies should examine how variation in life history and natural history explains variation in proneness to extinction, tolerance to human disturbance, and tolerance to invasive species at local scales.

Although we are excited that some wildlife managers and conservation biologists appreciate the expertise offered by behavioral biologists, we believe that many still fail to appreciate the importance of proximate mechanisms. Some pigeonhole behavioral research as too species-specific and fail to recognize that behavioral biologists can produce generalized and predictive models. With an increased appreciation of what conservation behaviorists have to offer, and a fundamental integration of them into recovery teams, we believe that future recovery efforts will be more successful and cost-effective, and that the exciting integrative field of conservation behavior will prosper.

BIBLIOGRAPHY

Afonso P, Fontes J, Holland KN, Santos RS. 2008. Social status determines behavior and habitat usage in a temperate parrotfish: implications for marine reserve design. *Mar Ecol Prog Ser.* **359**:215–227.

Alberts AC. 2007. Behavioral considerations of headstarting as a conservation strategy for endangered Caribbean rock iguanas. *Appl Anim Behav Sci.* **102:**380–391.

Angeloni, L, Schlaepfer, MA, Lawler, JJ, Crooks, KR. 2008. A reassessment of the interface between conservation and behavior. *Anim Behav.* **75**:731–737.

Bar-David S, Saltz D, Dayan T. 2005. Predicting the spatial dynamics of a reintroduced population: the Persian fallow deer. *Ecol Appl.* **15**:1833–1846.

Beck B, Casgro M, Stoinski T, Ballou J. 2002.The efects of prerelease environ- ments and postrelease management on survivorship in reintroduced golden lion tamarins. In: D. Kleiman, A. Rylands, editors, *Lion Tamarins Biol Conserv.* Washington, DC: Smithsonian Institution Press; p. 283–300.

Blumstein DT. 2001. Book review: behaviour and conservation (Gosling LM, Sutherland WJ, editors). *J Wildlife Manag.* **65**:601–603.

Blumstein DT, Berger-Tal O. 2015. Understanding sensory mechanisms to develop effective conservation and management tools. *Curr Opin Behav Sci.* **6**:13–18.

Blumstein DT, Fernández-Juricic E. 2004. Lead editorial: the emergence of conservation behavior. *Conserv Biol.* **18**:1175–1177.

Blumstein DT, Fernández-Juricic E. 2010. *A Primer of Conservation Behavior.* Sinauer Associates, Sunderland, MA.

Berger-Tal O, Polak T, Oron A, Lubin Y, Kotler BP, Saltz D. 2011. Integrating animal behavior and conservation biology: a conceptual framework. *Behav Ecol.* **22**:236–239.

Berger-Tal O, Saltz D. 2014. Using the movement patterns of reintroduced animals to improve reintroduction success. *Curr Zool.* **60**:515–526.

Buchholz R. 2007. Behavioral biology: an effective and relevant conservation tool. *Trends Ecol Evol.* **22**:401–407.

Caro T. 1998. The significance of behavioral ecology for conservation biology. In: T. Caro, editor, *Behavioral Ecology and Conservation Biology.* Oxford University Press, Oxford; p. 3–26.

Caro T. 1999. The behaviour-conservation interface. *Trends Ecol Evol.* **14**(9):366–369. http://dx.doi.org/10.1016/S0169-5347(99)01663-8

Caro T. 2007. Behavior and conservation: A bridge too far? *Trends Ecol Evol.* **22**:394–400.

Clemmons JR, Buchholz R. 1997. *Behavioral Approaches to Conservation in the Wild.* Cambridge: Cambridge University Press; p. 3–26.

Craft ME, Volz E, Packer C, Meyers LA. 2009. Distinguishing epidemic waves from disease spillover in a wildlife population. *Proc R Soc B Biol Sci.* **276**:1777–1785.

Davis GE, David MG, Steven AA. 2004. National parks as scientific benchmark standards for the biosphere; or, how are you going to tell how it used to be, when there's nothing left to see? *George Wright Forum* **21**(2):34–44.

Dicks LV, Walsh JC, Sutherland WJ. 2014. Organising evidence for environmental management decisions: a '4S' hierarchy. *Trends Ecol Evol.* **29**:607–613.

Ehrlich PR. 2001. Intervening in evolution: ethics and actions. *Proc Natl Acad Sci U S A.* **98**:5477–5480.

Greggor AL, Berger-Tal O, Blumstein DT, Angeloni L, Bessa-Gomes C, Blackwell BF, et al. 2016. Research priorities from animal behaviour for maximis- ing conservation progress. *Trends Ecol Evol.* **31**:953–964.

Griffin AS, Blumstein DT, Evans CS. 2000. Training captive-bred or translocated animals to avoid predators. *Conserv Biol.* **14**:1317–1326.

Harcourt, AH. 1999. The behavior-conservation interface. *Trends Ecol Evol.* **14**:490.

Heithaus MR, Wirsing AJ, Frid A, Dill LM. 2007. Behavioral indicators in marine conservation: lessons from a pristine seagrass ecosystem. *Isr J Ecol Evol.* **53**:355–370.

Festa-Bianchet M, Apollonio M, editors. 2003. *Animal Behavior and Wildlife Conservation.* Island Press, Washington, DC.

Geist V, Walther F. 1974. *The Behavior of Ungulates and Its Relation to Management.* IUCN, Morges.

Greggor AL, Clayton NS, Phalan B, Thornton A. 2014. Comparative cognition for conservationists. *Trends Ecol Evol.* **29**:489–495.

Groom MJ, Meffe GK, Carroll CR. 2006. *Principles of Conservation Biology.* 4th edition. Sinauer Associates, Sunderland, MA.

Hurley BP, Garnas J, Cooperband MF. 2015. Assessing trap and lure efectiveness for the monitoring of Sirex noctilio. *Agric For Entomol.* **17**:64–70.

Ikuta LA, Blumstein DT. 2003. Do fences protect birds from human disturbance? *Biol Conserv.* **112**:447–452.

Indermaur L, Schmidt BR. 2011. Quantitative recommendations for amphibian terrestrial habitat conservation derived from habitat selection behavior. *Ecol Appl.* **21**(7):2548–2554. http://dx.doi.org/10.1890/10-2047.1

Knight J. 2001. If they could talk to the animals. *Nature* **414**:246–247.

Kotler BP, Morris DW, Brown JS. 2007. Behavioral indicators and conservation: wielding "the biologist's tricorder." *Isr J Ecol Evol.* **53**:237–244.

Krebs JR, Davies NB. 1981. *An Introduction to Behavioural Ecology.* Blackwell Scientific Press, Oxford.

Krebs JR, Davies NB. 1997. *Behavioural Ecology: An Evolutionary Approach.* Blackwell Scientific Press, Oxford.

Kuhn TS. 1970. *The Structure of Scientific Revolutions.* 2nd ed. The University of Chicago Press, Chicago, IL.

Lindell CA. 2008. The value of animal behavior in evaluations of restoration success. *Restor Ecol.* **16**:197–203.

Linklater WL. 2004. Wanted for conservation research: behavioral ecologists with a broader perspective. *Bioscience.* **54**:352–360.

Malo JE, Acebes P, Traba J. 2011. Measuring ungulate tolerance to human with flight distance: a reliable visitor management tool? *Biodiv Conserv.* **20**(14):3477–3488. http://dx.doi.org/10.1007/s10531-011-0136-7

Manor R, Saltz D. 2003. Impact of human nuisance disturbance on vigilance and group size of a social ungulate. *Ecol Appl.* **13**:1830–1834.

McLean IG, Lundie-Jenkins G, Jarman PJ. 1996. Teaching an endangered mammal to recognize predators. *Biol Conserv.* **75**:51–62.

Moore JA, Bell BD, Linklater WL. 2008. The debate on behavior in conservation: New Zealand integrates theory with practice. *Bioscience.* **58**:454–459.

Mori Y, Mitani Y, Watanabe Y, Sato K. 2007. A behavioral indicator of prey patch richness derived from diving behavior: the proportion of residence time to the standard time. *Isr J Ecol Evol.* **53**:347–354.

Nelson, XJ. 2014. Animal behavior can inform conservation policy, we just need to get on with the job – or can it? *Curr Zoology* **60**: 479–485.

Norris K. 2004. Managing threatened species: the ecological toolbox, evolutionary theory and declining-population paradigm. *J Appl Ecol.* **41**:413–426.

Owens IPF. 2006. Where is behavioural ecology going? *Trends Ecol Evol.* **21**:356–361.
Owen-Smith N, Cain JW 3rd. 2007. Indicators of adaptive responses in home range utilization and movement patterns by a large mammalian herbivore. *Isr J Ecol Evol.* **53**:423–438.
Parsons DJ. 2004. Supporting basic ecological research in U.S. national parks: challenges and opportunities. *Ecol Appl.* **14**(1):5–13. http://dx.doi.org/10.1890/03-5085
Parsons MH, Apfelbach R, Banks PB, Cameron EZ, Dickman CR, Frank ASK, et al. 2018. Biologically meaningful scents: a framework for understanding predator–prey research across disciplines. *Biol Rev.* **93**:98–114.
Pe'er G, Saltz D, Thulke HH, Motro U. 2004. Response to topography in a hilltopping butterfly and implications for modeling nonrandom dispersal. *Anim Behav.* **68**:825–839.
Price JC, Banks PB, Greggor AL. 2019. What evidence exists on the efectiveness of diferent types of olfactory lures as attractants for invasive mammalian predators? A systematic map protocol. *Environ Evid.* https://doi.org/10.1186/s13750-019-0156-4.
Putman BJ, Blumstein DT. 2019. What is the efectiveness of using conspecifc or heterospecifc acoustic playbacks for the attraction of animals for wildlife management? A systematic review protocol. *Environ Evid.* https://doi.org/10.1186/s13750-019-0149-3.
Ramirez A. 2004. Ecological research and the Costa Rican park system. *Ecol Appl.* **14**(1):25–27. http://dx.doi.org/10.1890/03-5086
Reed JM. 2002. Animal behavior as a tool in conservation biology. In: A. A. Aguirre, R. S. Ostfeld, C. A. House, and G. M. Tabor, editors, *Conservation Medicine: Ecological Health in Practice.* Oxford University Press, New York; p. 145–163.
Robertson BA, Hutto RL. 2006. A framework for understanding ecological traps and an evaluation of existing evidence. *Ecology.* **87**:1075–1085.
Robertson BC, Elliott GP, Eason DK, Clout MN, Gemmell NJ. 2006. Sex allocation theory aids species conservation. *Biol Lett.* **2**:229–231. https://doi.org/10.1098/rsbl.2005.0430.
Saltz D, Rowen M, Rubenstein DI. 2000. The effect of space-use patterns of reintroduced Asiatic wild ass on effective population size. *Conserv Biol.* **14**:1852–1861.
Searle KR, Hobbs NT, Gordon IJ. 2007. It's the "foodscape", not the landscape: using foraging behavior to make functional assessments of landscape condition. *Isr J Ecol Evol.* **53**:297–316.
Schakner ZA, Petelle MB, Berger-Tal O, Owen MA, Blumstein DT. 2014. Developing effective tools for conservation behaviorists: reply to Greggor et al. *Trends Ecol Evol.* **29**:651–652.
Schultz CB. 1998. Dispersal behavior and its implications for reserve design in a rare Oregon butterfly. *Conserv Biol.* **12**:284–292.
Shier DM. 2006. Effect of family support on the success of translocated blacktailed prairie dogs. *Conserv Biol.* **20**:1780–1790.
Shier DM, Owings D. 2006. Effects of predator training on behavior and postrelease survival of captive prairie dogs. *Biol Conserv.* **132**:126–135.
Snijders L, Greggor AL, Hilderink F, Doran C. 2019. Effectiveness of animal conditioning interventions in reducing Human-wildlife conflict. A systematic map protocol. *Environ Evid.* https://doi.org/10.1186/s1375 0-019-0153-7
Sutherland WJ. 1998. The importance of behavioural studies in conservation biology. *Anim Behav.* **56**(4):801–809. http://dx.doi.org/10.1006/anbe.1998.0896
Swaisgood R. 2007. Current status and future directions of applied behavioral research for animal welfare and conservation. *Appl Anim Behav Sci.* **102**(3–4):139–162. http://dx.doi.org/10.1016/j.applanim.2006.05.027
Swaisgood, R. 2010. Conservation and animal behavior. In: *Encyclopedia of Animal Behavior.* Oxford: Academic Press. http://dx.doi.org/10.1016/B978-0-08-045337-8.00163-7
van Gils JA, Kraan C, Dekinga A, Koolhaas A, Drent J, de Goeij P, Piersma T. 2009. Reversed optimality and predictive ecology: burrowing depth forecasts population change in a bivalve. *Biol Lett.* **5**:5–8.
Vargas A, Anderson SH. 1999. Efects of experience and cage enrichment on predatory skills of black-footed ferrets (Mustela nigripes). *J Mammal.* **80**:263–9.
Williams R, Ashe E. 2007. Killer whale evasive tactics vary with boat number. *J. Zoology.* **272**(4):390–7. http://dx.doi.org/10.1111/j.1469-7998.2006.00280.x
Wallace MT, Buchholz R. 2001. Translocation of red-cockaded woodpeckers by reciprocal fostering of nestlings. *J Wildl Manage.* **65**:327–333.
Whelan CJ, Jedlicka DM. 2007. Augmenting population monitoring programs with behavioral indicators during ecological restorations. *Isr J Ecol Evol.* **53**:279–295.

Whiteside MA, Sage R, Madden JR. 2015. Diet complexity in early life afects survival in released pheasants by altering foraging efciency, food choice, handling skills and gut morphology. *J Anim Ecol.* **84**:1480–1489.
Wright JT, Byers JE, Koukoumaftsis LP, Ralph PJ, Gribben PE. 2010. Native species behavior mitigates the impact of habitat-forming invasive seaweed. *Oecologia.* **163**:527–534.
Zidon R, Saltz D, Shore LS, Motro U. 2009. Behavioral changes, stress, and survival following reintroduction of Persian fallow deer from two breeding facilities. *Conserv Biol.* **23**:1026–1035.

12 Impact of Hormones and Drugs on Animal Behavior

Elohozino Oghale Benneth, Isaac Babatunde Fasipe, Oluwashina Owolabi and Emmanuel Onche

12.1 INTRODUCTION

The complex interplay between hormones, drugs, and animal behavior, which is a multifaceted field of research and has the digestive system acting as a crucial intermediary, highlights that hormones such as cortisol and serotonin are significantly affected by nutrient absorption efficiency and gut health, which are essential for stress and mood regulation (Mayer, 2011). Similarly, drug metabolism varies greatly among species with different types of digestive systems, impacting the drugs' effectiveness and the resulting behavioral responses, as noted by Mikkelsen et al. (2023) clearly indicating that such variation necessitates species-specific considerations in veterinary pharmacology to optimize therapeutic outcomes and minimize adverse effects. For example, ruminants and monogastric animals display different responses to identical pharmacological treatments due to variations in gut microbiota and digestive processes (Gänzle, 2015). Recognizing these interactions is crucial for enhancing veterinary care and promoting animal well-being effectively. This study explores how hormonal changes and pharmacological substances impact behavior through the perspective of various digestive systems, providing valuable insights into species-specific factors and treatment approaches.

12.2 BACKGROUND ON THE RELATIONSHIP BETWEEN HORMONES, DRUGS, AND ANIMAL BEHAVIOR

The connection between hormones, drugs, and animal behavior is an intricate and multifaceted field of research, closely linked to the biological processes of animals' physiology. Hormones are chemical messengers secreted by endocrine glands and fulfill crucial functions in controlling behavioral responses, e.g., cortisol, which is a hormone released during stress, is capable of greatly modifying an animal's response to environmental stimuli, resulting in heightened alertness or aggressive behavior (Dickens et al., 2011). Serotonin, a neurotransmitter produced from dietary tryptophan, impacts mood and social interactions by influencing aggression, anxiety, and reproductive behaviors, while drugs that include pharmaceuticals used in veterinary medicine have been observed to affect hormonal pathways through regulated and modulated behavior; specifically, antidepressants such as selective serotonin reuptake inhibitors (SSRIs) modify serotonin levels, thereby influencing mood and anxiety-related behaviors (Peremans et al., 2003). Moreover, anti-inflammatory drugs can influence animal behavior by alleviating pain and discomfort, leading to subsequent changes in physical activity and social interaction patterns (Lascelles et al., 2008) (Figure 12.1).

The digestive system has a vital function in this interplay, processing nutrients that become neurotransmitters and hormones, and hosting a microbiome that interacts with the gut-brain axis and eventually shapes behavior (Mayer et al., 2014). Variations in the digestive system influence the metabolism and utilization of these substances like those between monogastric animals (pigs) and ruminants (cows) (Stevens & Hume, 1998). This metabolic variation can result in species-specific

 DOI: 10.1201/9781003710271-12

FIGURE 12.1 Hormonal trigger in dogs (photo credit: Oreva Ogwah, Orangeslice).

differences in drug effectiveness and hormonal reactions, consequently impacting behavior in distinct ways characteristic of each species. Recognizing these relationships is essential for creating effective veterinary treatments and enhancing animal well-being significantly. By examining the digestive system's function in hormone and drug processing, researchers can more accurately forecast behavioral responses and adapt treatments correspondingly.

12.3 OVERVIEW OF DIFFERENT ANIMAL DIGESTIVE SYSTEMS (MONOGASTRIC, AVIAN, RUMINANT, AND PSEUDO-RUMINANT)

Animal digestive systems exhibit substantial variation across species, mainly classified into monogastric, avian, ruminant, and pseudo-ruminant systems, each tailored to meet specific dietary requirements and occupy unique ecological niches.

Monogastric systems: Monogastric animals, including humans, pigs, and dogs, possess single-chambered stomachs only. Their digestive process starts in the mouth with mechanical and enzymatic food breakdown, proceeds to the stomach for acid and enzyme treatment, and finishes in the small intestine where nutrients are mostly absorbed (McDonald et al., 2020). The large intestine absorbs water and forms feces, enabling efficient processing of omnivorous diets that consist of both plant and animal material contents.

Avian systems: Birds have a distinctive digestive system, featuring a crop for storing food, a proventriculus for enzymatic digestion, and a gizzard for mechanically breaking down ingested materials efficiently (Klasing, 2015). The small intestine is the main location for absorbing essential nutrients, and waste is eliminated through the cloaca. This system efficiently processes food to meet the high energy demands required for flight.

Ruminant systems: Ruminants, including cows, sheep, and goats, possess a complex, four-chambered stomach comprising the rumen, reticulum, omasum, and abomasum sections, enabling them to break down fibrous plant material through microbial fermentation in the rumen, thereby yielding volatile fatty acids as a source of energy (Van Soest, 2018). The reticulum captures foreign objects, the omasum absorbs water and essential nutrients, and the abomasum functions similarly to a monogastric stomach with acidic and enzymatic digestion processes.

Pseudo-ruminant systems: Pseudo-ruminants, including camels and alpacas, possess a three-chambered stomach and a digestive process analogous to true ruminants, but with a relatively less complex mechanism, effectively decomposing fibrous plant matter and conserving water, a crucial adaptation for survival in harsh, water-scarce arid environments (Dittmann et al., 2020). These digestive adaptations exemplify evolutionary responses to dietary requirements and environmental pressures, highlighting the intricate diversity and complex nature of animal nutritional strategies.

12.4 IMPORTANCE OF UNDERSTANDING THE DIGESTIVE SYSTEM'S ROLE IN MEDIATING THESE EFFECTS

Understanding the animal digestive system's crucial role in mediating the effects of hormones and drugs on behavior is vital due to the complex interplay between physiological processes and behavioral responses.

The digestive system substantially affects the bioavailability and metabolism of hormones and drugs, ultimately influencing their effectiveness and behavioral impact, with various animal species exhibiting diverse digestive mechanisms, such as the intricate fermentation process in ruminants versus the simple digestion in monogastric animals, resulting in distinct pharmacokinetic and pharmacodynamic outcomes (Van Soest, 2018). Hormones like ghrelin and cortisol, which regulate hunger and stress responses respectively, are produced and modulated by the digestive system's intricate processes (Mayer et al., 2014) (Figure 12.2).

Ghrelin, secreted by the stomach, regulates feeding behavior and energy balance, according to research (Cummings & Overduin, 2007). Differences in digestive anatomy and physiology can significantly impact hormone levels, leading to variations in behavioral patterns, including feeding, mating, and stress response mechanisms (Müller et al., 2015). Drugs given to animals interact differently based on their digestive system characteristics, with the gut microbiota—which exhibits significant variation across different species—playing a crucial role in drug metabolism, thereby influencing drug absorption and overall efficacy (Gautam et al., 2019). For example, antibiotics can disrupt the balance of gut microbiota, leading to behavioral changes connected to the gut-brain axis (Dinan & Cryan, 2017). Moreover, elucidating the complex interactions between the gut microbiome and host digestive systems can inform evidence-based veterinary practices,

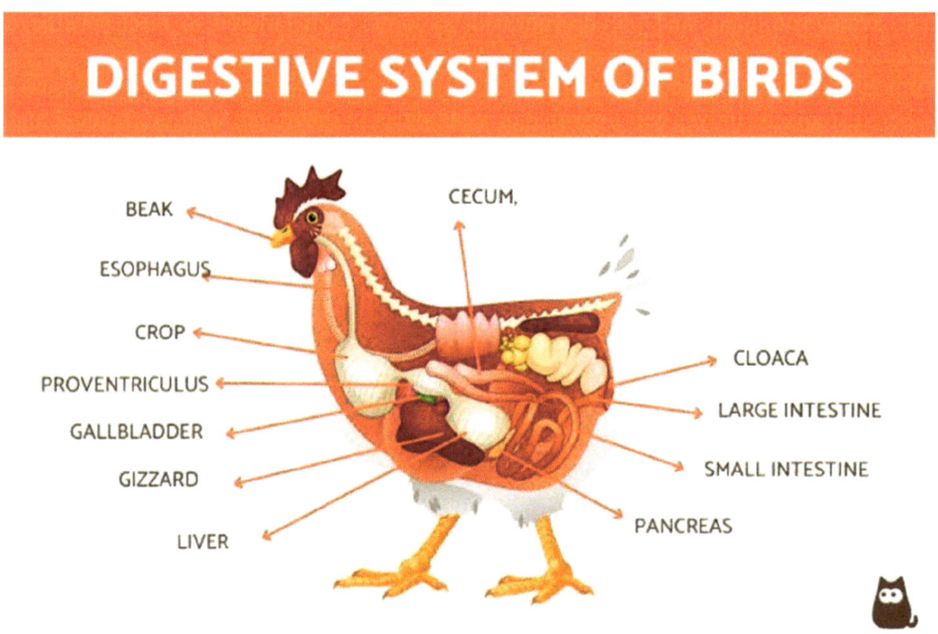

FIGURE 12.2 Avian digestive system (photo credit: Nick A. Romero, Animal Wised).

leading to enhanced animal welfare through targeted treatments that maximize therapeutic efficacy while minimizing deleterious effects (Martínez-Miró et al., 2016). Therefore, a thorough grasp of the digestive system's function is crucial for precisely forecasting and controlling the behavioral impacts of hormones and drugs across different animal species.

12.5 HORMONAL INFLUENCES ON BEHAVIOR THROUGH DIGESTIVE SYSTEM INTERACTIONS

12.5.1 Overview of Key Hormones Affecting Behavior (e.g., Cortisol, Serotonin, Ghrelin)

Hormones play vital roles in modulating animal behavior, serving as biochemical messengers that shape mood, energy levels, and stress responses. Three hormones with a profound impact on animal behavior are cortisol, serotonin, and ghrelin, significantly influencing behavioral patterns.

Cortisol is a steroid hormone secreted by the adrenal glands in response to stress and low blood glucose levels and often called the "stress hormone" due to its role in the body's fight-or-flight reaction, is associated with enhanced alertness and energy activation but also with anxiety and aggressive behavior in animals; for example, looking at chronic stress in livestock, it can be seen that prolonged exposure to handling or environmental changes is capable of leading to elevated cortisol levels thereby resulting in altered feeding behaviors and reduced reproductive success (Moberg & Mench, 2000). In wildlife, fluctuations in cortisol levels, a key stress hormone, exert a profound influence on territoriality and mating behaviors among individuals, thereby exerting a consequential impact on population dynamics, reproductive success, and overall ecosystem interactions (Boonstra, 2013) (Figure 12.3).

Serotonin is a neurotransmitter that is pivotal in governing feelings of happiness and well-being and is primarily distributed in the gastrointestinal tract, blood platelets, and central nervous system, and its regulatory role extends to mood modulation, profoundly influencing social behaviors,

FIGURE 12.3 Avian animal (photo credit: Martking).

aggression, and anxiety in animals, with heightened serotonin levels correlating with decreased aggression and the establishment of more stable social hierarchies, particularly observed in species like primates and rodents (Peremans et al., 2003). Dietary factors, including the availability of tryptophan, play a crucial role in modulating serotonin production within the body, consequently exerting a significant influence on behavioral patterns and responses (Meyer et al., 2010).

Ghrelin is a peptide hormone primarily produced in the stomach, often referred to as the "hunger hormone" due to its appetite-stimulating properties and effects. According to Asakawa et al. (2001), ghrelin levels rise before meals, thereby triggering feeding behavior in the brain, subsequently decreasing post-meal. Besides regulating energy homeostasis, it also impacts reward-related behaviors and stress response systems, which is exemplified by studies on rodents showing how its administration boosts food-seeking behavior and motivation for food rewards consistently. The authors further revealed that while also interacting with stress pathways, it potentially elevates anxiety-related behaviors in food-deprived animals, thereby resulting in heightened stress responses consistently.

The intricate interplay between hormones and animal behavior is complex, with each hormone potentially influencing and being influenced by various environmental and physiological factors simultaneously. Understanding these hormonal relationships is essential for enhancing animal welfare and informed management decisions. For instance, effective stress management and optimal dietary strategies can help regulate hormonal balance, promoting healthier and more stable behavioral patterns in both domestic and wild animal populations (Peremans et al., 2003).

12.5.2 Mechanisms of Hormone Production and Regulation in Different Digestive Systems

Hormone production and regulation in animals are intricately connected to their digestive systems, exhibiting significant variations across different species, where hormones like ghrelin, leptin, and insulin play a vital role in regulating energy balance, appetite, and metabolism, having their production and activity heavily influenced by the complex structure and function of the digestive system (Dickens et al., 2011).

Monogastric animals: In monogastric animals, such as humans and pigs, the stomach and intestines play a crucial role in hormone production and regulation. Ghrelin, also known as the "hunger hormone", is primarily produced in the stomach and stimulates appetite when the stomach is empty, triggering food intake (Kojima & Kangawa, 2005). Insulin, produced by the pancreas, regulates blood glucose levels and is significantly influenced by nutrient absorption in the small intestine, maintaining energy homeostasis (Ahren, 2005).

Ruminants: Ruminants, such as cows and sheep, possess a complex stomach comprising four compartments, enabling them to efficiently ferment and digest fibrous plant material through a unique digestive process. The fermentation process in the rumen influences the production of short-chain fatty acids, which in turn affect the release of gut hormones like GLP-1, playing a crucial role in glucose homeostasis and satiety regulation (Drucker, 2006). The intricate microbial ecosystem in the rumen significantly impacts the production of hormones that regulate energy balance, metabolic functions, and overall physiological processes in ruminant animals (Cummings & Overduin, 2007).

Hindgut fermenters: Hindgut fermenters, exemplified by species such as horses and rabbits, exclusively depend on their large intestines and cecum to facilitate fermentation processes, wherein microbial activity generates short-chain fatty acids that play a pivotal role in modulating gut hormones like PYY and GLP-1, thereby exerting profound effects on appetite regulation and mechanisms governing energy expenditure (Müller et al., 2015).

Carnivores: Carnivorous animals, such as cats and dogs, possess relatively simple and short digestive tracts compared to others. Their hormone regulation is heavily influenced by protein and fat intake, impacting the release of hormones like CCK, which promotes satiety, digestion, and overall nutrient absorption (Rehfeld, 2004).

In summary, the structure and function of different animal digestive systems significantly impact the complex mechanisms of hormone production, regulation, and metabolism, affecting overall appetite, energy balance, and species-specific nutritional needs. Understanding these intricate mechanisms provides valuable insights into species-specific nutritional requirements and pharmacological needs for optimal health and well-being.

12.5.3 Case Studies of Hormonal Impacts on Behavior in Various Animals (e.g., Stress Response in Ruminants vs. Monogastric Animals)

Hormones significantly influence animal behavior, with notable variations observed between ruminant and monogastric digestive systems, reflecting adaptability to different diets, environments, and ecological niches. Understanding these hormonal differences provides valuable insights into animal adaptability and behavioral responses.

12.5.3.1 Ruminants: Cortisol and Stress Response

In ruminant animals, such as cattle and sheep, cortisol serves as a primary indicator of stress, and its elevated levels can result from various environmental stressors, including transportation and handling. For instance, a study conducted by González et al. (2012) revealed that transporting cattle induces a substantial elevation in cortisol levels, precipitating discernible alterations in behavior such as heightened aggression and restlessness, underlining the imperative of effective stress management in agricultural practices, as prolonged stress exerts detrimental effects on growth rates and milk production, thereby impacting animal welfare and productivity significantly. The rumen, a specialized compartment of the stomach crucial for mediating stress responses, exhibits significant influence over nutrient absorption, overall health, and behavioral responses in ruminant animals, wherein stress-induced alterations in rumen microbial populations can disrupt fermentation patterns, impair digestive efficiency, shorten feeding times, and evoke behavioral signs of discomfort such as heightened vocalizations, panting, and restlessness, thereby exerting profound implications on animal well-being (Fisher et al., 2002).

12.5.3.2 Monogastric Animals: Serotonin and Behavioral Regulation

Monogastric animals such as pigs and dogs depend on serotonin (5 HT) as a pivotal hormone governing mood and behavior, synthesized and produced both in the gut and the brain, where the gut microbiota substantially contribute to its synthesis and production processes. Schmidt et al.'s (2015) study revealed that dietary changes altering gut microbiota composition significantly impacted serotonin levels and associated behaviors in pigs. Elevated serotonin levels correlated with reduced aggression and improved social interactions, underscoring the gut-brain axis's role in behavioral regulation. Dogs often experience anxiety-related behaviors due to serotonin dysregulation. Flint et al.'s (2013) research demonstrated that administering SSRIs to dogs with anxiety disorders led to substantial reductions in separation anxiety and obsessive-compulsive behaviors. Targeting hormonal pathways offers therapeutic potential for modulating behavior in monogastric animals, presenting new treatment opportunities.

12.5.3.3 Comparative Analysis

A comparison of ruminants and monogastric animals reveals distinct mechanisms and outcomes in hormonal influences on behavior, despite shared experiences. Ruminants' stress responses are closely linked to their unique digestive processes, with cortisol impacting behavior and digestive efficiency. In contrast, monogastric animals' behaviors are shaped by gut-brain interactions involving serotonin, with significant implications for anxiety and social behavior management. These case studies highlight the importance of considering digestive system differences when examining hormonal effects on animal behavior. Tailoring management practices to species-specific digestive differences can improve animal welfare and productivity.

12.6 DRUG METABOLISM IN DIFFERENT DIGESTIVE SYSTEMS

12.6.1 Description of How Different Digestive Systems Process Drugs (Absorption, Metabolism, Excretion)

Different animal digestive systems process drugs through distinct mechanisms of absorption, metabolism, and excretion, influenced by various anatomical and physiological factors.

Absorption: Craigmill et al. (2020) puts forward that monogastric animals, such as humans and pigs, exhibit primary drug absorption in the small intestine due to its expansive surface area and high vascularization, with the stomach's acidic environment influencing drug stability and solubility, thereby affecting absorption rates and highlighting the importance of gastric considerations in pharmacology. The author points out that ruminant animals like cows, sheep, and goats, in contrast, possess a complex, multi-chambered stomach, where the rumen's microbial population significantly degrades drugs before they reach the small intestine, often resulting in reduced bioavailability of orally administered medications.

Metabolism: The liver serves as the primary site for drug metabolism across all animal species, with varying efficiency and pathways. Monogastric animals, such as humans and pigs, heavily rely on hepatic enzymes like cytochrome P450 for drug biotransformation processes (Klein et al., 2020). In contrast, ruminants utilize ruminal microbes for pre-systemic metabolism, leading to the formation of distinct metabolites not found in monogastric systems, as reported by Dorne et al. (2021). For example, the extensive microbial fermentation process in ruminants can significantly alter the chemical structure of certain drugs before they enter the bloodstream for absorption (Figure 12.4).

Excretion: The kidneys and liver are the primary organs responsible for drug excretion in the body. In monogastric animals, drugs typically undergo hepatic metabolism before being excreted through urine or bile (Sweeney et al., 2019). Ruminants may experience prolonged drug excretion times due to enterohepatic recirculation, where drugs and metabolites are secreted into bile, reabsorbed from the intestine, and returned to the liver, extending their systemic presence (Toutain & Lees, 2020).

Hindgut fermenters: Hindgut fermenters, such as horses, have a large cecum and colon where microbial activity impacts drug metabolism and excretion patterns, although to a lesser extent than

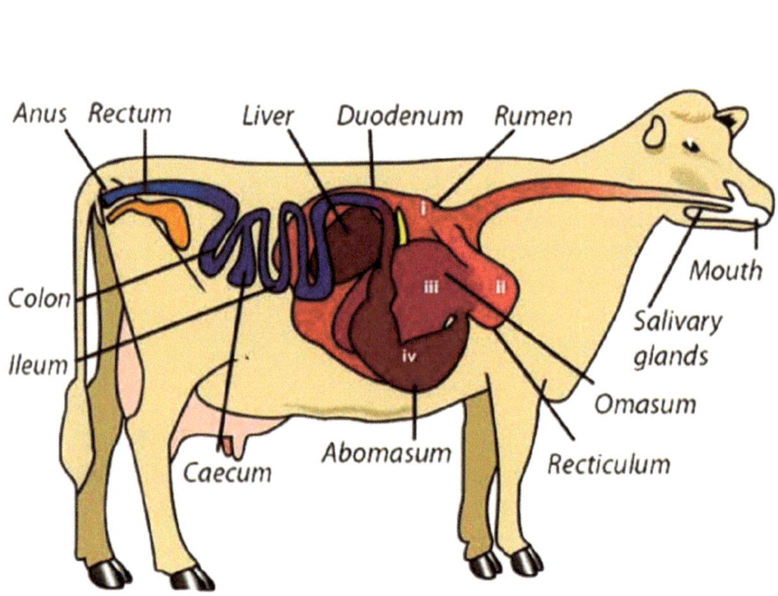

FIGURE 12.4 Ruminant digestive system (photo credit: Williams Considine, *Irish Farmers Journal*).

in ruminants (Riviere & Papich, 2018). These differences highlight the importance of species-specific pharmacological studies to ensure drug therapy efficacy and safety across various animal species. Understanding these processes enables the development of tailored veterinary treatments and optimized drug formulations for different digestive systems.

12.6.2 Impact of Gut Microbiota on Drug Metabolism and Behavior

The gut microbiota plays a crucial role in drug metabolism and behavior modulation in animals, influencing various physiological processes and impacting drug efficacy and safety. Microorganisms in the gut alter drug bioavailability, efficacy, and toxicity through biotransformation, gut environment modification, and host metabolic pathway interaction, leading to significant effects. Gut microbiota-mediated drug metabolism involves enzymatic activities that activate or deactivate pharmaceutical compounds. Certain bacteria produce enzymes like beta-glucuronidase, deconjugating glucuronidated drugs and potentially reactivating them within the gut, impacting drug pharmacokinetics.

This process significantly affects drug pharmacokinetics and pharmacodynamics, leading to variations in drug efficacy and safety among different animals (Gautam et al., 2019). The gut microbiota influences the gut-brain axis, which is a bidirectional communication system between the gastrointestinal tract and central nervous system, impacting behavior and cognitive functions. Alterations in gut microbiota composition impact neuroactive compound production, including short-chain fatty acids, neurotransmitters, and inflammatory cytokines, affecting animal behavior and neurological processes. Germ-free mice exhibit altered anxiety-like behavior and cognitive functions, which can be normalized by microbiota transplantation, highlighting the gut microbiota's role in behavior modulation (Dinan & Cryan, 2017)

Diet and environmental factors shape the gut microbiota, influencing drug metabolism and behavior. A high-fiber diet promotes SCFA-producing bacteria growth, enhancing gut barrier function and modulating systemic inflammation, potentially impacting drug absorption and neurobehavioral outcomes (Schmidt et al., 2015). Research by Gautam et al. (2019) indicates significant variation in gut microbiota composition and function among species, necessitating species-specific approaches where this interspecies variation demands tailored approaches to drug administration and behavioral interventions, considering the gut microbiota's role in drug metabolism and behavior.

The gut microbiota plays a crucial role in drug metabolism and behavior in animals, influencing enzymatic activities and gut-brain axis modulation processes. Understanding the complex interactions between the gut microbiota and host is essential for developing effective pharmacological treatments and enhancing animal well-being across species.

12.6.3 Comparative Analysis of Drug Efficacy and Side Effects in Monogastric Animals vs. Ruminants

A comparative analysis of drug efficacy and side effects in monogastric animals, such as pigs, dogs, and humans, versus ruminants, including cattle, sheep, and goats, reveals notable differences influenced by their distinct digestive physiologies and metabolic pathways. Monogastric animals have a simple stomach structure with a single compartment for food digestion, whereas ruminants have a complex, multi-compartmental stomach that facilitates microbial fermentation (Craigmill et al., 2020). These anatomical differences significantly impact drug metabolism, absorption, and efficacy. In monogastric animals, drugs are primarily absorbed in the small intestine and undergo hepatic metabolism before reaching systemic circulation, resulting in predictable as well as rapid drug responses (Mayer, 2011). However, variations in gastric pH, transit time, and enzymatic activity can affect drug bioavailability and metabolism, leading to unpredictable outcomes which emphasize the fact that differences in gastrointestinal transit times between species may influence the onset and duration of drug action, necessitating species-specific considerations (Mikkelsen et al., 2023). For instance, dogs have longer gastrointestinal transit times compared to humans, potentially altering drug absorption kinetics and efficacy in canine patients (Figure 12.5).

Ruminants, as highlighted by Van Soest (2018), have been observed to exhibit unique digestive physiology characterized by microbial fermentation in the forestomach, followed by passage to the abomasum and small intestine, which significantly impacts drug absorption and metabolism. As a result, certain drugs may undergo extensive microbial degradation in the rumen, leading to reduced systemic bioavailability and efficacy. In addition, the presence of feed particles and rumen contents can bind drugs, reducing absorption rates and potentially leading to decreased efficacy.

Consequently, drug efficacy in ruminants can be compromised compared to monogastric species, necessitating tailored pharmacological approaches. Differences in metabolic rates and drug clearance mechanisms between monogastric and ruminant species contribute to variations in drug efficacy and side effects, highlighting the need for species-specific considerations. For instance, hepatic enzyme systems responsible for drug metabolism exhibit species-specific differences, affecting drug clearance rates and potential toxicity in various species. Variations in renal function and drug excretion pathways can influence the duration of drug action and susceptibility to adverse effects in different species.

In summary, the comparative analysis underscores the importance of considering digestive system physiology and metabolic pathways when assessing drug efficacy and safety across animal species, highlighting the need for tailored pharmacological approaches in veterinary medicine to ensure optimal treatment outcomes.

12.7 BEHAVIORAL CHANGES INDUCED BY HORMONAL FLUCTUATIONS

12.7.1 Examples of Hormonal Cycles (e.g., Estrous Cycle) and Their Behavioral Effects

Mesen and Young (2015) explain how the estrous cycle in female mammals is a vital reproductive process that consists of recurring phases—proestrus, estrus, metestrus, and diestrus—which control reproductive readiness and fertility. The hypothalamic-pituitary-gonadal (HPG) axis regulates hormonal fluctuations, distinguishing each phase and orchestrating reproductive processes,

FIGURE 12.5 Ruminant animal (photo credit: Isaac Fasipe).

including mating behaviors and fertility. According to their findings, proestrus begins with rising estradiol levels that trigger increased activity and receptivity to mating cues, signaling reproductive preparedness and readiness for breeding; estrus, which is the peak fertility phase, is characterized by maximum estradiol levels that induce overt mating behaviors, vocalizations, and attraction of potential mates, ensuring successful breeding; metestrus, which follows ovulation, is marked by declining estradiol and rising progesterone that leads to decreased sexual receptivity and preparation for potential pregnancy or pseudo-pregnancy behaviors; and diestrus, which is the luteal phase, is dominated by high progesterone levels that promote maternal behaviors, nest-building, and reproductive adaptability, highlighting the flexibility of reproductive strategies. Perry et al. (2016) point out that the behavioral expressions of estrous cycle phases vary across species due to ecological and social factors, reflecting adaptive responses to environmental pressures and reproductive strategies exemplified by polygynous species like domestic sheep, which exhibit pronounced estrus behaviors to attract multiple mates, while monogamous species like prairie voles strengthen pair bonds through estrus behaviors, demonstrating diverse reproductive strategies.

Environmental cues, social dynamics, and stressors also influence estrous behaviors, shaping reproductive outcomes and fertility. In mice, exposure to male pheromones accelerates puberty onset and synchronizes estrous cycles among females housed together, which illustrates social regulation of reproductive processes (Oboti et al., 2014). Similarly, stressors like social isolation or predator presence disrupt the estrous cycle and alter behavioral responses, underscoring the impact of environmental factors on reproductive health and fertility (Woodruff et al., 2010). Having a comprehension of the interplay between hormonal fluctuations and behavioral changes across estrous cycle phases provides valuable insights into reproductive strategies, social dynamics, stress responses, and conservation efforts in diverse animal species. This knowledge is crucial for developing effective reproductive management strategies, conserving endangered species, elucidating the evolutionary underpinnings of animal behavior, and promoting biodiversity and ecological health.

12.7.2 Influence of Digestive Efficiency and Nutrient Absorption on Hormonal Balance

The complex interplay between digestive efficiency, nutrient absorption, and hormonal balance has a profound impact on animal behavior and overall health, influencing various physiological processes. Efficient nutrient digestion and absorption are crucial for maintaining homeostasis and regulating hormonal balance, ensuring proper bodily functions and overall well-being as observed in recent studies that have illuminated the intricate mechanisms by which the digestive system influences hormonal balance, shedding light on the complex relationships between gut health and endocrine function (Mayer et al., 2014). A comprehensive understanding of the intricate relationships between digestive processes, nutrient absorption, and hormonal balance is essential for elucidating the underlying mechanisms of behavioral changes and developing targeted interventions to maintain hormonal homeostasis and promote animal health and well-being.

12.7.3 Behavioral Studies in Species with Distinct Digestive Systems (e.g., Carnivores vs. Herbivores)

Studies examining species with distinct digestive systems reveal significant behavioral differences shaped by dietary needs and digestive physiology. Carnivores, such as lions and wolves, have evolved simple stomachs adapted for meat digestion, resulting in shorter digestive tracts that facilitate rapid protein and fat absorption (Miller and Zachary, 2017). Their behaviors, including hunting strategies and social structures, are closely linked to their high-protein diet requirements and efficient predation techniques (Pesta & Samuel, 2014). Herbivores have evolved complex digestive systems to process plant material. Ruminants, like cows and sheep, have multi-chambered stomachs that enable cellulose fermentation by symbiotic microorganisms, leading to prolonged digestion times (Dittmann et al., 2020). This extended digestive process influences their behavior, manifesting in extensive grazing periods, rumination, and a significant portion of their daily activities dedicated to these processes (Van Soest, 2018).

Digestive physiology differences also impact social behaviors and stress responses. Ruminants exhibit predictable, repetitive behaviors associated with feeding and digestive patterns, influencing their social structures and group foraging behaviors (Rands et al., 2008). Herbivores' social structures are influenced by resource availability, while carnivores, which requires fewer but more sporadic meals, demonstrate behaviors aligned with territoriality, solitary hunting, or complex pack dynamics (Sells et al., 2021). The gut-brain axis plays a crucial role in mediating behavioral responses to digestive processes. Studies have shown that gut microbiota composition, influenced by diet and digestion, affects mood, stress levels, and behavior in both carnivores and herbivores (Dinan & Cryan, 2017).

Variations in gut flora lead to differences in aggression, social interactions, and coping mechanisms under stress, as seen from recent research which highlights the importance of considering digestive system differences when evaluating animal behavior (Gacia-Cabrerizo et al., 2024). This approach provides a comprehensive understanding of how physiological adaptations to diet influence physical health, behavioral ecology, and social dynamics across species (Kohl et al., 2020). Future studies integrating digestive physiology and behavioral science will further elucidate complex interactions and improve animal welfare and management practices.

12.8 PHARMACOLOGICAL INTERVENTIONS AND BEHAVIORAL OUTCOMES

12.8.1 Overview of Common Drugs Used in Veterinary Practice and Their Behavioral Effects

Veterinarians utilize various pharmaceuticals to manage animal health conditions, each with potential effects on behavior, highlighting the importance of understanding these impacts. Analgesics like NSAIDs are frequently employed to alleviate pain, subsequently improving behavior by reducing stress and aggression. These medications lead to more normal, relaxed behaviors by mitigating pain-induced stress and aggression, promoting overall well-being (Queremel Milani et al., 2023) (Figure 12.6).

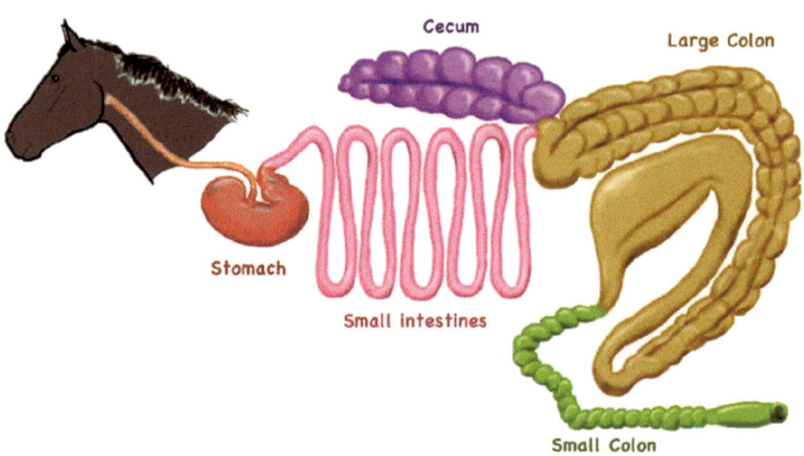

FIGURE 12.6 Pseudo-ruminant digestive system (photo credit: Doug Mayo, UGA Extension).

Amoxicillin and enrofloxacin are antibiotics that effectively treat bacterial infections, and irrespective of their importance for infection control, antibiotics can also disrupt gut microbiota and lead to behavioral changes like irritability, decreased activity, and potentially even long-term psychological effects (Patangia et al., 2022). Probiotics are co-administered to mitigate adverse effects, support gut health, and promote a balanced microbiome. Anxiolytics and sedatives, like benzodiazepines and alpha-2 agonists, manage anxiety and aggression, particularly during stressful situations like veterinary visits or transportation, and have been noted to induce calmness, reduce hyperactive behaviors, and may cause sedation and lethargy as side effects, which requires careful monitoring (Edinoff et al., 2021).

Hormonal therapies, including synthetic hormones, treat conditions like inappropriate sexual behaviors or aggression linked to hormonal imbalances, restoring balance and promoting healthy behaviors. These therapies stabilize mood, reduce unwanted behaviors, and may lead to weight gain and changes in activity levels due to systemic effects, emphasizing the importance of close observation (Annesi & Gorjala, 2010). Anticonvulsants like phenobarbital and levetiracetam effectively manage epilepsy, preventing seizure-induced stress and aggression. However, they may cause side effects like sedation and ataxia, altering normal behavior and necessitating careful management (Fiorillo et al., 2015).

Behavioral drugs like SSRIs treat disorders like obsessive-compulsive disorder and separation anxiety, modulating neurotransmitter levels to promote healthy behaviors. By modulating neurotransmitter levels, these drugs improve mood and behavior, though they may take weeks to achieve full efficacy and potentially cause initial gastrointestinal upset (Moraczewski et al., 2023). Understanding drugs' behavioral impacts is crucial for veterinarians to balance therapeutic benefits with potential side effects, ensuring optimal animal welfare and informed treatment decisions. Ultimately, veterinarians must consider the complex interplay between drugs, behavior, and overall health to provide comprehensive care and promote the well-being of animals in their care.

12.8.2 Interaction between Drugs and Digestive System Types (e.g., Antibiotics, Anti-Inflammatories)

The interaction between drugs and digestive system types has a profound impact on pharmacokinetics and pharmacodynamics, which influences medication efficacy and side effects significantly; therefore, understanding these interactions is crucial for optimizing therapeutic outcomes and minimizing adverse effects. Different digestive systems, including monogastric, ruminant, and hindgut fermenters, process drugs uniquely due to variations in anatomy and microbiota composition. Ruminants like cows possess a complex, multi-chambered stomach system where drugs undergo extensive microbial fermentation before reaching the bloodstream, unlike monogastric animals, such as humans and pigs, which have a single-chambered stomach where drug absorption primarily occurs in the small intestine (Kaur et al., 2023). This leads to potential drug degradation or modification, altering efficacy and half-life. Antibiotics are significantly affected by digestive system types such that monogastric animals absorb antibiotics like tetracycline efficiently in the small intestine, and on the other hand, ruminants' microbial activity can inactivate certain antibiotics, requiring higher doses or alternative administration routes to achieve therapeutic levels (Tomanic et al., 2023) (Figure 12.7).

Antibiotics can disrupt gut microbiota, leading to dysbiosis, which varies in severity depending on the digestive system. For example, hindgut fermenters like horses are sensitive to antibiotics disrupting their cecal and colonic flora, potentially causing severe colitis (Chaucheyras-Durand et al., 2022). Anti-inflammatories including NSAIDs, interact differently with various digestive systems and are absorbed in the stomach and small intestine of monogastric animals, although their use is limited by gastrointestinal irritation and ulceration, while in ruminants, NSAIDs are less likely to cause direct gastric damage due to the buffering effect of the rumen but can still cause ulcerations

FIGURE 12.7 Pseudo-ruminant animal (photo credit: Bob Baffert, Belmont Stakes).

if overused (Takeuchi, 2012). The prolonged presence of NSAIDs in the rumen can affect rumen fermentation and overall digestive efficiency, which further emphasizes that the type of digestive system plays a pivotal role in the determination of drug absorption, metabolism, and side effects. Tailoring drug administration protocols to specific digestive systems will therefore enhance therapeutic efficacy and reduce adverse outcomes, highlighting the need for species-specific veterinary pharmacology (Souto et al., 2023).

12.8.3 Behavioral Case Studies Post-Drug Administration in Animals with Different Digestive Systems

Behavioral case studies post-drug administration reveal significant variations across animals with different digestive systems, highlighting the critical role of digestive physiology in mediating drug effects. For instance, in monogastric animals like dogs, the administration of psychotropic drugs such as fluoxetine (an SSRI) has been shown to alleviate symptoms of separation anxiety, resulting in reduced destructive behavior and vocalization (Simpson et al., 2007).

Conversely, in ruminants like cows, the works of Laven et al. (2012) reveal that the intricate multi-chambered stomach can alter drug metabolism and bioavailability observed in the administration of non-steroidal anti-inflammatory drugs (NSAIDs) in cattle, which often requires higher doses compared to monogastric animals to achieve therapeutic effects due to the prolonged absorption process in the rumen. The authors believe that behavioral changes post-NSAID administration in ruminants typically include improved mobility and reduced signs of pain-related distress, indicating effective analgesia.

In hindgut fermenters such as horses, the impact of drugs like antibiotics on behavior can be profound due to the crucial role of the cecum and colon in fermentation and nutrient absorption. Antibiotic treatment can disrupt the gut microbiota, leading to gastrointestinal disturbances and associated behavioral changes such as colic symptoms and reduced appetite (Ramirez et al., 2020). Post-administration behavior in horses often necessitates close monitoring to mitigate adverse effects on gut health and behavior.

These case studies underscore the necessity for species-specific considerations in veterinary pharmacology, acknowledging that digestive system differences significantly influence drug absorption, metabolism, and consequent behavioral outcomes. Therefore, tailored pharmacological strategies are imperative to optimize therapeutic efficacy and minimize adverse behavioral responses in animals with diverse digestive physiologies.

12.9 ENVIRONMENTAL AND NUTRITIONAL FACTORS

12.9.1 How Diet Influences Hormone Levels and Drug Efficacy

Diet plays a crucial role in modulating hormone levels and drug efficacy in animals through various mechanisms including nutrient absorption, gut microbiota composition, and metabolic processes. Different dietary components can either enhance or inhibit hormone production. For instance, dietary proteins and amino acids are essential for synthesizing hormones like insulin and growth hormones, while fatty acids serve as precursors for steroid hormones (Jahan-Mihan et al., 2011). Carbohydrate intake significantly impacts insulin secretion and sensitivity such that high-carbohydrate diets can lead to hyperinsulinemia, potentially altering the balance of other hormones such as glucagon and cortisol, resulting in this imbalance which may influence behaviors related to stress and energy regulation (Schneiderman et al., 2005). Moreover, dietary fiber affects the secretion of gut hormones like ghrelin and peptide YY, which regulate appetite and satiety, thus indirectly influencing feeding behaviors and metabolic health (Efimtseva et al., 2021). Gut microbiota, profoundly influenced by diet, also plays a pivotal role in hormone regulation. Fermentable fibers and prebiotics enhance the growth of beneficial gut bacteria capable of producing short-chain fatty acids (SCFAs) that have systemic hormonal effects, and these SCFAs can modulate the release of gut hormones as well as influence the brain-gut axis, affecting animal behavior and stress responses (Davani-Davari et al., 2019) (Figure 12.8).

Diet also affects drug metabolism, primarily through its impact on the liver and gut microbiota. The liver's ability to metabolize drugs is influenced by the availability of specific nutrients and dietary components, e.g. a diet high in antioxidants and certain vitamins can enhance hepatic enzyme activity thereby improving drug metabolism (Lobo et al., 2010). Conversely, Ji et al. (2023) are of the notion that while nutrient deficiencies or high-fat diets can impair liver function and reduce drug efficacy leading to increased risk of adverse effects, gut microbiota shaped by diet can alter drug bioavailability and efficacy. Furthermore, the authors state that microbial enzymes can metabolize drugs before they reach systemic circulation, either activating or deactivating them, e.g.

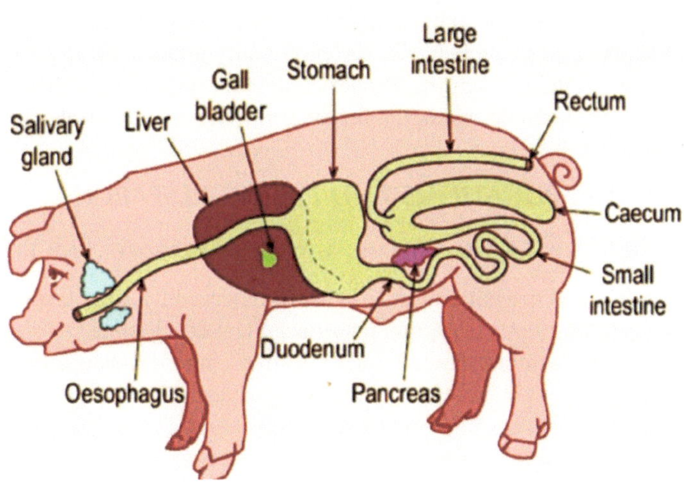

FIGURE 12.8 Monogastric digestive system.

dietary fibers that promote beneficial bacteria can enhance the bioavailability of certain drugs by modifying gut pH and enzyme activity.

In conclusion, diet significantly influences hormone levels and drug efficacy through its influence on nutrient availability, gut microbiota, and metabolic pathways. Understanding these interactions is crucial for optimizing dietary strategies to improve animal health and therapeutic outcomes.

12.9.2 Comparative Analysis of Dietary Impacts on Monogastric vs. Ruminant Behavior

Diet has a profound impact on animal behavior, with varying effects between monogastric and ruminant species due to their unique digestive system characteristics (Ranabir & Reetu, 2011). Monogastric animals, like pigs and chickens, have a single-chambered stomach for quick food digestion while ruminants such as cattle and sheep have a complex stomach designed for fermenting plant material over extended periods (Appleton, 2018).

One of the primary differences in dietary impact is related to nutrient absorption and its influence on energy levels and overall well-being. Monogastric animals tend to exhibit immediate behavioral responses to dietary changes due to rapid digestion and absorption rates always (Efimtseva & Chelpanova, 2021). High-energy diets, for example, can lead to increased activity and aggressive behavior in pigs; conversely, low-quality diets result in lethargy and reduced social interaction among animals as demonstrated by Meer et al. (2017). Ruminants rely heavily on microbial fermentation in their rumen to break down complex carbohydrates efficiently which affects energy availability, influences production of volatile fatty acids, and impacts metabolic processes significantly as seen from the studies of Varesi et al. (2022). The authors reveal that diets high in fiber promote calm behavior in ruminants due to steady release of VFAs and sustained energy while diets rich in easily fermentable carbohydrates cause rapid fermentation, leading to acidosis, discomfort, restlessness, aggression, and potentially harmful health issues.

Additionally, dietary composition significantly affects the gut microbiota, influencing the gut-brain axis and impacting behavior patterns. In monogastric animals, shifts in gut microbiota due to dietary changes can substantially impact neurotransmitter levels, affecting mood and behavior regulation (Azad et al., 2020). In ruminants, a balanced diet supports a stable microbial population, promoting normal rumination behavior and reducing stress-related behaviors effectively. Feeding strategies differ between monogastric and ruminant groups. Monogastrics receive controlled meal portions, inducing competitive behaviors when resources are scarce and in contrast, ruminants have continuous access to forage which aligns with natural grazing behavior and reduces competitive interactions, thereby promoting social harmony (Van Soest, 2018) (Figure 12.9).

In summary, dietary impacts on behavior are profoundly influenced by monogastric and ruminant animals' digestive system structures. Proper comprehension of these differences is necessary for optimizing animal welfare and managing behavior through dietary strategies.

12.10 PRACTICAL IMPLICATIONS AND FUTURE RESEARCH

12.10.1 Implications for Animal Welfare, Veterinary Practices, and Pharmaceuticals

The study of hormones and drugs on animal behavior is crucial for understanding their impacts on animal welfare, veterinary practices, and pharmaceutical development. Hormones like cortisol, serotonin, and ghrelin play pivotal roles in regulating stress, mood, and appetite, influencing behaviors such as aggression, social interaction, and feeding patterns. For example, high cortisol levels, which is often a response to stress, can lead to increased aggression and anxiety in animals.

Drugs given to animals have significant behavioral effects and include antibiotics, anti-inflammatories, and psychotropic medications which have substances that can alter brain chemistry and neurotransmitter levels, potentially leading to changes in mood, cognition, and overall behavior.

FIGURE 12.9 Monogastric animal (photo credit: De Heus, Animal Nutrition).

The digestive system plays a crucial role in how these drugs are processed and their subsequent impact on behavior. For instance, the gut microbiota's composition can influence drug efficacy and side effects, highlighting the need for tailored veterinary interventions based on an animal's specific digestive system. Pharmaceutical companies are also impacted by these findings, and in the course of the development of drugs, they must consider variations in digestive systems across different species to ensure efficacy and safety. Research on the gut-brain axis and its role in behavior modulation is particularly relevant for drug development aimed at improving mental health and behavior in animals. A good knowledge of the interplay between hormones, drugs, and behavior is essential for improving animal welfare in veterinary practices. Effective management of stress and anxiety through hormonal regulation or appropriate medication can enhance animal well-being, reducing stress-related behaviors and improving their overall quality of life. Veterinarians must consider the animal's digestive system when prescribing medications, as differences in absorption and metabolism can significantly affect the drug's behavioral impact.

In conclusion, the intricate relationship between hormones, drugs, and animal behavior requires a comprehensive approach in veterinary practices and pharmaceutical development. By considering the digestive system's role, practitioners and researchers can better address behavioral issues and improve animal welfare. Future research should continue to explore these dynamics to provide deeper insights into effective treatments and interventions tailored to various species.

12.10.2 Recommendations for Managing Hormonal and Drug Impacts Based on Digestive System Types

Understanding the complex interplay between hormones, drugs, and animal behavior requires a thoughtful and refined approach, particularly when considering the varied digestive systems across different animal species, which could conversely lead to effective management strategies that also consider the distinct physiological and metabolic processes of different digestive systems.

Monogastric animals: With an insight that monogastric animals possess a single-chambered stomach that results in direct and rapid drug absorption, it is necessary to note that these animals require carefully tailored drug dosages to prevent overdose due to efficient absorption rates; additionally, hormonal treatments such as cortisol for stress need close monitoring for potential rapid systemic effects (Kusov et al., 2023). Nutritional support including probiotics enhances gut health, modulating the gut-brain axis and positively influencing behavior.

Ruminants: These animals have a complex, multi-chambered stomach, significantly impacting drug metabolism and hormone regulation. Drugs given to ruminants must account for microbial fermentation processes in the rumen which affect drug efficacy and behavior (Cammack et al., 2018). Hormonal therapies should consider slower absorption rates and prolonged effects. For example, long-acting formulations can maintain consistent hormonal levels, stabilizing behavior over time.

Hindgut fermenters: In hindgut fermenters, animals experience significant fermentation in the cecum and colon. Drugs and hormones in these animals may have delayed absorption and prolonged effects, which makes the use of slow-release drugs that align with the digestive processes really crucial (Song et al., 2023). Moreover, managing dietary fiber is vital, as it impacts the microbial ecosystem and synthesis of behavioral compounds like short-chain fatty acids.

12.10.2.1 General Recommendations

Personalized medicine: Tailoring treatments to the specific digestive physiology of the animal optimizes drug efficacy and minimizes adverse behavioral effects.

Nutritional support: Incorporating dietary strategies that support gut health can enhance the effectiveness of hormonal and pharmacological interventions significantly (Berding et al., 2021).

Monitoring and adjustment: Regularly tracking behavioral responses and tailoring treatment plans to individual- and species-specific needs are crucial.

Holistic approach: Integrating pharmacological treatments with environmental and management practices can lead to improved behavioral outcomes. For instance, minimizing environmental stress and providing enrichment activities can enhance the effectiveness of hormonal therapies.

Successfully implementing strategies requires a profound understanding of the animal's digestive system, ensuring that interventions are both effective and supportive of optimal behavioral well-being. This comprehensive approach leads to more compassionate, evidence-based practices in veterinary care, animal management, and overall animal welfare.

12.10.3 Identification of Gaps in Current Research and Suggestions for Future Studies

The study of hormones and drugs on animal behavior through the lens of varying digestive systems reveals several research gaps that require further investigation. Current literature has primarily focused on either the hormonal impacts or drug effects on animal behavior, often overlooking the complex interplay between these factors, digestive systems, and species-specific differences.

12.10.3.1 Gaps in Current Research

Comparative studies on digestive systems: There is a lack of comparative studies investigating how different digestive systems, such as monogastric, ruminant, and hindgut fermenters, affect hormone and drug metabolism and subsequent animal behavior. Most research focuses on a single species or digestive system, limiting understanding of cross-species variations and hindering a comprehensive approach (Karasov and Douglas, 2013).

Gut-brain axis investigations: While the gut-brain axis is recognized as a crucial mediator of behavior, detailed studies investigating links between digestive system variations and changes in this axis are scarce. The mechanisms by which gut microbiota, influenced by different digestive processes, impact brain function and behavior remain underexplored, requiring further investigation (Cryan & Dinan, 2012).

Longitudinal behavioral studies: Many studies only provide short-term observations of behavioral changes post-drug or hormone administration. Longitudinal studies that assess long-term behavioral impacts and adaptations over time are limited, restricting understanding of chronic effects, habituation, and sensitization possibilities.

Holistic environmental interactions: Research often isolates hormonal or pharmacological effects without considering the complex, holistic environmental interactions, including diet, stressors, social dynamics, and other factors that significantly influence behavior. Comprehensive studies that integrate these various factors are urgently needed to provide a more accurate and comprehensive understanding of behavior in naturalistic settings, as demonstrated by Harris and McDade (2018).

12.10.3.2 Suggestions for Future Studies

Enhanced comparative research: Future studies should adopt a comparative approach, investigating how diverse digestive systems affect the pharmacokinetics and pharmacodynamics of various drugs and hormones. This approach will elucidate species-specific responses, enabling informed decisions and tailored veterinary practices that improve animal health outcomes.

Integrated gut-brain axis research: There is a pressing need for comprehensive, integrated studies that focus on the gut-brain axis across diverse species, exploring how variations in gut microbiota composition and digestive processes impact behavior. Advanced techniques like metagenomics, neuroimaging, and bioinformatics can be employed to uncover the complex relationships between gut microbiota, brain function, and behavior, as demonstrated by Zhu et al. (2020).

Longitudinal and multidisciplinary studies: Conducting longitudinal studies that track behavioral changes over extended periods will provide valuable insights into the chronic effects of hormonal and pharmacological interventions on behavior. Incorporating multidisciplinary approaches, combining behavioral science and microbiology, will offer a comprehensive understanding of complex biological processes.

Environmental and nutritional ocntexts: Future research should consider the broader environmental and nutritional contexts, investigating how diet and environmental stressors interact with hormonal and drug effects on animal behavior and welfare. This approach will help develop holistic management strategies for animal welfare and health, incorporating insights from other relevant research (Mkwanazi et al., 2019).

By addressing these gaps, future research can provide a more nuanced and comprehensive understanding of how hormones and drugs influence animal behavior, considering the critical role of different digestive systems and nutritional factors. This knowledge is essential for improving animal welfare, optimizing veterinary interventions, and advancing the field of behavioral pharmacology, ultimately enhancing the well-being of animals in various settings.

12.11 CONCLUSION

The digestive system profoundly impacts the intricate relationship between hormones, drugs, and animal behavior. This chapter has explored how different digestive systems—monogastric, ruminant, and hindgut fermenters—modulate hormonal fluctuations and pharmacological interventions on behavior. Key findings highlight the critical role of nutrient absorption efficiency, gut microbiota diversity, and the gut-brain axis in this relationship.

Hormones such as cortisol, serotonin, and ghrelin, integral to stress response, mood regulation, and hunger signals, exhibit varied impacts on behavior depending on the digestive system. For instance, ruminants display distinct stress responses compared to monogastric animals due to differences in gut microbiota composition and fermentation processes. Similarly, the metabolism of drugs such as antibiotics and anti-inflammatories is significantly influenced by the digestive system, affecting efficacy and behavioral outcomes. The gut-brain axis, a vital component of this interplay, underscores bidirectional communication between the gastrointestinal tract and central nervous system. Variations in digestive systems modulate this communication, influencing behavioral responses to hormonal fluctuations and drug interventions. Diet and environmental factors further compound these effects, with dietary habits influencing hormone levels and drug metabolism, impacting behavior.

Despite these insights, significant gaps persist in our understanding. Firstly, comprehensive studies comparing hormonal and pharmacological impacts across diverse species with different digestive systems are lacking. Most research focuses on a limited number of model organisms, failing to capture the full spectrum of digestive and behavioral diversity. Secondly, the mechanisms underlying the gut-brain axis in different digestive systems require further elucidation, particularly regarding the influence of specific gut microbiota profiles on hormonal and drug effects. Future research should aim to address these gaps by conducting comparative studies across diverse species with varied digestive systems. Investigating the role of specific gut microbiota in mediating hormonal and drug effects on behavior will provide deeper insights into the gut-brain axis. Additionally, longitudinal studies examining long-term impacts of hormonal fluctuations and drug administration on behavior, considering dietary and environmental variables, are essential.

Understanding these dynamics will have significant implications for animal welfare, veterinary practices, and pharmaceutical development. Tailoring treatments based on digestive system type can enhance pharmacological intervention efficacy and mitigate adverse behavioral outcomes. Ultimately, integrating digestive system considerations into behavioral research will lead to more holistic and effective approaches in managing animal health and behavior.

REFERENCES

Ahren, B. (2005). Autonomic regulation of islet hormone secretion—Implications for health and disease. *Diabetologia*, 48(3), 393–395.

Annesi, J. J., & Gorjala, S. (2010). Relations of self-regulation and self-efficacy for exercise and eating and BMI change: A field investigation. *BioPsychoSocial Medicine*, 4(1), 10. https://doi.org/10.1186/1751-0759-4-10

Appleton, J. (2018). The gut-brain axis: Influence of microbiota on mood and mental health. *Integrative Medicine (Encinitas)*, 17(4), 28–32.

Asakawa, A., Inui, A., Kaga, T., Yuzuriha, H., Nagata, T., Fujimiya, M., ... & Kasuga, M. (2001). A role of ghrelin in neuroendocrine and behavioral responses to stress in mice. *Neuroendocrinology*, 74(3), 143–147.

Azad, M. A. K., Gao, J., Ma, J., Li, T., Tan, B., Huang, X., & Jie, Y. (2020). Opportunities of prebiotics for the intestinal health of monogastric animals. *Animal Nutrition*. https://doi.org/10.1016/j.aninu.2020.08.001.

Berding, K., Vlckova, K., Marx, W., Schellekens, H., Stanton, C., Clarke, G., Jacka, F., Dinan, T. G., & Cryan, J. F. (2021). Diet and the Microbiota–gut–brain axis: Sowing the seeds of good mental health. *Advances in Nutrition (Bethesda, Md.)*, 12(4), 1239–1285. https://doi.org/10.1093/advances/nmaa181

Boonstra, R. (2013). Reality as the leading cause of stress: Rethinking the impact of chronic stress in nature. *Functional Ecology*, 27(1), 11–23.

Cammack, K. M., Austin, K. J., Lamberson, W. R., Conant, G. C., & Cunningham, H. C. (2018). Tiny but mighty: The role of the rumen microbes in livestock production. *Journal of Animal Science*. https://doi.org/10.1093/jas/skx053.

Chaucheyras-Durand, F., Sacy, A., Karges, K., & Apper, E. (2022). Gastro-intestinal Microbiota in equines and its role in health and disease: The black box opens. *Microorganisms*, 10(12), 2517. https://doi.org/10.3390/microorganisms10122517.

Craigmill, A. L., Riviere, J. E., & Webb, A. I. (2020). *Handbook of comparative pharmacokinetics and residues of veterinary antimicrobials*. CRC Press..

Cryan, J. F., & Dinan, T. G. (2012). Mind-altering microorganisms: the impact of the gut microbiota on brain and behaviour. *Nature Reviews. Neuroscience*, 13(10), 701–712. https://doi.org/10.1038/nrn3346.

Cummings, D. E., & Overduin, J. (2007). Gastrointestinal regulation of food intake. *Journal of Clinical Investigation*, 117(1), 13–23.

Davani-Davari, D., Negahdaripour, M., Karimzadeh, I., Seifan, M., Mohkam, M., Masoumi, S., Berenjian, A., & Ghasemi, Y. (2019). Prebiotics: Definition, types, sources, mechanisms, and clinical applications. *Foods (Basel, Switzerland)*, 8(3), 92. https://doi.org/10.3390/foods8030092.

Dickens, M. J., Delehanty, D. J., & Romero, L. M. (2011). Stress: An inevitable component of animal translocation. *Biological Conservation*, 141(9), 2349.

Dinan, T. G., & Cryan, J. F. (2017). The microbiome-gut-brain axis in health and disease. *Gastroenterology Clinics of North America*, 46(1), 77–89.

Dittmann, M. T., Frey, J., & Kreuzer, M. (2020). Comparative digestive physiology in ruminants and camelids. *Animal Nutrition and Feed Technology*, 20(3), 297–309.
Dorne, J. L., et al. (2021). Comparative pharmacokinetics and metabolism of veterinary drugs. *Journal of Veterinary Pharmacology and Therapeutics*, 44(1), 110–123.
Drucker, D. J. (2006). The biology of incretin hormones. *Cell Metabolism*, 3(3), 153–165.
Edinoff, A. N., Nix, C. A., Hollier, J., Sagrera, C. E., Delacroix, B. M., Abubakar, T., Cornett, E. M., Kaye, A. M., & Kaye, A. D. (2021). Benzodiazepines: Uses, dangers, and clinical considerations. *Neurology International*, 13(4), 594–607. https://doi.org/10.3390/neurolint13040059.
Efimtseva, E. A., Institute of Physiology of K omi Science Centre of the Ural Branch of the Russian Academy of Sciences, Federal Research Centre "K omi Science Centre of the Ural Branch of the Russian Academy of Sciences," Chelpanova, T. I., & Institute of Physiology of K omi Science Centre of the Ural Branch of the Russian Academy of Sciences, Federal Research Centre "K omi Science Centre of the Ural Branch of the Russian Academy of Sciences." (2021). Dietary fiber as modulators of gastrointestinal hormonal peptide secretion. *Voprosy Pitaniia*, 90(4), 20–35. https://doi.org/10.33029/0042-8833-2021-90-4-20-35.
Fiorillo, A., Dell'Osso, B., Albert, U., Atti, A. R., Carmassi, C., Carrà, G., Cosci, F., Del Vecchio, V., Di Nicola, M., Ferrari, S., Goracci, A., Iasevoli, F., Luciano, M., Martinotti, G., Nanni, M. G., Nivoli, A., Pinna, F., Poloni, N., Pompili, M., … Volpe, U. (2015). Bridging the gap between education and appropriate use of benzodiazepines in psychiatric clinical practice. *Neuropsychiatric Disease and Treatment*, 1885. https://doi.org/10.2147/ndt.s83130
Fisher, A. D., Stewart, M., Verkerk, G. A., Morrow, C. J., & Matthews, L. R. (2002). The effects of feed restriction and genotype on adrenocortical responsiveness, behaviour, and growth in cattle. *Livestock Production Science*, 73(1), 33–44.
Flint, G., Gray, C., Kogan, L., & Young, K. (2013). Efficacy of fluoxetine for the treatment of anxiety-related behaviors in dogs. *Journal of Veterinary Behavior*, 8(6), 420–428.
Gänzle, M. G. (2015). Lactic metabolism revisited: metabolism of lactic acid bacteria in food fermentations and food spoilage. *Current Opinion in Food Science*, 2, 106–117.
García-Cabrerizo, R, Cryan, J. F. (2024). A gut (microbiome) feeling about addiction: Interactions with stress and social systems. *Neurobiology of Stress*, 30, 100629. https://doi.org/10.1016/j.ynstr.2024.100629.
Gautam, A., Dixit, S., & Srinivasan, S. (2019). Role of gut microbiota in modulating metabolism and neurobehavioral outcomes. *Journal of Biosciences*, 44(2), 101–112.
González, L. A., Schwartzkopf-Genswein, K. S., Bryan, M., Silasi, R., & Brown, F. (2012). Relationships between transport conditions and welfare outcomes during commercial long haul transport of cattle in North America. *Journal of Animal Science*, 90(10), 3640–3651.
Harris, K. M., & McDade, T. W. (2018). The biosocial approach to human development, behavior, and health across the life course. *The Russell Sage Foundation Journal of the Social Sciences: RSF*, 4(4), 2–26. https://doi.org/10.7758/rsf.2018.4.4.01
Jahan-Mihan, A., Luhovyy, B. L., El Khoury, D., & Anderson, G. H. (2011). Dietary proteins as determinants of metabolic and physiologic functions of the gastrointestinal tract. *Nutrients*, 3(5), 574–603. https://doi.org/10.3390/nu3050574
Ji, J., Jin, W., Liu, S.-J., Jiao, Z., & Li, X. (2023). Probiotics, prebiotics, and postbiotics in health and disease. *MedComm*, 4(6). https://doi.org/10.1002/mco2.420
Karasov, H. W., & Douglas, A. E. (2013). Comparative digestive physiology. In *Comprehensive physiology* (pp. 741–783). Wiley. https://doi.org/10.1002/cphy.c110054
Kaur, H., Kaur, G., Gupta, T., Mittal, D., & Ali, S. A. (2023). Integrating omics technologies for a comprehensive understanding of the microbiome and its impact on cattle production. *Biology*, 12(9), 1200. https://doi.org/10.3390/biology12091200
Klasing, K. C. (2015). *Comparative Avian nutrition*. CABI Library.
Klein, H. J., Zierler, R., & Kreutzer, C. (2020). Comparative digestive physiology in animals. *Animal Physiology Reviews*, 49(3), 204–217.
Kohl, K. D., Coogan, S. C. P., & Raubenheimer, D. (2020). The nutritional ecology of wild carnivores. *Annual Review of Animal Biosciences*, 8, 145–165.
Kojima, M., & Kangawa, K. (2005). Ghrelin: Structure and function. *Physiological Reviews*, 85(2), 495–522.
Kusov, P. A., Kotelevtsev, Y. V., & Drachev, V. P. (2023). Cortisol monitoring devices toward implementation for clinically relevant biosensing in vivo. *Molecules (Basel, Switzerland)*, 28(5), 2353. https://doi.org/10.3390/molecules28052353.
Lascelles, B. D. X., Gaynor, J. S., & Smith, E. S. (2008). Amelioration of pain in dogs and cats. *Veterinary Clinics: Small Animal Practice*, 38(6), 1337–1355.

Laven, R., Chambers, P., & Stafford, K. (2012). Using non-steroidal anti-inflammatory drugs around calving: Maximizing comfort, productivity and fertility. *Veterinary Journal (London, England: 1997)*, 192(1), 8–12. https://doi.org/10.1016/j.tvjl.2011.10.023.

Lobo, V., Patil, A., Phatak, A., & Chandra, N. (2010). Free radicals, antioxidants and functional foods: Impact on human health. *Pharmacognosy Reviews*, 4(8), 118. https://doi.org/10.4103/0973-7847.70902

Martínez-Miró, S., Tecles, F., Ramón, M., Escribano, D., Hernández, F., Madrid, J., ... & Cerón, J. J. (2016). Causes, consequences and biomarkers of stress in swine: An update. *BMC Veterinary Research*, 12(1), 171.

Mayer, E. A. (2011). Gut feelings: the emerging biology of gut–brain communication. *Nature Reviews Neuroscience*, 12(8), 453–466.

Mayer, E. A., Tillisch, K., & Gupta, A. (2014). Gut/brain axis and the microbiota. *The Journal of Clinical Investigation*, 125(3), 926–938.

McDonald, P., Edwards, R. A., Greenhalgh, J. F. D., Morgan, C. A., Sinclair, L. A., & Wilkinson, R. G. (2020). *Animal nutrition* (8th ed.). Pearson.

Meer, Y. van der, Gerrits, W. J. J., Jansman, A. J. M., Kemp, B., & Bolhuis, J. E. (2017). A link between damaging behaviour in pigs, sanitary conditions, and dietary protein and amino acid supply. *PLoS One*, 12(5), e0174688. https://doi.org/10.1371/journal.pone.0174688.

Meyer, J. S., Novak, M. A., Bowman, R. E., & Harlow, H. F. (2010). Behavioral responses to stress in rhesus monkeys (Macaca mulatta). *Psychosomatic Medicine*, 42(3), 357–367.

Mesen, T. B., & Young, S. L. (2015). Progesterone and the luteal phase. *Obstetrics and Gynecology Clinics of North America*, 42(1), 135–151. https://doi.org/10.1016/j.ogc.2014.10.003.

Mikkelsen, L. F., Van Cruchten, S., & Makin, A. (2023). The use of Göttingen minipigs in juvenile studies. In *Drug discovery and evaluation: Safety and pharmacokinetic assays* (pp. 1–19). Cham: Springer International Publishing.

Miller, A. D., & Zachary, J. F. (2017). Nervous system. In *Pathologic basis of veterinary disease* (pp. 805–907. e1). Elsevier.

Mkwanazi, M. V., Ncobela, C. N., Kanengoni, A. T., & Chimonyo, M. (2019). Effects of environmental enrichment on behaviour, physiology and performance of pigs — A review. *Asian-Australasian Journal of Animal Sciences*, 32(1), 1–13. https://doi.org/10.5713/ajas.17.0138.

Moberg, G. P., & Mench, J. A. (2000). *The biology of animal stress: Basic principles and implications for animal welfare*. CABI.

Moraczewski, J., Awosika, A. O., & Aedma, K. K. (2023). *Tricyclic antidepressants*. StatPearls Publishing. https://www.ncbi.nlm.nih.gov/books/NBK557791/

Müller, T. D., Nogueiras, R., Andermann, M. L., Andrews, Z. B., Anker, S. D., Argente, J., ... & Horvath, T. L. (2015). Ghrelin. *Molecular Metabolism*, 4(6), 437–460.

Oboti, L., Pérez-Gómez, A., Keller, M., Jacobi, E., Birnbaumer, L., Leinders-Zufall, T., Zufall, F., & Chamero, P. (2014). A wide range of pheromone-stimulated sexual and reproductive behaviors in female mice depend on G protein Gαo. *BMC Biology*, 12(1). https://doi.org/10.1186/1741-7007-12-31

Patangia, D. V., Anthony Ryan, C., Dempsey, E., Paul Ross, R., & Stanton, C. (2022). Impact of antibiotics on the human microbiome and consequences for host health. *Microbiology Open*, 11(1). https://doi.org/10.1002/mbo3.1260

Peremans, K., Audenaert, K., De Cort, W., Blanckaert, P., & van Bree, H. (2003). Effects of serotonin in animal behavior and cognition. *European Journal of Pharmacology*, 478(3), 243–259.

Perry, A. N., Carter, C. S., & Cushing, B. S. (2016). Chronic social isolation enhances reproduction in the monogamous prairie vole (Microtus ochrogaster). *Psychoneuroendocrinology*, 68, 20–28. https://doi.org/10.1016/j.psyneuen.2016.02.016

Pesta, D. H., & Samuel, V. T. (2014). A high-protein diet for reducing body fat: mechanisms and possible caveats. *Nutrition & Metabolism*, 11(1), 53. https://doi.org/10.1186/1743-7075-11-53.

Queremel Milani, D. A., & Davis, D. D. (2023). *Pain management medications*. StatPearls Publishing. https://www.ncbi.nlm.nih.gov/books/NBK560692/

Ramirez, J., Guarner, F., Bustos Fernandez, L., Maruy, A., Sdepanian, V. L., & Cohen, H. (2020). Antibiotics as major disruptors of gut Microbiota. *Frontiers in Cellular and Infection Microbiology*, 10. https://doi.org/10.3389/fcimb.2020.572912.

Ranabir, S., & Reetu, K. (2011). Stress and hormones. Indian *Journal of Endocrinology and Metabolism*, 15(1), 18. https://doi.org/10.4103/2230-8210.77573

Rands, S. A., Cowlishaw, G., Pettifor, R. A., Rowcliffe, J. M., & Johnstone, R. A. (2008). The emergence of leaders and followers in foraging pairs when the qualities of individuals differ. *BMC Evolutionary Biology*, 8(1), 51. https://doi.org/10.1186/1471-2148-8-51

Rehfeld, J. F. (2004). A centenary of gastrointestinal endocrinology. *Hormone and Metabolic Research*, 36(11/12), 735–741.

Riviere, J. E., & Papich, M. G. (2018). *Veterinary pharmacology and therapeutics*. Wiley-Blackwell.

Schmidt, T., Koop, K., Altherr, B. J., & Belik, V. (2015). Influence of dietary composition on the gut microbiota and serotonin levels in pigs. *Journal of Animal Science*, 93(3), 1431–1440.

Schneiderman, N., Ironson, G., & Siegel, S. D. (2005). Stress and health: Psychological, behavioral, and biological determinants. *Annual Review of Clinical Psychology*, 1(1), 607–628. https://doi.org/10.1146/annurev.clinpsy.1.102803.144141

Sells, S. N., Mitchell, M. S., Podruzny, K. M., Gude, J. A., Keever, A. C., Boyd, D. K., Smucker, T. D., Nelson, A. A., Parks, T. W., Lance, N. J., Ross, M. S., & Inman, R. M. (2021). Evidence of economical territory selection in a cooperative carnivore. *Proceedings. Biological Sciences*, 288(1946), 20210108. https://doi.org/10.1098/rspb.2021.0108

Song, Y., Day, C. M., Afinjuomo, F., Tan, J.-Q. E., Page, S. W., & Garg, S. (2023). Advanced strategies of drug delivery via oral, topical, and parenteral administration routes: Where do equine medications stand? *Pharmaceutics*, 15(1), 186. https://doi.org/10.3390/pharmaceutics15010186

Souto, E. B., Santini, A., Barbosa, C. I., Baldim, I., Campos, J. R., Fernandes, A. R., Mazzola, P. G., Andreani, T., Dias, I. R., Durazzo, A., Lucarini, M., Atanasov, A. G., & Silva, A. M. (2023). Modified drug delivery systems for veterinary use: Pharmaceutical development and applications. *Current Bioactive Compounds*, 19(1), 69–76. https://doi.org/10.2174/1573407218666220404110837

Simpson, B. S., Landsberg, G. M., Reisner, I. R., Ciribassi, J. J., Horwitz, D., Houpt, K. A., Kroll, T. L., Luescher, A., Moffat, K. S., Douglass, G., Robertson-Plouch, C., Veenhuizen, M. F., Zimmerman, A., & Clark, T. P. (2007). Effects of reconcile (fluoxetine) chewable tablets plus behavior management for canine separation anxiety. *Veterinary Therapeutics: Research in Applied Veterinary Medicine*, 8(1), 18–31.

Stevens, C. E., & Hume, I. D. (1998). *Comparative physiology of the vertebrate digestive system*. Cambridge University Press.

Sweeney, M. J., & Papich, M. G. (2019). Influence of food and diet on veterinary drug pharmacokinetics. *Journal of Veterinary Pharmacology and Therapeutics*, 42(3), 213–223.

Takeuchi, K. (2012). Pathogenesis of NSAID-induced gastric damage: Importance of cyclooxygenase inhibition and gastric hypermotility. *World Journal of Gastroenterology: WJG*, 18(18), 2147. https://doi.org/10.3748/wjg.v18.i18.2147

Toutain, P. L., & Lees, P. (2020). Integration and modeling of pharmacokinetic and pharmacodynamic data. *Journal of Veterinary Pharmacology and Therapeutics*, 43(1), 24–30.

Van Soest, P. J. (2018). *Nutritional ecology of the ruminant* (2nd ed.). Cornell University Press.

Varesi, A., Chirumbolo, S., Campagnoli, L. I. M., Pierella, E., Piccini, G. B., Carrara, A., Ricevuti, G., Scassellati, C., Bonvicini, C., & Pascale, A. (2022). The role of antioxidants in the interplay between oxidative stress and senescence. *Antioxidants (Basel, Switzerland)*, 11(7), 1224. https://doi.org/10.3390/antiox11071224

Woodruff, T. J., Janssen, S. J., Guillette, L. J., Jr, & Giudice, L. C. (Eds.). (2010). *Environmental impacts on reproductive health and fertility*. Cambridge University Press. https://doi.org/10.1017/cbo9780511674686

Zhu, S., Jiang, Y., Xu, K., Cui, M., Ye, W., Zhao, G., Jin, L., & Chen, X. (2020). The progress of gut microbiome research related to brain disorders. *Journal of Neuroinflammation*, 17(1). https://doi.org/10.1186/s12974-020-1705-z

13 Behavior of Courtship

*Rajesh Kumar, Jeetendra Singh Rajoriya,
A. P. Singh, Pramod Kumar and Kabir Alam*

13.1 INTRODUCTION

A courtship is defined as a set of display behaviors in which an animal attempts to attract a mate and exhibit their desire to copulate. These behaviors often include ritualized movement ("dances"), vocalizations, mechanical sound production, or displays of beauty, strength, or agonistic ability. The term courtship refers to the behavioral interaction that occurs between males and females before, during, and just after the act of mating. The main function of courtship behavior is to ensure that the two individuals are of the same species. It also involves stereotyped behaviors. Courtship behavior informs a potential mate that the intention is breeding and not aggression. The reproductive behavior of animals plays an important role in reproduction, affecting both the success of mating and survival of the young (Hafez, 1992). Behavioral patterns associated with courtship and copulation, birth, maternal care, and suckling attempts of the newborn have a dramatic quality that has attracted students of mammalian behavior (Phillips, 1993).

13.2 PSYCHOSOCIAL ASPECTS OF BEHAVIOR

The encounter of sexual partners is the first step of reproductive behavior. In free-living animals, this occurs largely under the influence of pre-existing social structure and the territorial or home-range behavior of males and females, and leads to an organized pattern of reproduction that varies with the socio-spatial or territorial characteristics of the species. The territories of males and females are overlapping, with a permanent association between potential sexual partners. In other species, such as in the wild rabbit and beaver, the territory is occupied by a permanent couple or harem, and the male avoids any encounter outside his territory. This pattern persists under artificial environments. For example, male rabbits breeding in cages display sexual behavior toward receptive females only after the male occupies the cage for a sufficiently long time to consider it his territory. Territorial behavior is intensified during the season of reproduction, and in fact, in many species, such as the seal, it exists only at that time. Under feral conditions, farm animals do not defend defined territories against intruders, but herds and flocks tend to occupy a "home range". The basic unit is matriarchal, consisting of a female, her adult female offspring, and their immature young. Such a matriarchal herd is remarkably stable. It persists after a temporary dispersion of its members or mixing in large groups of several hundred individuals (African antelopes, bison). This stability is the consequence of strong inter-individual bonds resulting from contacts occurring during infancy. Experimentally, cows reared together from birth form such a stable group even in the middle of a large herd. Such a bond, limited to the dams and their female offspring, could be the basis of the social organization in ungulates. Increasing population density may lead to abnormal behavior such as tail biting, cannibalism, and an increased level of aggression. Such abnormal behavior may be detrimental in that the performance of individual animals with low social status in the group would decline because of stress (Randolph *et al.*, 1981).

 DOI: 10.1201/9781003710271-13

13.3 MECHANISMS OF SEXUAL BEHAVIOR

The physiologic signal that originates sexual motivation is the gonadal steroid balance. Transmitted by the blood flow, the hormones activate the central nervous system. The humoral signal is transformed into sexual motivation, or sex drive. The motor patterns of copulatory activity are programmed according to pre-existing species-specific neuronal circuits. The behavioral interactions leading to copulation can be divided into four major phases: mutual searching for the sexual partner; identification of the physiologic state of the partner; the sequence of behavioral interactions resulting in the adoption of the mating posture by the female; and the mounting reaction of the male leading to copulation. The "mating stance" of the sow—clear, long-lasting, and easy to release by an experimenter—is especially suitable for a study of the releasing mechanism. During the "standing reaction", the receptive sow is absolutely immobile, arches her back, and cocks the ears, and this reaction may be exhibited when an estrous female is touched on the back. Only 48% of estrous gilts, however, will "stand" in the absence of the male. Of those gilts in estrus that do not "stand" in the absence of a male, 60% will show the immobility if presented with the odor of a boar. Thus, the stimuli emitted during pre-copulatory interactions facilitate the release of the female's postural response. Females in the mating stance are mounted immediately, and this reaction seems to be released mainly by visual and tactile cues. A restrained female, although not in estrus, is immediately mounted even by a sexually experienced bull or ram. Similarly, a ram does not copulate selectively with an estrous ewe when presented with two restrained anestrous females. Sexual reactions of the male toward stimuli other than those emanating from the female are common. For example, the bull or the boar reacts rapidly to a restrained male or to a dummy. The sexual releaser for mounting may be the overall shape of the female and her immobility. The other visual, olfactory, or acoustic information from the estrous female may be of minor but complementary importance.

13.4 CORTEX AND SENSORY CAPACITIES

Deprivation of sensory capacity can inhibit sexual behavior, reduce the ability to detect the partner, and/or impair orientation. Inexperienced males are impaired to a greater degree than experienced ones. If one sense is inhibited, another sense that is ordinarily used to a lesser degree may be augmented. Thus, elimination of the stimuli to visual receptors in males results in the use of tactile and olfactory receptors. Copulation in domestic mammals is not suppressed with the elimination of vision, smell, or hearing, provided contact with the partner has been established. Tactile stimuli are involved in the organization of postural responses of copulation (e.g., immobilization of the estrous sow and lordosis in the estrous rat in response to flank palpation). Sensory input from the penis is important for the normal development of mounting and thrusting behaviors in bulls and goats.

13.5 NEURAL MECHANISMS OF ERECTION AND EJACULATION

Erection is predominantly under the influence of the parasympathetic system. The parasympathetic nerves in the bull, which supply the external genitalia, arise from the sacral segments of the spinal cord. The copulatory patterns of the male are primarily governed by the neuromuscular anatomy and blood supply of the penis. The bull, ram, and boar have a fibroelastic penis that is relatively small in diameter and rigid when nonerect. Although the penis becomes more rigid on rapid erection, it enlarges little, and the amount of contractile tissue is limited. Protrusion is effected mainly by straightening the "S"-shaped flexure and relaxation of the retractor muscle. On the other hand, the stallion has a typical vascular penis with no sigmoid flexure. The function of the penis as an organ of intromission depends on the power of erection as a result of sexual excitement. The size, shape, and length of the penis vary greatly between the flaccid and the erect state.

Intromission and ejaculation are elicited by tactile stimuli (warmth of vagina and slipperiness of mucus) acting on the penile receptors. The penis of the bull and the ram is sensitive to temperature, whereas that of the stallion is more sensitive to pressure exerted by the contractions of the vaginal walls. In the boar, the corkscrew-shaped tip of the penis is engaged in the cervix during mating. The pressure exerted by this is sufficient to elicit ejaculation even without any thermal stimulation.

13.6 SEXUAL BEHAVIOR

Sexual behavior of males and females, stimuli eliciting male sexual behaviors, sex drive (libido), patterns of female sexual behavior, and intensity of estrus have been reviewed (Qureshi *et al.*, 2000a, 2000b). Various patterns of courtship, display, motor activities, and postures are directed to bring the male and female gametes together to ensure fertilization, pregnancy, and propagation of the species (Hafez, 1992). The encounter of sexual partners is the first step that occurs largely under the influence of pre-existing social structures and the territorial or home-range behavior of males and females (Mloszewski, 1983). The behavioral interactions leading to copulation can be divided into four major phases: mutual searching for the sexual partner; identification of the physiologic state of these partners; the sequence of behavioral interaction resulting in the adoption of the mating posture by the female; and the mounting reaction of the male leading to copulation (Gordon, 1996).

13.7 MALE SEXUAL BEHAVIOR

Male sexual behavior in buffaloes is similar to but less intense than that in cattle bulls. Libido is suppressed during the hotter period of the day, particularly in the swamp buffalo (Jainudeen and Hafez, 1992a). Normal copulation encompasses a sequence of the following behavioral elements: courtship, erection and protrusion, mounting, intromission, ejaculatory thrust and ejaculation, and dismounting. Each of the elements in the sequence has certain unique characteristics (Hafez, 1992).

13.7.1 Pre-Copulatory Behavior (Courtship)

Courtship is more evident in open range rather than under the restricted conditions on smaller farms (Jainudeen, 1986). The experienced bull detects the proestrous female two days before estrus and remains in her general vicinity. This activity is termed "guarding behavior" which is defined as a female coming into estrus being closely guarded by a bull who attempts to keep other bulls (low in social rank) at a distance; but females often prove evasive, thereby attracting the attention of other bulls and leading to a series of displacements by males of successively high rank. The most dominant male in the herd is usually in attendance (Estes, 1992).

Nudging can be seen in courtship; it prompts the female to move forward (Vale *et al.*, 1990) after nudging by the male and provides reciprocal stimulation, and once mating has occurred, nudging will be stopped. Nosing the perineum and hind quarters of the females is a rather male activity. Nudging and nosing activities are called "prompting behavior" (Gordon, 1996). Males regularly monitor females in estrus and can prompt them to urinate by licking and nosing the vulva and perineum; in response to prompting, females in heat start to urinate. At this time, the male may take one or two sips of urine and may also lick the vulva (Pathak, 1992); simultaneously the male starts some soft penile movements and protrudes a few centimeters of penis.

Tending behavior is a common sexual behavior displayed by males during the pre-copulatory period (courtship). It means that the male maintains close bodily contact and associates with the female in estrus while being near her (Estes, 1992), and both sexes contribute to this temporary bond, thereby facilitating repeated mating and ensuring optimum conditions for fertilization. If she does not bolt, he mounts, and a receptive cow stands still with her tail arched. The tending male may lose interest; other bulls that visually recognize the tending behavior may come up and replace the first animal (Mloszewski, 1983). It is a behavior often shown by the male while consorting with the female before and during estrus (Figure 13.1).

Sniffing and licking the female's genitalia are the most frequent patterns, suggesting an important function of chemical communication through olfaction (Figure 31.1). The male smells the female's urine, which contains pheromones (Fraser and Broom, 1990). Also there is tactile stimulation is made by nuzzling and licking the perineal region of the female. The male may sniff the grass on which the female may have lain or urinated. The early response of the bull to the estrus scent is "flehmen", which is widespread and prominent in ungulates, including females (Houpt *et al.*, 1993). This behavior consists of a forward-extended neck and muzzle, the upper lip curled up exposing the gums and teeth (Hafez, 1992) with constricted nares (Gordon, 1996) or closed nasal apertures (Sule *et al.*, 2001), and the head is elevated (Jainudeen and Hafez, 1992a) (Figure 13.2).

This noticeable flehmen behavior appears to be related to chemo-sensory evaluation by the vomeronasal (Jacobson's) organ, which is an additional olfactory receiver connected to the roof of the mouth and nasal cavity and is instrumental in eliciting a reflex act that is expressed in the flehmen phenomenon (Estes, 1992). Flehmen is also performed by males in response to urine from other

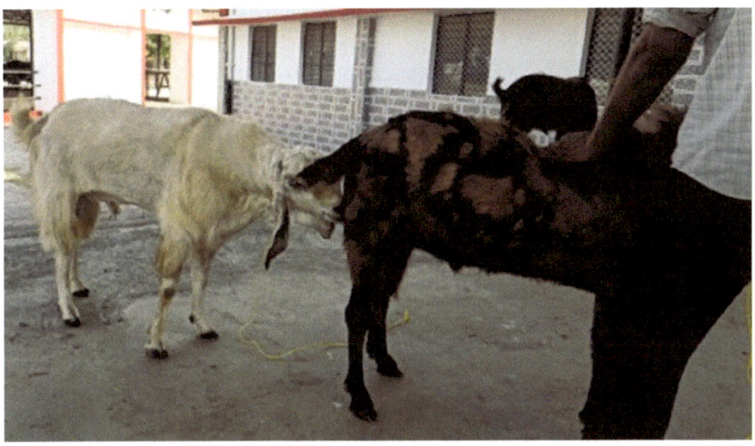

FIGURE 13.1 Sniffing behavior shown by bucks during semen collection (courtesy: Kabir Alam).

FIGURE 13.2 Flehmen reaction shown by buffalo bull during semen collection (courtesy: Kabir Alam).

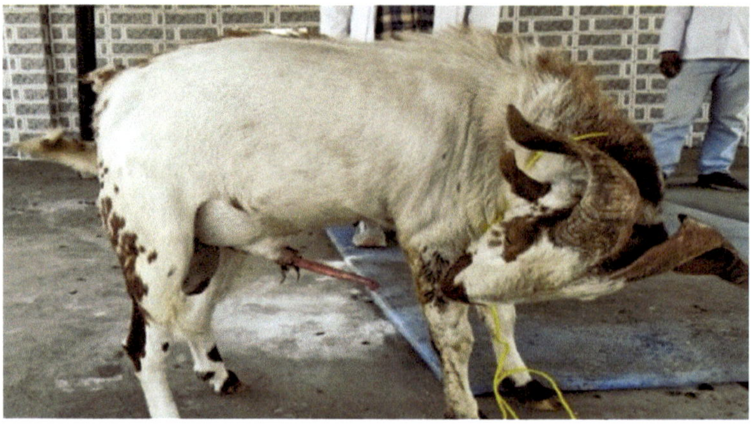

FIGURE 13.3 Self-enurinaring behavior shown by male goat (courtesy: Kabir Alam).

males or to the male's own urine; when males approach each other at pasture, they may threaten each other. Novel odors from inorganic compounds or clothing from handlers will occasionally elicit the flehmen response (Albright and Arave, 1997). Flehmen is rarely performed by females during sexual encounters with males; however, buffalo cows will perform flehmen when sniffing other cows that are either in estrus or proestrus. It may also be performed by females towards the birth fluids of newborn animals (Gordon, 1996).

A tending male tests the female's readiness to stand and be served by licking and sniffing. Then, when the female becomes well established as one approaching estrus, the "chin-resting behavior" becomes more frequent (Hill *et al.*, 1992). Chin-resting behavior often appears before mounting as the bull positions himself behind the cow and raises his head so that his chin and throat are in contact with the cow's rump. Estrous cows respond to chin-resting pressure by standing to be mounted. The female either moves away from under his chin, which rests on her rump, or stands, but without leaning into the male (Mloszewski, 1983). Chin resting can also be observed between males as a form of dominance or appeasement gesture (Vale, 1990). The male goat exhibits self-enurination behavior (Figure 13.3), which encompasses downward turning of head and shoulders while urinating onto the face and front legs before or after primary courtship and copulation (Price *et al.*, 1986) (Figure 13.3).

13.7.2 Mounting

In the presence of proestrous female, the bull attempts several mounts (unsuccessful); the penis becomes partially erected and protrudes from the prepuce. During this activity, the bull excretes dribblings of accessory fluid, derived from the Cowper's gland and differing from the seminal plasma emitted from the vesicular glands during ejaculation (Hafez, 1992). The female buffalo in estrus emits a strong scent and attracts the bull. The male tends the estrous cow and rests his chin on the female "chinning behavior", and she in turn responds by "standing". The buffalo bull mounts her; i.e. the withers above the female's pelvis, fixes his forelegs around the female's back, grasps her firmly, and performs rhythmic pelvic thrusts (Mloszewski, 1983).

In mounting, the bull quickly shifts his weight to the hind legs, lifts his shoulder and forelegs off the ground, and straddles the cow near the middle of her back (Figure 13.4). The force of muscular contractions during mounting is so strong that the bull's hind legs are often lifted completely off the ground (Albright and Arave, 1997). The sexual releaser for mounting may be the overall shape of the female and her immobility. The other visual, olfactory, or acoustic information from the estrous female may be of minor but complementary importance (Gupta *et al.*, 1994). The copulatory mounting lasts only some tens of seconds in buffaloes (Pathak, 1992) (Figure 13.4).

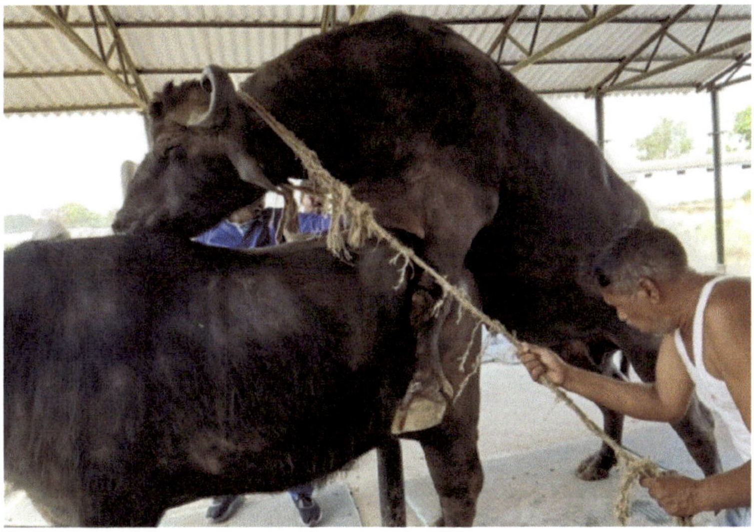

FIGURE 13.4 Mounting of buffalo bulls, foreleg clasp on female's back during semen collection (courtesy: Kabir Alam).

13.7.3 Intromission

At mounting, the abdominal muscles of the bull, particularly the rectus abdominis muscles, contract suddenly. As a result, the pelvic region of the bull is quickly brought into direct apposition to the external genitalia of the cow (Hafez, 1992). Intromission is performed quickly, the ejaculatory thrust is given with maximum vigor, and semen is ejaculated.

13.7.4 Ejaculation and Dismounting

After intromission, the ejaculatory thrust is quick, and semen is ejaculated as a single violent rush intra-vaginally near the os-cervix (Barkawi *et al.*, 1993). In buffalo bulls, an intense generalized muscular contraction takes place at ejaculation (Albright and Arave, 1997). After ejaculation, the abdominal muscles then relax and the bull dismounts slowly; the penis is soon retracted into the prepuce (Hafez, 1992). After ejaculation, the bull head is pressed down on the female body (Figure 13.5).

13.7.5 Post-Copulatory Behavior

This period is called "refractory period". Most animals show no sexual activity immediately following copulation. The duration of the refractory period is highly variable and increases gradually when several copulations are allowed successively with the same female (Houpt *et al.*, 1993). Following ejaculation and dismounting, buffalo bulls show a refractory period, which is a state of sexual exhaustion. This state is not principally a physical one; however, it refers mainly to the loss of stimulus by the female. A quick return to mounting behavior is shown by males when they are given an opportunity to mate a new estrous female (Hafez, 1992).

In a study by Mloszewski (1983), it was noticed that, after the male has dismounted, he usually continues to tend the female buffalo and mounts her again, perhaps in 10 minutes or so, but the interval and the number of mountings vary with the male. The frequency of copulation varies with the species, the breed, the ratio of males and females, available space, and the nature of sexual

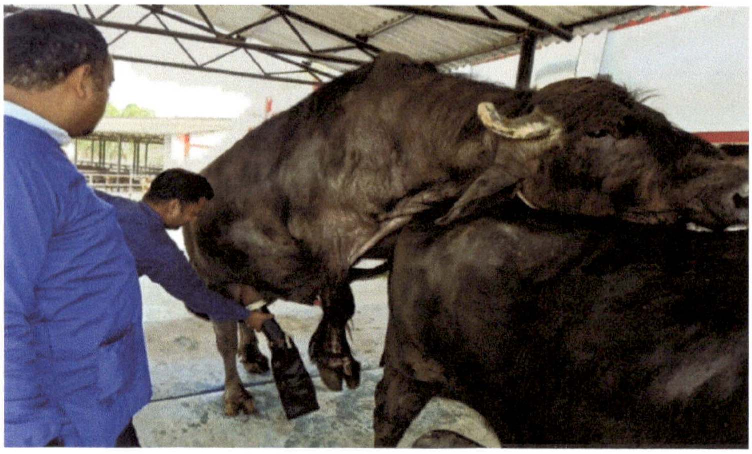

FIGURE 13.5 After ejaculation, the bull head is pressed down on the female body (courtesy Kabir Alam).

stimuli (Albright and Arave, 1997). Some bulls have been observed to copulate over 80 times within 24 hours or 60 times within 6 hours, with an average of 21 copulations before exhaustion was observed (Gordon, 1996). After ejaculation, the buffalo bull presses his head on the female's body; moreover, this behavior was also shown during semen collection using an artificial vagina.

13.7.6 Frequency of Copulation

The frequency of copulation varies with the species, the breed, and the ratio of males and females present, available space, period of sexual rest, climate, and nature of sexual stimuli. The maximal number of ejaculations is higher in bulls and rams than in stallions and boars. After a long sexual rest, a ram may copulate up to 50 times on the first day after joining with the ewes, but this frequency is greatly reduced on subsequent days. The goat, the stallion, and the boar reach exhaustion after a lesser number of ejaculations than in the ram and the bull.

13.8 DURATION OF ESTRUS

The duration of estrus is influenced by species, breed, climate, and management. Estrus is limited to about a day in sheep and cattle but to longer periods in the sow and the mare. In species in which the period of sexual receptivity is short, ovulation takes place after its end, but in species that remain receptive for long periods, ovulation occurs during estrus.

13.9 GENERAL FACTORS AFFECTING MALE SEXUAL BEHAVIOR

13.9.1 Neural Mechanism

Erection is under the influence of the parasympathetic system. The parasympathetic nerves in the bull that supply the external genitalia arise from the sacral segments of the spinal cord. The copulatory patterns of males are primarily governed by the neuromuscular anatomy and blood supply of the penis. The bull has a fibroelastic penis that is relatively small in diameter and rigid when non-erect, while it becomes more rigid on rapid erection. Visual and tactile cues elicit sexual desire in the bull. Intromission and ejaculation are elicited by tactile stimuli (warmth of vagina and slipperiness of mucus) acting on the penile receptors. The penis of the bull is sensitive to temperature.

13.9.2 Genetic Factors

Breed and strain differences in sexual performance are frequently observed. Males of dairy breeds are more active than beef males, whereas Brahman bulls are sluggish. Yorkshire boars are easier to train for semen collections than Durocs. More differences in the pattern of sexual behavior occur among pairs of identical twin bulls than between members of the pair. Breed differences in the duration of estrus in sheep and pigs may be partly due to differences in ovulation rates. Individual differences in the amount of sexual stimulation required to elicit "immobilization reaction" in the sow are independent of sexual experience.

13.9.3 Environmental Factors

The effect of external stimulation on sexual behavior is more pronounced in the male than in the female.

a. **Effect of novelty of stimulus females (Coolidge effect)**
Sexual activity of the male increases when new females in the herd become receptive. If four receptive ewes are available, the ram mates three times as much as he does when only one ewe is in estrus. The enhancing effect of a new stimulus animal should be kept in mind under modern husbandry conditions. For instance, changing the teaser bull is an effective way to increase sexual behavior of a sluggish male.

b. **Non-specific stimuli**
Non-specific external stimuli may lead to sexual activity in males of low libido. Changing the location of semen collection by moving the teaser animal or encouraging the bull are all effective in sluggish bulls.

c. **Presence of other animals**
The presence of other males while teasing a female or copulating improves sexual libido of the male. Social hierarchy, however, may interfere with sexual activity when several males compete for one receptive female. The dominant male performs most of the copulations and restricts the sexual performance of his subordinates. When females are in excess, however, dominant males cannot effectively control the activity of their inferiors. Adult rams usually dominate or "boss" yearling rams, and the degree of their dominance is greater than the dominance of yearling rams over other yearlings. Unfortunately, social dominance is not correlated with the fertility of males; thus, a "bossy" male who is infertile or diseased may depress the conception rate of the entire breed. The size of the pasture also affects the competition among males and the number of copulations per female. The sexual performance of males of some species can be improved by providing the animals an opportunity to view other males engaged in sexual activity. Although rams are not stimulated by observing sexual activity, they are aroused when provided the opportunity to smell another male who has just copulated.

d. **Seasonal effect**
Seasonal variations in sexual behavior of sheep, goats, and horses are mostly due to seasonality of hypothalamic/pituitary function controlling the secretion of gonadal hormones. Seasonal changes are also reported in the responsiveness of ovariectomized ewes, does, and sows to exogenous hormones, showing a direct effect of seasons on the response of the CNS. For example, in the ovariectomized female goat (doe), an injection of 30 μg of estradiol will stimulate the full expression of estrous behavior during the fall, the breeding season. During the spring, however, this dose of estrogen is not effective. This finding can

be replicated in environmentally controlled rooms, providing evidence for the importance of photoperiod on the expression of sexual behaviors. The intensity of sexual behavior is reduced in hot climates. The plane of nutrition per se does not seem to affect sexual behavior. However, any physical trouble may seriously affect sexual expression (e.g., inflammation of the hooves or joints, change in teeth, eczema, pains from accidents, or certain diseases).

13.10 EFFECT OF EXPERIENCE

The efficiency of copulation of males and females is improved by experience. Social deprivation during infancy drastically impairs adult sexual behavior. Individual contacts before puberty can have an organizing effect on subsequent sexual performance. Young, inexperienced males are usually awkward during their first contact with a receptive female; they approach hesitantly, spend a long time exploring the genitalia, mount with erection, and ejaculation is weak. Watching another male copulate enhances sexual desire in the spectator when he is subsequently allowed to copulate.

13.11 UTERINE MOTILITY AND SPERM TRANSPORT

Sterile matings with a vasectomized male stimulate sperm transport in the rabbit but not in sheep. Stimulation of the genitalia or pre-coital stimuli causes contraction of the cervix and uterus of the ewe and the cow, as a result of the release of oxytocin. Vaginal distention and pre-coital stimulation cause maximal oxytocin release in sheep and goats. Oxytocin release often occurs before actual coitus has taken place.

13.12 EFFECT OF MALE ON ANESTROUS FEMALES

Many domestic females undergo periods of anestrus. By the end of seasonal anestrus in sheep and goats, the introduction of the male results in an earlier and synchronized appearance of estrous cycles. Even the Merino sheep, which is reputed to breed throughout the year, ovulates and breeds spontaneously for only a restricted period during autumn. The peak of estrus observed 17 to 18 days following the introduction of the male represents, in fact, the second cycle, and this allows normal fertilization. An androgen-dependent pheromone from the male is responsible for similar synchronization of estrus in mice (Bronson, 1984) and possibly sheep. Neither sight nor contact is necessary for the synchronization of the first estrus of the breeding season in sheep; nursing the young delays estrus compared with milking in cows and ewes. The presence of the ram results in an earlier postpartum estrus in nursing ewes, making the postpartum estrus similar in dry and milked females. Male goats exhibit a unique "self-marking" behavior: self-urination that likely enhances the odor signal detected by females during the breeding season. The introduction of the boar shortly before spontaneous puberty in a group of previously isolated gilts results in earlier onset of estrus. This phenomenon is caused by odor emanating from the boar. The pig's characteristic body odor is produced by skin gland secretions. The carpal glands are well differentiated on the front legs. In the male, secretions of the preputial pouch, which give off its "boar" scent, are involved in sexual behavior.

13.13 ABNORMALITIES IN SEXUAL BEHAVIOR

13.13.1 Male Animals

13.13.1.1 Inability to Copulate (Impotence)

Physical disabilities may impede or prevent mating by causing failure in copulatory behavior i.e., mounting, intromission, or ejaculation.

13.13.1.2 Impotence to Mount

This is defined by Houpt (1991) as a loss of libido and considered a clinical problem. Inability to mount is a common disorder encountered in older buffalo bulls. It is associated with locomotor dysfunction or musculoskeletal diseases such as fractures, sprains, and osteoarthritic lesions of the hind limbs causing the bulls to become lame and hesitant to mount (Jainudeen and Hafez, 1992a). Obese bulls can be difficult to arouse sexually, so the diet is closely controlled. Some bulls become unwilling to mount when placed on slippery floors; bulls trained to mount other males, as practised in artificial insemination, may lose interest and refuse to mount. This inhibition may be overcome by changing the sexual stimulus to a new steer or bull (Houpt *et al.*, 1993) or restraining the bull from the teaser animal or allowing him to watch other bulls, which may sexually arouse him (Vale *et al.*, 1990).

13.13.1.3 Failure to Achieve Intromission or Ejaculation

This is a condition in which the penis fails to enter the vagina. It may result from insufficient protrusion of the penis from the sheath or deviation of the penis. Another serious cause of inability to protrude the penis is hematoma of the penis as a result of rupture of the corpus cavernosum of the penis (Houpt *et al.*, 1993). It commonly occurs in bulls during coitus when the penis is thrust against the perineum of the cow. Abnormal venous drainage of the corpus cavernosum in the bull could result in a penis that is too flaccid for intromission despite good libido (Fraser and Broom, 1990). Tumor of the glans penis may occasionally prevent protrusion of the penis. Affected bulls are reluctant to serve or are incapable of achieving intromission. A series of pelvic thrusts may occur after a rectal intromission, but the penis is most often withdrawn without ejaculation. Abortive ejaculations may occur if the female refuses the intromission or the penis fails to penetrate the vulva (Hafez, 1992).

13.13.1.4 Buller-Steer Syndrome

Male cattle or buffaloes, "riders", in intensive housing are prone to mount herdmates willing to stand to be mounted, "bullers". This occurs mainly in feedlot herds. The buller may be ridden to the point of exhaustion or even death by the herdmates. Separating the buller from the group is the only corrective measure with a sure result (Badawey *et al.*, 1972; Houpt, 1991).

13.13.1.5 Homosexuality

This refers to sexual behavior among males, particularly at puberty and when young males are housed together. The stimuli eliciting the male's sexual response are essentially visual. The homosexual males in sex-aggregated groups become heterosexual when placed with females and again homosexual when segregated (Hafez, 1992).

13.13.1.6 Hyper- and Hypo-Sexuality

In males, hypersexuality consists of increased sexual excitement, increased frequency of copulation, and attempted copulations with young males and females of the same or different species. Hyposexuality is characterized by abnormalities in the ejaculatory pattern. Certain males may fail to ejaculate in spite of protrusion or erection of the penis, whereas others cannot mount or exhibit no sexual desire for varying periods of time (Houpt *et al.*, 1993).

13.13.1.7 Auto-Erotic Behavior

This refers to self-arousal of sexual response, which is called "masturbation (Houpt, 1991) in males, and it is most common among bulls on a high protein ration. All bulls masturbate, especially at times of inactivity. Since all domestic animals have been observed to masturbate, it would be difficult to consider this as abnormal behavior. The bull performs pelvic thrusts, with an arched back and a partially erect penis which moves in and out of the sheath until ejaculation occurs.

13.13.2 Female Animals

13.13.2.1 Silent Heat (Quiescent Estrus)

Silent heat is the failure to indicate the signs of estrus even though the reproductive tract is at the height of influence by estrogen and follicular stimulating hormone (FSH) and ovulation is imminent (Hafez, 1992). It is also called "silent ovulation" (Albright and Arave, 1997) or quiet ovulation (Jainudeen and Hafez, 1992b), and is defined as the occurrence of ovulation without overt estrus signs. Silent heat is the condition of subestrus in buffaloes and occurs when corpus luteum is present in one of the ovaries in the absence of visible signs of estrus (Gordon, 1996).

The estrus behavior ceases during heavy rain or snowfalls. Extreme cold or heat will result in much diminished sexual activity. Submissive cows may avoid mounting dominant cows in estrus (Pathak, 1992). A good number of buffaloes remain unbred during summer months because of silent heat. Poor expression of heat or even completely silent estrus has been attributed to the low profile of gonadotropic hormones or disturbance in the ratio of certain micro-minerals (Mudgal, 1992). Maximum concentration of FSH and lowest concentration of LH are indicative of the peak breeding period.

Albright and Arave (1997) summerized the methods used for heat detection as visual observations; use of teaser or vasectomized bulls fitted with a marking harness; training dogs to sniff out cows in estrus; progesterone levels in the milk (Qureshi *et al*; 2000a); painting the tail and rump each day and noting when the paint is scruffed off after riding; androgenized females, i.e., cows treated with androgen (male hormone); pedometers to measure the distance walked, which indicates that the use of activity monitors may be effective in improving reproductive performance; and automatic detection. These methods can be added to the three major approaches designed to monitor changes in physical activity: walking distance (pedometry); changes in the electrical resistance of the reproductive tract tissues (impedance monitoring); and pressure to the tailhead region accompanying standing to be mounted. All may be electrically coupled to a personal computer for analysis.

13.13.2.2 Nymphomania

This is also called "bulling" (Albright and Arave, 1997). It is a follicular cystic ovary disease of cattle and buffaloes (Gordon, 1996). It is a condition characterized by female cows reacting like males; the cause of the condition is usually an ovarian cyst that produces excessive estrogens for a prolonged period of time (Kreul, 1991). It has a higher occurrence in dairy than in beef herds, probably due to the more intensive housing and management. Milk production can drop noticeably because these cows are often high producers. Most cases occur within 3–8 weeks post-partum, coinciding with the development of the first follicle following calving. Ovulation fails, and in the absence of adequate luteinizing hormone, several follicles may grow to form multiple cysts (Qiu Huai and Leo Jun, 1995). Affected cows act like bulls, they become more like bulls vocally and in body conformation. Each cow actively seeks out other cows and mounts them, but does not stand for other cows (Houpt, 1991).

Nymphomaniac cows show continuous intense estrus behavior for long periods. Affected cows may show frequent, intermittent estrus, stand for mounting, and aggressively pursue monosexual "bulling" behavior as pawing the ground, bellowing, and attempting to mount herdmates (Edwards, 1995). Musculinization of the head and neck and a prominent tailhead are characteristic of the chronically nymphomaniac cow. Manual rupture of the cyst is one of the oldest treatments, though it has risks of hemorrhage and scar tissue formation. Veterinary treatment using a source of luteinizing hormone such as pituitary extracts or chronic gonadotropin (HCG) or gonadotropin-releasing hormone (GnRH) can be successful (Houpt, 1991).

13.13.2.3 Post-Partum Anestrus

The resumption of estrus cyclicity after calving, an important factor in obtaining satisfactory reproductive performance, remains a significant problem (Gordon, 1996). It is a condition of ovarian inactivity that occurs during lactation in many mammalian species. Anestrus is more common in suckled species (Suthar and Kavani, 1992). Anestrus has been classified as either "pre-service" or "post-service". Pre-service anestrus refers to the absence of observed estrus in the immediate postpartum period, while post-service anestrus is defined as the absence of observed estrus after an unsuccessful insemination. Jainudeen (1986), attributed the anestrus to the maintenance of the corpus luteum of the previous pregnancy or the persistence of the corpus luteum of the first postpartum estrus cycle, which may be non-ovulatory.

The possible underlying reasons for anestrus may be physiopathological in origin or they may be due to management practices (Janakirman, 1988). Physiopathological factors include lactation, suckling, nutrition, infections, hormonal imbalance, parity, and breed of the animal. Management practices are mainly responsible for inadequate estrus detection. Barkawi *et al.* (1993) found that estrus behavior was more pronounced in the cold season than in the hot season, as also noted by several other authors (Janakiman, 1988; Suthar and Kavani, 1992; Gordon, 1996).

13.13.2.4 Mounting Humans

Sexually active cows may lose their inhibitions toward humans and attempt to mount them. Adult cows or heifers in estrus and young bulls after puberty are most likely candidates for such aberrant behavior. Caution should be taken when walking among groups of loose-housed buffalo cows (El-Kaschab, 1994).

13.13.2.5 Delayed Puberty

Delayed puberty is a reproductive problem. Even though the animals achieve the desired chronological age of puberty as well as a normal weight, estrus does not occur.

13.14 CONCLUSION

The term courtship refers to the behavioral interaction that occurs between males and females before, during, and just after the act of mating. The main function of courtship behavior is to ensure that the two individuals are of the same species. It also involves stereotyped behaviors. Courtship behavior informs a potential mate that the intention is breeding and not aggression. The reproductive behavior of animals plays an important role in reproduction, affecting both the success of mating and survival of the young, but under some circumstances, the copulatory behavior of males and females can be extremely low. This might be improved by preventing males from slipping while mounting and by not using females that are large in comparison with the males. However, it is possible that the frequent mounting events by males stimulate the receptivity of females.

Courtship is ambiguous. However, it has generally been treated within the context of copulatory behavior. It is argued herein that by at least temporarily considering certain patterns as courtship, they can be analyzed within both the classical ethological and sociobiological/behavioral-ecological traditions. Such analysis can yield improved insight into the functioning of these behavioral patterns in particular and courtship communication in general. Nonejaculatory intromissions are evolved signals in tactile modality, and display many of the properties of other courtship signals. They synchronize, persuade, orient, and isolate at the same time as they manipulate and function with some dishonesty. These characteristics of "copulatory" patterns can be effectively analyzed as courtship behavior and courtship communication.

In species in which social learning influences the development of courtship behaviors and preferences, understanding the social transmission of these traits is fundamental to an understanding of the evolutionary processes affecting those populations. Unlike strict genetic transmission, behaviors that are transmitted socially can spread very quickly through populations. The relatively rapid diffusion of social traditions, by affecting the social and potentially physical environment of populations, can alter selection pressures and subsequent evolution.

REFERENCES

Albright, J.L. and Arave, C.W. (1997). *The behaviour of cattle.* 1st ed, CAB International. Wallingford, UK.

Badawey, A.B.A., El-Sawaf, S.A. and El-Wishy, A.B. (1972). Sexual behavior of buffalo bulls. *J. Egyptian Vet. Med. Assoc.*, **32**:81–99.

Barkawi, A.K., Bedeir, L.H. and El- Wardani, M.A. (1993). Sexual behaviour of Egyptian buffaloes in post-partum perio. *Buffalo J.*, **9**:225–236 .

Bronson, F.H. (1984). Energy allocation and reproductive development in wild and domestic house mice. *Biol. Reprod.*, **31**:83–88.

Edwards, T.A. (1995). Buller syndrome. What's behind this abnormal sexual behaviour? *Large Anim. Veterin.*, **50**:6.

El-Kaschab, S. (1994). Reproductive efficiency of Egyptian fuffalo bulls raised under farmers, conditions. *Indian J. Anim. Sci.*, **64**:1097–1103.

Estes, R.D. (1992). *The behaviour guide to African mammals.* 1st ed. University of California Press, Berkeley, CA, pp. 199–200.

Fraser, A.F. and Broom, D.M. (1990). *Farm animal behaviour and welfare.* 3rd ed., Bailliere Tindall, London, pp. 156–161.

Gordon, I. (1996). *Controlled reproduction in cattle and buffaloes.* 1st ed., CAB International, Willingford, pp. 432–466.

Gupta, B.D., Kaushik, S.N. and Mishra, R.R. (1994). Study on reproduction efficiency parameters of murrah buffaloes. *Indian J. of Dairy Sci.*, **47**:257–264.

Hafez, E.S.E. (1992). *Reproduction in farm animals.* 6th ed., Lea & Febiger Press, Philadelphia, PA.

Hill, F.I., Knight, T.W., Death, A.F., Wyeth, T.K. and Ridland, M. (1992). Oestrus synchronization and oestrus detection in swamp buffaloes (Bubalusbubalis). *Proc. NewZealand Soc. Anim. Prod.*, **52**:25–27.

Houpt, K.A. (1991). *Domestic animal behaviour for veterinarians and animal scientists.* 2nd ed., Iowa State University, Ames, IA, p. 408.

Houpt, K.A., Ah, K., Maher-Calabria, V. and Klebs, N. (1993). The welfare of dairy bulls. *Appl. Anim. Hehav. Sci.*, **37**:81.

Jainudeen, M.R. (1986). Reproduction in water buffalo. In: *Current therapy in theriogenalogy*, by Morrow, P.A. W.B. Saunders, Philadelphia, PA, pp. 443–449.

Jainudeen, M.R. and Hafez, E.S.E. (1992a). Cattle and buffalo. In: *Reproduction in fartm animals,* by Hafez, E.S.E. 6th ed., Lea & Febiger, Philadelphia, PA, p. 315.

Jainudeen, M.R. and Hafez, E.S.E. (1992b). Reproductive failure in females. In: *Reproduction in farm animals.* 6th ed., by hafez, E.S.E. Lea & Febiger. Philadelphia, U.S.A., pp. 261–311.

Janakiman, K. (1988). Some aspects of reproductive problems in buffaloes. In: P*roceedings II World Buffalo Congress.* Indian Council of Agricultural Research, New Delhi, India, pp. 264–270.

Kreul, W. (1991). Water Buffaloes- exotic animals with a future. *Tierzuchter*, **43**:518–519.

Mloszewski, M.I. (1983). *The behaviour and ecology of the African buffalo.* 1st ed., Cambridge University Press, Cambridge, pp. 180–194.

Mudgal, V.D. (1992). Reproduction in River buffaloes. In: *Buffalo production,* by Tulloh, N.M. and Holmes, J.H.G. 1st ed., Elsiever Science Publisher, Amsterdam, pp. 171–181; 23 ref.

Pathak, N.N. (1992). Behaviour and training of River buffaloes. In: *World animal science. 6. Buffalo production.* 1st ed., by Tulloh, N.M. and Holmes, J.H.G., Elsevier Science Publisher, Amsterdam, p. 223.

Phillips, C.J.C. (1993). *Cattle Behaviour Farming.* Press Books, Ipswich, UK, p. 212.

Price, E.O., Smith, V.M. and Katz, L.S. (1986). Stimulus condition influencing self-enurination, genital groomingand flehmen in male goat. *App Anim Behav Sci.*, **16**:371–381.

Qui Huai and Lco Jun (1995). *Water buffalo and Yak productin in China.* Animal Genetics Resources Information No. 15, FAO, Rome, pp. 83–99.

Qureshi, M.S., Samad, H.A., Habib, G., Siddiquce, M.M. and Nawab, G. (2000a). Milk progesterone profiles under various reproductive states in doing buffaloss. *Proc. National Sci., Council*, (B) China, **24**:70–75.

Qureshi, M.S., Siddiqui, M.M., Habib, G., Samed, H.A., Shamroz, J. and Bano, N. (2000b). Reproduction-nutrition interaction in dairy buffaloes. Effect of macro-minerals in take. *Egyptian J. Dairy Sci. (Cairo)*, **28**:109–119.

Randolph, H., Cromwell, G.L., Stahly, T.S. and Kratzer, D.O. (1981). Effects of group size and space allowance on performance and behavior of swine. *Anim. Sci.*, **53**:922–927.

Sule, S.R., Taparia, A.L., Jain, L.S. and Tailor, S.P. (2001). Reproductive status of surti buffaloes maintained under sub-humid conditions of Rajasthan. *Indian Vet. J.*, **78**:1049–1051.

Suthar, B.N. and Kavani, F.S. (1992). Occurrence and nature of first postpartum estrus in Mehsani buffaloes. *Indian J. of Anim. Reprod.*, **13**:161–164.

Vale, W.G., Ohashi, O.M., Sousay, J.S. and Ribeiro, H. F.L. (1990). Studies on the reproduction of water buffalo in the Amazon basin. In: *Livestock Reproduction in Latin America. FAO/IAEA Seminar*, Bogoto, pp. 204–210.

14 Conditioning and Habituation

Vasantha S. K. I., M. P. S. Tomar, S. S. R. Kona and R. S. Arya

14.1 INTRODUCTION

In a world where everything is unpredictable, organisms must adapt their behaviour to fit their ever-changing surroundings. Experience is the reason for this apparent shift in conduct. An organism can adapt and cope with changing environments due to this flexible learning process. More learned behaviours than intrinsic ones are displayed by highly intelligent animals. There are two types of learning: associative learning and non-associative learning (Figure 14.1).

14.2 NON-ASSOCIATIVE LEARNING

This is regarded as the most basic type of education. Instead of learning a new behaviour, an animal learns to adapt its current behaviour in response to environmental or stimulus changes. This modification of the current conduct is brought about by the stimulus that is frequently delivered. When there is no related event or stimulus (reward or punishment), these behavioural changes are seen. Put simply, there is no stimulus associated with the behaviour that is being displayed. As in associative learning, the response to a single stimulus determines the result regardless of whether reinforcement is positive or negative. The neural reflexes that underlie non-associative learning are split into habituation and sensitisation based on how strongly the subsequent consequence or reaction is diminished or strengthened.

a. Habituation.
b. Sensitisation.

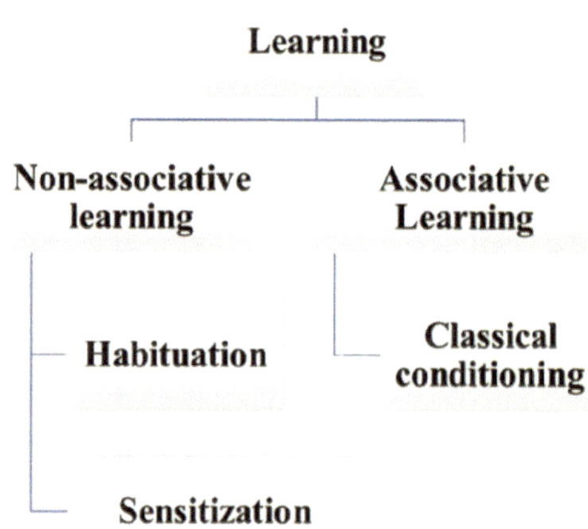

FIGURE 14.1 Classification of learning.

DOI: 10.1201/9781003710271-14

14.2.1 Habituation

This is the most basic type of learning that isn't associative. A behavioural reaction is reduced when a stimulus is presented repeatedly over time. When the stimulus is repeated, the first conditioned response displayed does not induce a second reaction. Throughout the animal kingdom, habituation is strongly conserved and aids in an animal's survival. An animal filters a lot of information it takes in from its surroundings during this process. It is able to concentrate on the pertinent data thanks to this filtering.

Through experimental study, the neural systems including the vertebrate spinal reflexes and the invertebrate mollusc Aplysia are used to understand the neurological processes behind habituation. Neither motor tiredness nor sensory receptor adaptation is the cause of the decrease in the motor response that underlies the behaviour being shown. But because there is less calcium ion influx into neuronal terminals, there is a decrease in the release of neurotransmitters (Table 14.1).

TABLE 14.1
Characteristics of Habituation

S.No	Original (Thompson and Spencer, 1966)	Modified (Rankin et al., 2009)
1.	"When a stimulus is applied repeatedly, the response to the stimulus decreases because the stimulus initially evokes a response (habituation). Usually, the decline is an exponential function that is negative in relation to the quantity of stimulus presentations."	When a stimulus is applied repeatedly, it causes a progressive decrease in some parameter of a response to an asymptotic level. This modification could include a decrease in the frequency and/or magnitude of the response. In many cases, the decrement is exponential, but it can also be linear; additionally, a response may exhibit facilitation prior to decrementing as a result of (or presumably as a result of) a concurrent process of sensitisation.
2.	"Withholding the stimulus causes the response to gradually improve (spontaneous recovery)."	"Withholding the stimulus causes the response to recover over time (spontaneous recovery)."
3.	"Habituation becomes progressively faster with repeated series of training sessions on habituation and spontaneous recovery (this phenomenon could be called potentiation of habituation)."	The response decay progressively gets faster and/or more noticeable after several series of stimulus repetitions and spontaneous recoveries (this phenomenon can be called potentiation of habituation).
4.	"When all else is equal, habituation occurs more quickly and/or prominently the more frequently stimulation occurs."	More frequent stimulation causes a more noticeable and/or swifter response decrement, as well as a faster spontaneous recovery (assuming the decrement has reached asymptotic levels), all other things being equal.
5.	"The rate and/or intensity of habituation increase with decreasing stimulus strength. Strong stimuli might not cause any appreciable habituation."	The behavioural response decrease within a stimulus modality is more rapid and/or pronounced the less intense the stimulus. Extremely strong stimuli might not cause a discernible decline in response.
6.	"Habituation training can have effects that go beyond the asymptotic or zero response level."	Even after the response reaches an asymptotic level – which could be zero or no response – the effects of repeated stimulation could still be felt. Exceeding asymptotic levels of stimulation can have the effect of changing subsequent behaviour, such as postponing the onset of spontaneous recovery.

(Continued)

TABLE 14.1 (Continued)
Characteristics of Habituation

S.No	Original (Thompson and Spencer, 1966)	Modified (Rankin et al., 2009)
7.	"Stimulus generalisation to other stimuli is demonstrated by the habituation of response to a particular stimulus."	The response decrement exhibits some stimulus specificity within the same stimulus modality. A second, novel stimulus is given, and the differences in the responses to the novel stimulus and the habituated stimulus are compared in order to assess stimulus specificity or stimulus generalisation. This test is properly called a stimulus generalisation test, but it has been incorrectly referred to as a dishabituation test in many paradigms (e.g., developmental studies of language acquisition).
8.	"When a different, typically powerful stimulus is presented, the habituated response (dishabituation) returns."	The decremented response to the original stimulus increases when a different stimulus is presented. We call this phenomenon "dishabituation." Note that an increase in response to the initial stimulus, rather than an increase in response to the dishabituating stimulus, is the appropriate way to test for dishabituation (see point #7 above). In fact, the response need not even be triggered by the dishabituating stimulus alone.
9.	"The amount of dishabituation produced becomes accustomed to the repeated application of the dishabituate stimulus; this phenomenon could be referred to as habituation of dishabituation."	The amount of dishabituation produced decreases with repeated application of the dishabituating stimulus (this phenomenon can be called habituation of dishabituation).
10.	—	Certain stimulus repetition protocols can produce response decrement properties that last for hours, days, or weeks, such as faster rehabilitation than baseline, smaller initial responses than baseline, smaller mean responses than baseline, and fewer frequent responses than baseline. Long-term habituation is the term used to describe this persistence of habitual features.

Source: As described by Thompson and Spencer (1966), Modified by Rankin et al. (2009).

Importance of habituation:

1. The temperaments of Indian cattle breeds and buffaloes differ from those of exotic and crossbred cattle. At the farmer level in India, habituation is a crucial mechanism (Bharti and Kamboj 2019) for:
    - Better acclimatisation of recently introduced cattle to their new surroundings.
    - Normal maternal behaviour in primiparous cattle, allowing nursing calves to consume their milk to ensure the health of the calves.
    - Adjustment to different farm tasks, such as routine milking.
    - Adaptation to new machinery or equipment intended for the welfare of animals.
    - Introducing primiparous cattle to milking techniques and enhancing milk let-down.
    - The ability to maintain a good milk temperament after weaning.
    - Reducing human-animal conflict in order to improve management techniques.
    - Reducing stress and ensuring the welfare of cattle; stopping erratic or abnormal behaviours in cattle; and improving healthcare through treatment and prophylaxis.
    - Completing all farm tasks without encountering any problems.
    - Creating plans for more effective cattle handling.

Habituation is necessary for humans to perform cognitive tasks because it helps them filter out stimuli that aren't relevant. Cognitive skills are disrupted by disruptions in habituation, as seen in patients with neuropsychiatric disorders.

Examples:

1. The animals that rely on alarm calls to alert them to the presence of predators in their surroundings are the best examples of habituation. When non-threatened species are present, they are unable to sound the alarm.
2. The squirrels' peaceful co-existence in the wild or in a park with the people who frequent it or live nearby.
3. A turtle's head retracts when it is touched; nevertheless, it cannot conceal or retract its head if it is repeatedly touched.
4. The demonstration of songbirds' decreased reactions upon being given a plush owl. They attempt to flee at first, but as time passes and the stuffed owl stays motionless inside the cage, fewer attempts are made to do so.
5. Ducks' habit of interacting with people all the time, and being fed by them without any threat.
6. The tendency of those residing close to an aerodrome or railway tracks becoming accustomed to the sounds of passing trains and planes.
7. Habituation to the fragrance of a perfume.

Factors affecting habituation:

1. Stimulation intensity: A variation in stimulation intensity causes the original response to be displayed. Less habituation will happen the higher the stimulus intensity.
2. Duration of stimulus: Shorter stimulus presentation times reduce the likelihood of habituation, which results in the display of the original response.
3. Stimulus frequency: The frequency of the stimulus affects the habituation process. Habituation will take place more quickly the higher the stimulus frequency.

14.2.2 Sensitisation

In this case, the response displayed is the result of repeated presentation of the stimuli. The response gradually intensifies, enhancing subsequent reactions to additional stressors or stimuli.

Characteristics of sensitisation (Abramson and Kieson 2016):

1. The strength of the stimulus has a direct correlation with the likelihood of sensitisation.
2. During sensitisation, generalisation may result from a stimulus's identical response to other stimuli that are similar.
3. If a sensitised stimulus is exposed repeatedly, desensitisation may occur.

Examples of sensitisation:

- Sensitisation to the sound of honey bees will protect elephants from sting bites.
- When an animal reacts to less painful sounds in the same way as louder ones, it is said to be sound-sensitive.
- Dogs that are frequently exposed to fireworks and thunderstorms develop a fear of other unexpected noises.

14.3 ASSOCIATIVE LEARNING

Classical conditioning (Abramson and Kieson 2016):

After associating with a normal stimulus that elicits a normal response from the animal, this learning involves the animal exhibiting a behavioural response to a neutral stimulus. When an animal is repeatedly exposed to both stimuli simultaneously, it learns to associate them and responds to the neutral stimulus alone in a way that is similar to the normal or unconditioned stimulus (US). Unconditioned response (UR) is the term used to describe the response elicited by an unconditioned stimulus (US). After being linked to a neutral stimulus, UR develops into a conditioned response (CR). It represents the most fundamental associative learning mechanism.

Characteristics of classical conditioning (Abramson and Kieson 2016):

1. There is a clear relationship between training efficacy and conditioned stimulus intensity.
2. There exists a direct proportionality between the intensity of the unconditioned stimulus and the effectiveness of training.
3. The time interval between a conditioned and an unconditioned stimulus determines how effective a training programme is.
4. The more pairings or associations there are between the stimuli, the more effective the training is.
5. When an unconditioned stimulus does not follow classical conditioning, the conditioned response eventually ends.
6. All similar stimuli elicit established responses to a specific conditioned stimulus.

Examples:

a. In the late 1800s and early 1900s, Russian scientist Ivan Pavlov trained dogs in classical conditioning experiments (Figure 14.2). Dogs salivate when they have food in their mouths or when they are exposed to the sight and smell of food. When their meal is ready to be served, the same thing takes place. The dogs were then trained to hear the ring of a bell prior to feeding time by Pavlov (Table 14.2). The dog eventually came to associate the sound with the meal. Then, the sound of the ringing bell caused the dog to begin drooling (Table 14.3).

b. Conditioned taste aversion of animals to certain feeds.

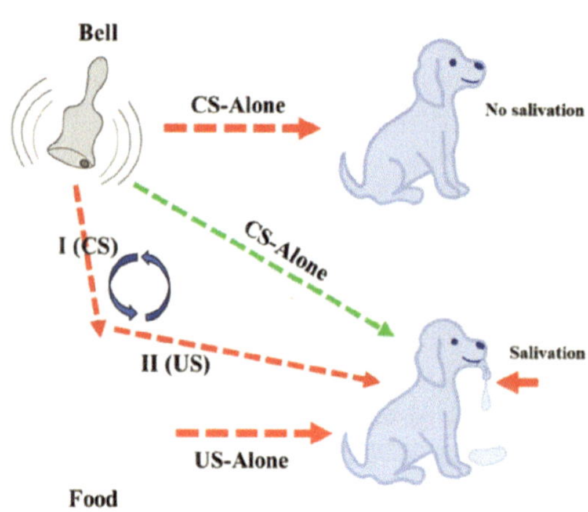

FIGURE 14.2 Diagrammatic presentation of Pavlov's conditioning.

TABLE 14.2
Sequence of Events in Pavlov's Classical Conditioning Experiment

1. Neutral stimulus (sound of a ringing a bell)	®	No response is elicited (Orientation response)
2. Unconditioned stimulus (Presence of food in the mouth)	®	Salivation (Unconditioned response – UR)
3. Conditioned stimulus – (sound of a ringing bell)	®	Salivation (Conditioned response)

TABLE 14.3
Term of Stimulus or Response, Acronym, and Actual Stimulus or Response of Dog

Term of Stimulus or Response	Acronym	Actual Stimulus or Response Exhibited by Dog
Conditioned stimulus	CS	Sound of a bell
Unconditioned stimulus	US	Food
Unconditioned response	UR	Salivation
Conditioned response	CR	Salivation

REFERENCES

Abramson, C. I. and E. Kieson. 2016. Conditioning methods for animals in agriculture: a review. *The journal Ciência Animal Brasileira* 17(3): 359–375.

Bharti, P. and M. L. Kamboj. 2019. Effect of habituation to milking and natural suckling on milking behaviour and milk production characteristics in primiparous Sahiwal cows. *Indian Journal of Dairy Science* 72(3): 312–317.

Rankin, C. H., T. Abrams, R. J. Barry, S. Bhatnagar, D. F. Clayton, J. Colombo, G. Coppola, M. A. Geyer, D. L. Glanzman, S. Marsland, F. K. McSweeney, D. A. Wilson, C. Wu and R. F. Thompson. 2009. Habituation revisited: An updated and revised description of the behavioral characteristics of habituation. *Neurobiology of Learning and Memory* 92: 135–138.

Thompson, R. F. and W. A. Spencer. 1966. Habituation: a model phenomenon for the study of neuronal substrates of behavior. *Psychological Review* 73:16–43.

15 Behavior in the Host-Parasite Interrelationship

Sirigireddy Sivajothi, Pattipati Malakondaiah, Bhavanam Sudhakara Reddy and Akkala Nagakeerthi

15.1 INTRODUCTION

Nearly every facet of an animal's behavior is intertwined with exposure to various types of parasites. Mating behavior, for instance, is crucial for the transmission of sexually transmitted bacteria, protozoa, and viruses, while foraging serves as a major route of infection for environmental and trophically transmitted bacteria and helminths. Additionally, social behavior contributes to the spread of various contact-transmitted infectious agents. Moreover, behavior plays a pivotal role in how hosts protect themselves against parasites, often acting as the primary defense against infection. Conversely, many behaviors undergo alterations when hosts become infected with parasites.

Behavioral changes due to infection can stem from various factors. Parasites might manipulate host behavior to bolster their own reproductive success, while alterations in host behavior can arise from immunological or pathological effects of parasite infection. Over longer periods, parasites exert selective pressures on hosts, potentially driving evolutionary shifts in behavior. Consequently, these changes in host behavior can influence parasite population dynamics and life history, impacting traits like virulence and transmission mode. Ultimately, behavior and parasitism are deeply interconnected, necessitating a holistic approach that considers both aspects to gain a comprehensive understanding.

15.2 HOST MANIPULATION BY PARASITES

The notion that a parasite can alter the phenotype of its host, whether by manipulating its behavior or changing its appearance, may have initially emerged in science fiction narratives, but it has since become a firmly established concept in the field of animal behavior. The first empirical evidence supporting this idea came from studies showing that amphipods infected with larval acanthocephalan parasites exhibited both unusual behavior and atypical coloration, rendering them more vulnerable to predation by the parasite's subsequent host. Since then, there has been continuous fascination and investigation into this phenomenon.

Consequently, the phenomenon of host manipulation by parasites has been observed in numerous host-parasite interactions, encompassing a wide array of living organisms across major phyla (Moore, 2002). While many documented cases involve subtle alterations in host behavior or appearance, some instances are particularly remarkable. Two trematode species have emerged as classical textbook examples of such manipulation. Both require the intermediate host, where the parasites develop as larvae, to be consumed by a definitive host that typically wouldn't prey on the intermediate host. The first example, *Dicrocoelium dendriticum* (Figure 15.1), relies on accidental ingestion from an ant by a sheep for transmission. Infected ants are compelled to climb to the tips of grass blades, where they remain, seemingly awaiting the arrival of a grazing sheep (Moore, 2002). The second trematode, *Leucochloridium spp.*, induces alterations in the size, shape, and coloration of the tentacles of its snail intermediate host. These tentacles pulsate vigorously in response to light,

 DOI: 10.1201/9781003710271-15

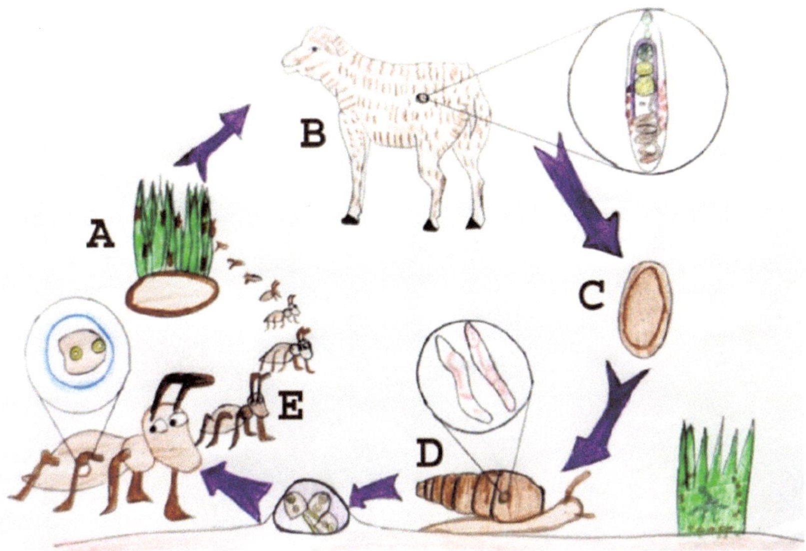

FIGURE 15.1 Phenomenon of host manipulation by parasites. (A) Accidental ingestion of an ant by a sheep, (B) sheep definitive host, (C) egg, (D) snail first intermediate host, (E) ant second intermediate host.

attracting the attention of birds, which serve as the next host for the parasite. This behavior presumably deceives birds into perceiving the colorful, pulsating tentacles as potential prey, such as caterpillars (Moore, 2002) (Figure 15.1).

Numerous other examples of host manipulation exhibit similarly peculiar behaviors. For instance, a nematode parasite induces a bright red coloration in the abdomen of its ant intermediate host and compels the ant to perch among patches of small red berries with its abdomen raised, awaiting the arrival of frugivorous birds, which serve as the definitive hosts for the nematode (Yanoviak et al., 2008). Another nematode, with a simpler life cycle, must release its eggs in water when its mayfly host returns to a stream to lay eggs. When this parasite infects a male mayfly, it feminizes the host, inducing morphological and behavioral changes that mimic those of a female mayfly, ensuring the host's return to the water, although only the parasite will be laying eggs (Vance, 1996).

Inside an orb-weaving spider, a parasitic wasp larva prompts its host to construct a peculiar pouch-like structure attached to its web shortly before the wasp emerges. This pouch serves as protection for the wasp larva against being washed away by heavy rain during its pupation after the spider's demise (Eberhard, 2000). Additionally, another parasitic wasp, utilizing caterpillars as hosts, can manipulate host behavior even after departing. Previously infected caterpillars, with a lifespan of only a few days, remain in close proximity to the newly emerged wasp larvae, shielding them from potential predators as the larvae undergo pupation (Grosman et al., 2008).

The widespread protozoan parasite *Toxoplasma gondii*, which relies on transmission (Figure 15.2) from a rat intermediate host to a cat definitive host, manipulates the innate aversion of its rat host to cat odor, reversing it into an attraction toward cat odor, with significant implications (Berdoy et al., 2000). Humans can also become infected by the parasite instead of rats. Although this does not facilitate parasite transmission, *T. gondii* still induces neurochemical changes in infected humans. Consequently, individuals with latent *T. gondii* infections exhibit personality traits and reaction times that differ from those of uninfected controls (Flegr et al., 2000; Havlicek et al., 2001) (Figure 15.2).

In essence, host manipulation by parasites refers to any change in the phenotype of a host, instigated by a parasite, that offers fitness advantages to the parasite. Specifically regarding altered host behavior, this typically entails infected hosts exhibiting behaviors that aid in the transmission or

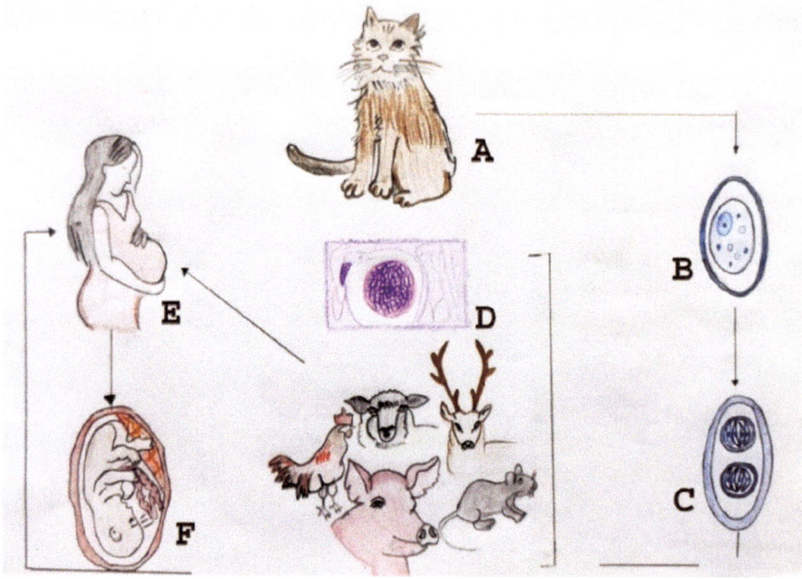

FIGURE 15.2 Altered host behavior aids in the transmission of the parasite. (A) Cat definitive host, (B) unsporulated oocyst, (C) sporulated oocyst, (D) herbivores and rodents intermediate hosts, (E) pregnant woman also acts as an intermediate host, and (F) congenital toxoplasmosis.

spread of the parasite, thereby enabling the fulfillment of its life cycle. This suggests that the phenotypic characteristics in the host altered by infection are influenced either directly or indirectly by genes in the parasite genome. In fact, host manipulation by parasites has been cited as a prominent example of extended phenotypes (Dawkins, 1982). A considerable portion of research on host manipulation by parasites has concentrated on its implications for the parasites and its mechanisms at the interface between host and parasite (Moore, 2002; Poulin, 2007; Thomas et al., 2005).

In recent times, broader implications of host manipulation have garnered increasing attention. For example, the presence of manipulative parasites within an animal population can influence host evolution (Poulin and Thomas, 1999), impact the ecology of other parasite species (Lafferty et al., 2000; Poulin, 2007), reshape the structure of the surrounding animal community (Mouritsen and Poulin, 2005; Thomas et al., 1998a), and may have veterinary significance (Lagrue and Poulin, 2010). These matters have been extensively reviewed elsewhere, with the present synthesis focusing solely on the original and foundational aspects of the phenomenon, providing an updated insight into our current comprehension of host manipulation as a parasite strategy.

15.3 ADAPTIVE MANIPULATION

However, many early and some recent studies have been inclined to hastily categorize any observed alteration in host phenotype as "adaptive manipulation." This prompted some authors to advocate for greater precision in the use of the term "adaptive" for presumed cases of manipulation (Poulin, 1995). In an effort to establish guidelines to prevent misuse of the adaptive label, Poulin (1995) proposed four basic criteria that a parasite-induced change in host behavior must fulfill to be considered a case of adaptive manipulation by parasites. In hindsight, only one criterion truly holds significance; nevertheless, let's briefly revisit these criteria. First and foremost, changes in host behavior subsequent to infection should align with expected outcomes based on their purported function. Alignment between a priori "design specifications" and the observed phenotypic alteration offers evidence of the influence of natural selection; conversely, post hoc efforts to attribute an adaptive function to an unforeseen behavioral change lack conviction. This criterion also encompasses the timing of events: the initiation of behavioral changes is anticipated to synchronize with the developmental stage of the parasite, at which it stands to gain from these alterations, for manipulation to be deemed adaptive.

For instance, the larval stages of parasitic worms typically commence inducing behavioral alterations in their intermediate hosts precisely when they are developmentally prepared to be transmitted through predation to their subsequent host (Hammerschmidt et al., 2009). Second, the complexity of parasite-induced behavioral changes can also indicate their adaptiveness. Simple traits are more prone to emerge by chance, perhaps as byproducts of pathology, whereas complex traits necessitate an organizing principle such as natural selection. The emergence of entirely new phenotypic traits appears to satisfy this criterion. Take, for example, ants infected by nematodes, whose abdomens transition from black to bright red, prompting them to perch among red berries. This phenotypic alteration appears both too intricate and too finely tuned to parasite transmission to be anything other than an adaptive manipulation (Yanoviak et al., 2008). Similarly, nematomorphs, or hairworms, must exit their terrestrial insect hosts and enter freshwater bodies to progress in their life cycle.

At the precise stage of the parasite's development, infected insects exhibit a wholly new behavior: they actively seek out water and immerse themselves in it (Thomas et al., 2002). Once again, this represents a complex behavioral shift that appears unlikely to occur randomly as a consequence of worm development within the host. In reality, what should be considered for this criterion is the complexity of the mechanism employed by the parasite to modify host behavior, rather than solely focusing on the eventual behavioral change. However, due to limited knowledge about the underlying mechanisms, we must evaluate the complexity of manipulation based solely on its observable phenotypic effects.

Third, when unrelated parasite lineages exhibit convergence in the type of behavioral alterations they induce in their hosts, it strongly suggests that these changes are cases of adaptive manipulation. Under similar selective pressures, we anticipate different taxa to address comparable challenges with analogous traits fulfilling similar functions. For instance, mermithid nematodes and hairworms, despite belonging to different and unrelated phyla (Nematoda and Nematomorpha, respectively), do so.

However, they have independently evolved remarkably similar life cycles: in both groups, during a late stage of their development within a terrestrial arthropod host, the parasite must emerge in water or water-saturated soil to continue its life cycle. As previously discussed, hairworms prompt their hosts to seek out water and immerse themselves in it (Hanelt et al., 2005; Thomas et al., 2002).

Remarkably, mermithids exhibit the same behavior manipulation in their hosts (Poulin and Latham, 2002). The independent emergence of nearly identical changes in host behavior following infection by unrelated parasite lineages strongly suggests adaptation. Fourth, for a trait to be considered adaptive, it must provide fitness advantages to its possessor. Therefore, parasites capable of manipulating the behavior of their hosts must achieve greater success in transmission compared to conspecifics that cannot alter host behavior.

Although this has only been verified in a small fraction of documented cases of parasite-induced alterations in host behavior, it remains the most compelling evidence available for adaptive manipulation. In the case of trophically transmitted parasites, this typically involves a predation test wherein equal numbers of parasitized (manipulated) and non-parasitized (unmanipulated) intermediate hosts are presented to a definitive host of the parasite in a semi-natural environment (Lagrue et al., 2007). Any preference toward greater consumption of manipulated prey is interpreted as enhanced transmission achieved through host manipulation. Looking back, Poulin's (1995) criteria as a whole were likely overly stringent, or at the very least excessively cautious.

Ultimately, it is only the fourth criterion, concerning the fitness effects for the parasite, that holds genuine significance. Regardless of whether the manipulation is simple or complex, appears well-suited to its function or not, and irrespective of its evolutionary origins, it will be favored and/or

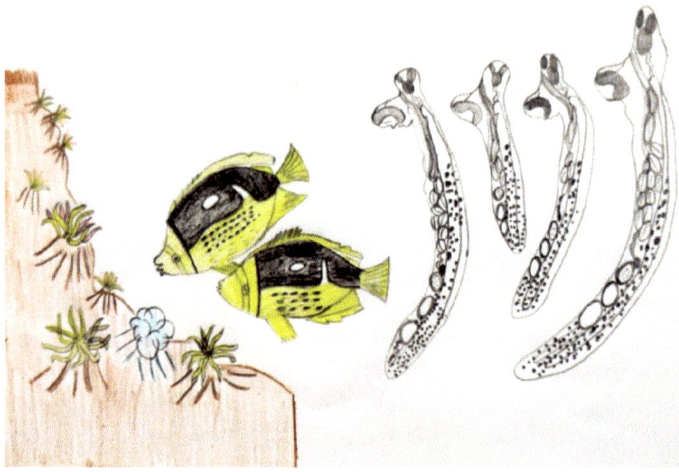

FIGURE 15.3 Example of adaptive manipulation.

maintained by selection if it enhances the fitness of the parasite, thus constituting a true adaptation. As emphasized by Thomas et al. (2005), if we elevate the coincidental byproduct scenario to the status of null hypothesis against which other interpretations must be assessed, then it should be feasible to test. However, in practice, there is no straightforward experimental method to differentiate between an advantageous byproduct and an advantageous direct outcome of selection.

Historically, traits in parasites may have evolved for other purposes that coincidentally also impact transmission, and these traits could have been co-opted for manipulation. The boundary between the original and the new function is often ambiguous. For example, consider the case of coral polyps infected by the trematode *Podocotyloides stenometra*. Infected polyps undergo a color change (Figure 15.3), turning bright pink, which makes them more visible and consequently preferred by the butterflyfish that act as the parasite's definitive hosts (Aeby, 2002).

Recently, researchers have identified the compound responsible for the pink coloration of infected polyps as a protein linked to the host's cytotoxic defense system (Palmer et al., 2009). Thus, the altered pigmentation of infected corals is a component of their typical immune response to infection. However, since this pigmentation change is harmless to trematodes and even advantageous for their transmission, natural selection should favor parasites that prompt more conspicuous color changes in their host. At a certain stage, beneficial side effects transition into adaptations. Therefore, to progress, it is likely time to settle the debate regarding adaptiveness: if a parasite-induced alteration in host behavior results in enhanced transmission of the parasite, then, provided there is a genetic foundation for this effect, whether it is incidental or not, it constitutes a case of adaptive manipulation.

15.4 VARIATION BETWEEN DIFFERENT SPECIES

Manipulating host behavior is an adaptation aimed at enhancing transmission and facilitating the completion of a parasite's life cycle. Similar to any other adaptation, the overall advantages it provides can vary based on the characteristics of the host-parasite relationship, especially the specific transmission hurdles and constraints encountered by a parasite. For certain parasite species, the advantages of manipulation may be considerable, leading to selection favoring this strategy. However, for others, the benefits may be minimal, and manipulation may not be favored.

At the core of all theoretical assessments regarding the net advantages of manipulation lies the concept of costs, which must be surpassed, in terms of fitness, by the gross benefits of manipulation for that strategy to be preferred. These costs of manipulation manifest in two ways. First, there may be induction costs, which refer to the physiological expenses associated with the mechanism employed by the parasite to induce a change in host behavior.

Some parasites may induce changes in host behavior without incurring any costs, simply by residing in the appropriate organ and coincidentally impairing tissue function, as long as residing in that organ does not lead to reduced parasite growth or survival compared to another location. However, most alterations in host behavior appear to arise from active interference with host neurochemistry, which may involve the secretion and release of substances by the parasite (Thomas et al., 2005). The development of specialized glands or tissues for the production of chemicals is likely to be costly.

It's important to note that such glands have never been identified in any parasite, and the induction costs have never been precisely quantified. Second, apart from any physiological costs linked to inducing manipulation, parasites that manipulate their hosts may also incur subsequent costs, which can be measured as an increased likelihood of premature death. These costs can be assessed in scenarios where manipulative parasites have conspecifics that benefit from manipulation without inducing it themselves. For example, the larval stages of the trematode *Microphallus papillorobustus*, which encyst in the cerebral region of their amphipod intermediate host, induce strong positive phototaxis and abnormal evasive responses in the host. This manipulation of host behavior leads to infected amphipods being more vulnerable to predation by aquatic birds, which act as definitive hosts for the parasite. However, not all *M. papillorobustus* larvae encyst in the head of amphipods; some also encyst in the abdomen. Amphipods possess the ability to mount an immune response against invading parasites, involving both encapsulation and melanization of the trematodes. Thomas et al. (2000) discovered that 17% of cerebral parasites are killed by encapsulation, whereas less than 1% of abdominal parasites meet this fate.

Another example involves the trematode (Figure 15.4) *Curtuteria australis*, which infects the cockle *Austrovenus stutchburyi* as an intermediate host; the parasites encyst in the foot of cockles

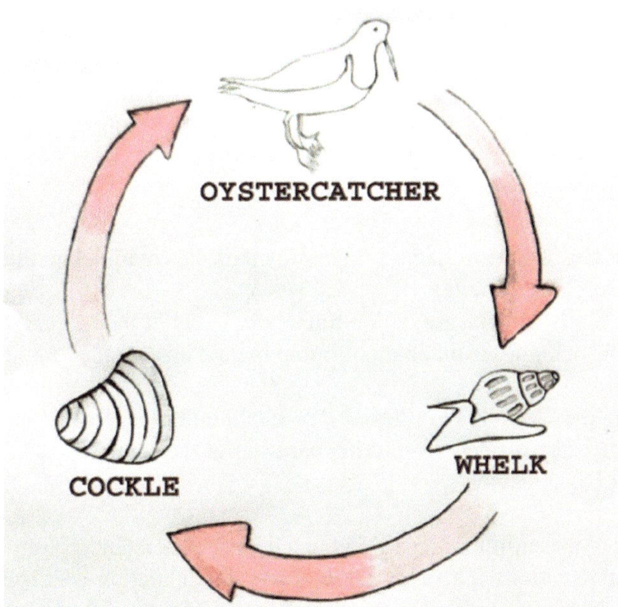

FIGURE 15.4 Trematodes typically encyst near the tip of a cockle's foot, with a detrimental impact on the host's burrowing ability.

and await predation by oystercatchers, their definitive host. As parasites accumulate in the foot of a cockle, they replace host muscle tissue and weaken the foot, causing heavily parasitized cockles to lose their ability to burrow. Consequently, they become stranded on the sediment surface of intertidal areas (Thomas and Poulin, 1998). Field experiments have demonstrated that manipulated cockles are approximately 5–7 times more likely to be consumed by bird definitive hosts compared to healthy, buried cockles (Mouritsen, 2002). The trematodes typically encyst near the tip of a cockle's foot, where their detrimental impact on the host's burrowing ability is most pronounced (Mouritsen, 2002). However, it is noteworthy that many *C. australis* parasites are also found in the middle of the foot or near its base. The advantages of host manipulation are enjoyed by all parasites: even though only those near the tip of the foot affect the host's ability to burrow, an oystercatcher consuming a cockle ingests all parasites along with host tissues.

However, an opportunistic predatory fish also preys on surface-stranded cockles, targeting exclusively the tip of the foot of those cockles attempting to burrow unsuccessfully. This fish predator is not a suitable definitive host, and any *C. australis* parasites ingested by it perish. Approximately one-third of all cockles display evidence that a portion of their foot has been cropped by the fish (Mouritsen and Poulin, 2003). In manipulated cockles stranded on the sediment surface, nearly one-fifth of parasites fall victim to fish predation (Mouritsen and Poulin, 2003). Interestingly, all of these parasites were encysted near the tip of the cockle's foot, rather than at its base. Consequently, in this system, parasites that induce host manipulation are exposed to a higher risk of mortality compared to conspecifics that do not manipulate their hosts. Thus, alterations in host behavior that benefit the parasite are improbable to be without cost, even if quantifying these costs is often challenging. Regarding induction costs, trade-offs will influence the evolution of host manipulation: any energy expended by the parasite on manipulating the host will be unavailable for growth, reproduction, or combating the host's immune system. These trade-offs, coupled with the potential risks of premature parasite mortality linked to manipulation, imply that selection may not consistently favor parasites that invest in host manipulation.

15.5 VARIATION WITHIN A SPECIES

Even among species known for inducing behavioral changes in their hosts, there exists variability among individual parasites in the extent of these induced changes. Therefore, not all manipulative organisms are identical. Factors such as age, size, sex, or abundance of parasites per host can account for some of this intraspecific diversity (Benesh et al., 2009). Such variation is anticipated, perhaps even mundane.

Certainly, let's break down those levels of variation:

1. **Variation in Host Characteristics:** The way parasites manipulate their hosts can differ based on individual host traits.
2. **Genetic Diversity in Parasites:** Within a population of parasites, there can be genetic variations influencing their inclination to manipulate hosts, reflecting different transmission strategies.
3. **Social Dynamics Among Parasites:** The likelihood of a parasite manipulating its host may be influenced by the actions of other parasites of the same species, indicating a social aspect to their behavior.

By examining each of these dimensions, we gain a deeper understanding of the complexity underlying parasite-host interactions. The expression of host manipulation isn't isolated from the host's initial phenotypic traits because it's essentially evaluated as a modification of those traits. Some hosts might be inherently more prone to parasite infection or more prone to significant phenotypic changes post-infection than others. Therefore, we anticipate a dynamic interplay between the inherent traits of the host and the parasite's ability to modify them. This perspective suggests potential selective advantages for the host as well, a concept explored by Blanchet et al. (2009a, 2009b).

In discussions about parasite-induced host manipulation, there's often an implied notion that hosts passively succumb to manipulation, akin to being mere puppets. However, in reality, natural selection should favor hosts capable of resisting attempts by parasites to alter their behavior, particularly if such alterations lead them to unfavorable environments or increase their vulnerability to predators. Nevertheless, the extent of selection pressure for hosts to resist manipulation may vary among individuals within a host population. Empirical evidence supports this notion. For instance, in freshwater fish acting as intermediate hosts for manipulative trematode parasites, older fish exhibit more significant declines in their antipredator behaviors post-infection compared to younger fish (Poulin, 1993). In species where hosts reproduce only once (semelparous), resisting manipulation would confer benefits to the infected host before reproduction, spanning most of its lifespan. Hence, there's a reasonable expectation for some hosts to resist parasite manipulation, offering a potential explanation for variations in the extent of behavioral changes induced by manipulative parasites within a species.

Certainly, host resistance can prompt parasites to adopt more aggressive strategies (Ponton et al., 2006a). Take, for instance, crickets hosting nematomorph worms. These parasites rely on water for their life cycle completion and induce crickets to seek water and jump in it (Hanelt et al., 2005). In a specific species, failure to comply with this manipulation incurs a greater fitness cost: compliant hosts retain reproductive capabilities post-parasite emergence, while non-compliant ones are typically killed or castrated (Biron et al., 2005). This showcases the complexity of host-parasite interactions and highlights the importance of considering the host's interests alongside parasite dynamics, rather than solely focusing on parasites in research endeavors.

15.6 HOW THE PARASITES MANIPULATE THE HOST BEHAVIOR

The mechanisms employed by parasites to modify host behavior post-infection are arguably the least understood aspect of host manipulation. Upon parasite acquisition, existing behavior patterns may undergo alterations, or entirely new behaviors may emerge, facilitated by either direct or indirect mechanisms (Lefèvre et al., 2009a; Thomas et al., 2005).

Parasites can induce changes in host behavior through direct secretion of neuroactive substances or indirectly by influencing or interfering with host biochemical pathways, coincidentally resulting in altered behavior (Table 15.1) However, distinguishing between these two mechanisms is often challenging.

Even in rare instances where secretions from a parasite with neurological impacts on the host have been identified, it remains uncertain whether these secretions are intended for manipulation or serve other functions, such as immune suppression (Thomas et al., 2005). For instance, the trematode Schistosoma mansoni secretes opioid peptides into its host, affecting both host immunity and neural function (Kavaliers et al., 1999). The original purpose of these secretions could have been immune suppression, unrelated to direct host manipulation. From the parasite's viewpoint, leveraging existing biochemical cascades and physiological processes within host cells and tissues might represent an optimal strategic approach to minimize the costs associated with inducing manipulation.

Parasites that target host compensatory responses might capitalize on pre-existing host mechanisms to fulfill their own transmission requirements without excessively impairing host fitness (Lefèvre et al., 2008, 2009a). For instance, vector-borne parasites are recognized for altering the feeding behavior of their vectors, such as increasing probing rates, thereby enhancing the likelihood of parasite transmission (Rogers and Bates, 2007).

Moreover, vectors typically experience a decrease in fecundity when they harbor parasites (Hurd, 2003). However, in the case of malaria-infected mosquitoes, when manipulation by the parasite leads to increased host biting rates per unit time, they regain their normal fecundity (Rossignol et al., 1986). This manipulation not only enhances parasite transmission but also enables the host

TABLE 15.1
Parasite Effects on Endocrine Hormones of Host

Name of the Parasite	Effect	Decreased Hormones	Increased Hormones
Ligula intestinalis	• Alteration of the gonadal development • Reproductive dysfunction	• Estrogen • Testosterone • Brain aromatase • Vitellogenin	
Taenia crassiceps *Taenia solium* *Taenia psiformis*	• Induction of feminization • Immune Th1/Th2 shift • Decreased weight of seminal vesicles • Increase production of IL-6 in splenocytes	• Testosterone • Dihydrotestosterone	• Estradiol
Schistoma mansoni	• Inhibition of development of the reproductive tract	• Gonadotropic hormones • Serotonin and dopamine secretion	
Toxoplasma gondii	• Increase in the testicular steroidogenesis and sexual behavior in rats • Decline in the innate fear of odor in cats • Hemorrhage • Hypocellularity of the decidua basalis • Apoptosis of placental cells • Decline in the numbers of mature natural killer cells in endometrium • Increased IFN- γ secretion • Increased indoleamine 2,3- • Dioxygenase (IDO) mRNA levels • Reduction of IL-15 mRNA levels		• Testosterone • Luteinizing hormone receptor • Estradiol
	• Not measured	Testosterone (women)	Testosterone (men)
	• Miscarriage • Enhancement of parasite reproduction		• Progesterone • Estrogen
Plasmodium	• Decreased in motility, count, and viability of sperms	• Testosterone	
	• Primary testicular failure • Prostate abnormalities	• Testosterone	• Estrogen and progesterone

to mitigate the fitness cost of infection, thus reducing the likelihood of host resistance (Lefèvre et al., 2008, 2009a). Utilizing the pre-existing physiological mechanisms of host compensatory responses may represent the most cost-effective method for a parasite to manipulate its host. At a proximate level, various studies have demonstrated that parasites achieve manipulation by directly or indirectly modifying the concentrations of hormones or neurotransmitters in their hosts. For example, the larval stages of numerous parasitic worms have been observed to induce alterations in the concentrations or activity of serotonin, dopamine, and/or other neurotransmitters in the brains of their intermediate hosts.

This phenomenon has been observed in various parasite-host interactions, including acanthocephalans within crustacean hosts (Tain et al., 2007), cestodes within fish hosts (Øverli et al., 2001), and trematodes within both crustacean and fish hosts (Shaw et al., 2009). Additionally, there is compelling evidence that many other types of parasites induce alterations in the neurochemistry of their host's brain as part of the changes they induce in host phenotype. These include insect parasitoids, nematomorphs, and the protozoan T. gondii (Thomas et al., 2002).

Indeed, manipulation doesn't always involve neurological pathways. For example, while nematomorphs and mermithid nematodes have developed similar life cycles and host manipulations, the latter seem to induce water-seeking behavior in their host through straightforward alterations in host hemolymph osmolality and ionic concentration (Williams et al., 2004). Regardless of whether alterations in brain chemistry occur during host manipulation, a commonality among most host-parasite systems where partial understanding of the mechanisms has been attained is the change in concentrations of substances synthesized by the host, which subsequently affect behavior following infection. These substances might either be produced by the parasite and released into the host or actively taken up from the host. However, there's scant evidence supporting the idea that parasites induce manipulation through direct production and release of these substances. Instead, they likely exploit existing biochemical pathways within host cells, either up- or downregulating them to modulate the concentrations of active neurochemicals or other products.

Detecting altered levels of serotonin in parasitized hosts is likely just one element of a more extensive biochemical cascade. It might be necessary to explore further up the biochemical chain to precisely understand how parasites induce changes in host behavior. A recent advancement in this direction involves the application of proteomic analyses to investigate host manipulation by parasites. The methodology involves comparing the levels of a wide array of proteins in specific tissues between manipulated and normal hosts, then investigating the functional roles of proteins that exhibit differences between the two host types. Studies have demonstrated that infection by manipulative parasites results in a modified profile of protein synthesis in parasitized hosts compared to nonparasitized conspecifics. Several candidate proteins have been identified as potentially associated with behavioral alterations in parasitized hosts (Lefèvre et al., 2009a). This approach can be especially enlightening when utilized within a comparative framework, as it allows for the identification of mechanistic differences and similarities between apparently similar host manipulations (Ponton et al., 2006b). In certain instances, proteomic studies indicate that parasites generate host mimetic proteins, which could potentially serve as the mechanistic basis for manipulation, though this hypothesis necessitates further validation (Lefèvre et al., 2009a).

Research on the proteomes of manipulated hosts is still in its early stages, leaving several gaps to be addressed. For instance, in the limited number of host-parasite systems examined so far, only a fraction of the host proteome has been explored, and there is a lack of information regarding host responses involving insoluble proteins or those with low molecular weight (Lefèvre et al., 2009a).

The most recent proteomic tools are now enabling exploration of previously uncharted regions of the proteome, providing researchers with the capability to delve deeper into parasite-induced host responses at this level. However, to fully understand these responses, it may be necessary to investigate further up the mechanistic chain. If protein synthesis indeed varies between manipulated and normal hosts, it likely stems from altered gene expression within the host genome.

Potentially, the most economical method for a parasite to initiate a biochemical cascade resulting in altered behavior would be to intervene in the host genome and partially suppress the expression of one or more host genes. Various environmental factors, such as components of the diet, can trigger the "turning off" of specific genes, leading to immediate phenotypic changes and/or transgenerational epigenetic effects (Jaenisch and Bird, 2003)

Numerous environmental factors, including those discussed by Jaenisch and Bird (2003), can induce DNA methylation. Given this, it's reasonable to consider whether parasites might also trigger DNA methylation in hosts (Poulin and Thomas, 2008). While several indirect lines of evidence support the notion that parasites could alter the expression of host genes, solid evidence is currently lacking. This gap exists primarily because no studies have yet examined parasite-induced changes in host methylation profiles and gene silencing within the context of host manipulation. Furthermore, even if parasites do suppress the expression of certain host genes, understanding the mechanism by which they accomplish this remains an unanswered question.

Once more, we find ourselves exploring the quest for specific secretory products or modulators released by the parasite into the host. The intricate sequence of events from infection to manipulation presents significant challenges for investigations at various levels, ranging from the genetic to the whole organism. While the altered biochemical and physiological processes of the host contribute to the manipulated host phenotype, the true source of manipulation likely resides within the parasite itself.

REFERENCES

Aeby, G.S., 2002. Trade-offs for the butterflyfish, Chaetodon multicinctus, when feeding on coral prey infected with trematode metacercariae. *Behav. Ecol. Sociobiol.* 52, 158–165.

Benesh, D.P., Seppa¨la¨, O., Valtonen, E.T., 2009. Acanthocephalan size and sex affect the modification of intermediate host colouration. *Parasitology* 136, 847–854.

Berdoy, M., Webster, J.P., Macdonald, D.W., 2000. Fatal attraction in rats infected with Toxoplasma gondii. *Proc. R. Soc. Lond. B* 267, 1591–1594.

Biron, D.G., Ponton, F., Joly, C., Menigoz, A., Hanelt, B., Thomas, F., 2005. Water-seeking behavior in insects harbouring hairworms: should the host collaborate? *Behav. Ecol.* 16, 656–660.

Blanchet, S., Me´jean, L., Bourque, J.-F., Lek, S., Thomas, F., Marcogliese, D.J., et al., 2009a. Why do parasitized hosts look different? Resolving the "chicken-egg" dilemma. *Oecologia* 160, 37–47.

Blanchet, S., Thomas, F., Loot, G., 2009b. Reciprocal effects between host phenotype and pathogens: new insights from an old problem. *Trends Parasitol.* 25, 364–369.

Dawkins, R., 1982. *The Extended Phenotype.* Oxford University Press, Oxford.

Eberhard, W.G., 2000. Spider manipulation by a wasp larva. *Nature* 406, 255–256.

Flegr, J., Kodym, P., Tolarova, V., 2000. Correlation of duration of latent Toxoplasma gondii infection with personality changes in women. *Biol. Psychol.* 53, 57–68.

Grosman, A.H., Janssen, A., de Brito, E.F., Cordeiro, E.G., Colares, F., Fonseca, J.O., et al., 2008. Parasitoid increases survival of its pupae by inducing hosts to fight predators. *PLoS One* 3, e2276.

Hammerschmidt, K., Koch, K., Milinski, M., Chubb, J.C., Parker, G.A., 2009. When to go: optimization of host switching in parasites with complex life cycles. *Evolution* 63, 1976–1986.

Hanelt, B., Thomas, F., Schmidt-Rhaesa, A., 2005. Biology of the phylum Nematomorpha. *Adv. Parasitol.* 59, 243–305.

Havlicek, J., Gasova, Z., Smith, A.P., Zvara, K., Flegr, J., 2001. Decrease of psychomotor performance in subjects with latent 'asymptomatic' toxoplasmosis. *Parasitology* 122, 515–520.

Hurd, H., 2003. Manipulation of medically important insect vectors by their parasites. *Annu. Rev. Entomol.* 48, 141–161.

Jaenisch, R., Bird, A., 2003. Epigenetic regulation of gene expression: how the genome integrates intrinsic and environmental signals. *Nat. Genet.* 33 (Suppl.), 245–254.

Kavaliers, M., Colwell, D., Choleris, E., 1999. Parasites and behaviour: an ethopharmacological analysis and biomedical implications. *Neurosci. Biobehav. Rev.* 23, 1037–1045.

Lafferty, K.D., Thomas, F., Poulin, R., 2000. Evolution of host phenotype manipulation by parasites and its consequences. In: Poulin, R., Morand, S., Skorping, A. (Eds.), *Evolutionary Biology of Host-Parasite Relationships: Theory Meets Reality.* Elsevier, Amsterdam, pp. 117–127.

Lagrue, C., and Poulin, R., 2010. Manipulative parasites in the world of veterinary science: implications for epidemiology and pathology. *Vet. J.* 184, 9–13.

Lagrue, C., Kaldonski, N., Perrot-Minnot, M.-J., Motreuil, B., Bollache, L., 2007. Modification of hosts' behavior by a parasite: field evidence for adaptive manipulation. *Ecology* 88, 2839–2847.

Lefèvre, T., Adamo, S.A., Biron, D.G., Misse´, D., Hughes, D., Thomas, F., 2009a. Invasion of the body snatchers: the diversity and evolution of manipulative strategies in host-parasite interactions. *Adv. Parasitol.* 68, 45–83.

Moore, J., 2002. *Parasites and the Behavior of Animals.* Oxford University Press, Oxford.

Mouritsen, K.N., 2002. The parasite-induced surfacing behaviour in the cockle Austrovenus stutchburyi: a test of an alternative hypothesis and identification of potential mechanisms. *Parasitology* 124, 521–528.

Mouritsen, K.N., Poulin, R., 2003. Parasite-induced trophic facilitation exploited by a non-host predator: a manipulator's nightmare. *Int. J. Parasitol.* 33, 1043–1050.

Mouritsen, K.N., Poulin, R., 2005. Parasites boost biodiversity and change animal community structure by trait-mediated indirect effects. *Oikos* 108, 344–350.

Øverli, Ø., Pa´ll, M., Borg, B., Jobling, M., Winberg, S., 2001. Effects of Schistocephalus solidus infection on brain monoaminergic activity in female three-spined sticklebacks Gasterosteus aculeatus. *Proc. R. Soc. Lond. B* 268, 1411–1415.

Palmer, C.V., Roth, M.S., Gates, R.D., 2009. Red fluorescent protein responsible for pigmentation in trematode-infected Porites compressa tissues. *Biol. Bull.* 216, 68–74.

Ponton, F., Biron, D.G., Moore, J., Møller, A.P., Thomas, F., 2006a. Facultative virulence as a strategy to manipulate hosts. *Behav. Processes* 72, 1–5.

Ponton, F., Lefe`vre, T., Lebarbenchon, C., Thomas, F., Loxdale, H.D., Marche, L., et al., 2006b. Do distantly related parasites rely on the same proximate factors to alter the behaviour of their hosts? *Proc. R. Soc. Lond. B* 273, 2869–2877.

Poulin, R., 1993. Age-dependent effects of parasites on anti-predator responses in two New Zealand freshwater fish. *Oecologia* 96, 431–438.

Poulin, R., 1995. "Adaptive" changes in the behaviour of parasitized animals: a critical review. *Int. J. Parasitol.* 25, 1371–1383.

Poulin, R., 2007. *Evolutionary Ecology of Parasites.* Princeton University Press, Princeton, NJ.

Poulin, R., Latham, A.D.M., 2002. Parasitism and the burrowing depth of the beach hopper Talorchestia quoyana (Amphipoda: Talitridae). *Anim. Behav.* 63, 269–275.

Poulin, R., Thomas, F., 1999. Phenotypic variability induced by parasites: extent and evolutionary implications. Parasitol. *Today* 15, 28–32.

Poulin, R., Thomas, F., 2008. Epigenetic effects of infection on the phenotype of host offspring: parasites reaching across host generations. *Oikos* 117, 331–335.

Rogers, M.E., Bates, P.A., 2007. Leishmania manipulation of sand fly feeding behavior results in enhanced transmission. *PLoS Pathog.* 3, 818–825.

Rossignol, P.A., Ribeiro, J.M.C., Spielman, A., 1986. Increased biting rate and reduced fertility in sporozoite-infected mosquitos. *Am. J. Trop. Med. Hyg.* 35, 277–279.

Shaw, J.C., Korzan, W.J., Carpenter, R.E., Kuris, A.M., Lafferty, K.D., Summers, C.H., et al., 2009. Parasite manipulation of brain monoamines in California killifish (Fundulus parvipinnis) by the trematode Euhaplorchis californiensis. *Proc. R. Soc. Lond. B* 276, 1137–1146.

Tain, L., Perrot-Minnot, M.-J., Ce´zilly, F., 2007. Differential influence of Pomphorhynchus laevis (Acanthocephala) on brain serotonergic activity in two congeneric host species. *Biol. Lett.* 3, 68–71.

Thomas, F., Adamo, S., Moore, J., 2005. Parasitic manipulation: where are we and where should we go? *Behav. Processes* 68, 185–199.

Thomas, F., Guldner, E., Renaud, F., 2000. Differential parasite (Trematoda) encapsulation in Gammarus aequicauda (Amphipoda). *J. Parasitol.* 86, 650–654.

Thomas, F., Poulin, R., 1998. Manipulation of a mollusc by a trophically transmitted parasite: convergent evolution or phylogenetic inheritance? *Parasitology* 116, 431–436.

Thomas, F., Renaud, F., de Meeu¨s, T., Poulin, R., 1998. Manipulation of host behaviour by parasites: ecosystem engineering in the intertidal zone? *Proc. R. Soc. Lond. B* 265, 1091–1096.

Thomas, F., Schmidt-Rhaesa, A., Martin, G., Manu, C., Durand, P., Renaud, F., 2002. Do hairworms (Nematomorpha) manipulate the water seeking behaviour of their terrestrial hosts? *J. Evol. Biol.* 15, 356–361.

Vance, S.A., 1996. Morphological and behavioural sex reversal in mermithid-infected mayflies. *Proc. R. Soc. Lond. B* 263, 907–912.

Williams, C.M., Poulin, R., Sinclair, B.J., 2004. Increased haemolymph osmolality suggests a new route for behavioural manipulation of Talorchestia quoyana (Amphipoda: Talitridae) by its mermithid parasite. *Funct. Ecol.* 18, 685–691.

Yanoviak, S.P., Kaspari, M., Dudley, R., Poinar, G., 2008. Parasite-induced fruit mimicry in a tropical canopy ant. *Am. Nat.* 171, 536–544.

16 Biological Altruism and Reciprocal Altruism in Behavior

Madhumita Roy

16.1 INTRODUCTION

Cooperative or helpful behavior frequently coexists with the existence of social activity. Many people interpret Darwin's (1859) theory of natural selection to suggest that individuals are subject to natural selection in order to maximize the number of offspring they can produce. However, Darwin brought up a criticism of this himself. What about other social animals like insects? This seems to call into doubt the validity of Darwin's theory. For example, only one of the ten thousand adult females in a normal honeybee colony actually reproduces! Once more, is the evolution of such worker sterility the product of natural selection? The phrase "cooperation" usually refers to an action that, despite possible personal sacrifices, benefits the group as a whole. Acting in a way that facilitates cooperation is the definition of cooperative behavior, even though without the cooperative behavior of other group members, the collaboration will not truly be accomplished. Animals exhibit cooperation in a wide range of behavioral settings, including mating, play, aggression, antipredatory behavior, foraging, and so on. Cooperation is found in many species in the natural world.

16.2 RANGE OF COOPERATIVE/ALTRUISTIC BEHAVIORS

a) **Helping in the birthing process:** Kunz et al. (1994) noted an intriguing instance of collaboration in the birthing process of a Roderick's fruit bat (*Pteropus rodricensis*). In these bats, non-pregnant females help pregnant females give birth by approaching the pregnant female and encircling the expectant mother with their wings. Peculiarly, the mothers don't adopt the customary feet-down stance until assistants assume it and provide them with instruction (Brown 1987). Other animals that have been documented to assist with childbirth include marmosets (*Callithrix jacchus*), bottlenose dolphins, Indian elephants (*Elephas maximus*), dogs (*Lycaon pictus*), and raccoon dogs (*Nyctereutes procyonoides*).
b) **Social grooming:** Primate researchers most commonly witness a cooperative behavior called social grooming, also known as allo-grooming, in which one member grooms another. Numerous species of primates have been observed engaging in a variety of grooming behaviors. Such behavior may accomplish multiple goals at once, including eliminating ectoparasites, lowering stress, forming coalitions, reducing aggression from other individuals, especially from dominant group members, joining new groups, warding off potential predators, and forming future relationships with individuals who have unique abilities that others lack.

   But as Maestripieri (1993) noted, there are drawbacks to grooming other individuals as well. For instance, it was discovered that adult female macaques engaged in social grooming were less watchful of their offspring's surroundings, exposing the young to higher rates of aggression from other group members.

DOI: 10.1201/9781003710271-16

c) **Group hunting:** Numerous species have been documented engaging in group hunting, including Harris's hawks (*Parabuteo unicinctus*). Among the various hunting techniques employed are: i) "Surprise pounce," in which multiple hawks approach from various directions and converge on a rabbit. ii) Using a "flash and ambush" tactic, in which the hawks circle and shift their observation to the spot where the rabbit vanishes. While one or more hawks pounce on the victim and make the kill, one or two hawks attempt to break through the cover. iii) "Really attack," in which party members take turns chasing a rabbit for several minutes in a continuous phase. Cooperative hunting has been observed almost exclusively in social species, such as lions, killer whales, chimpanzees, and wolves (Bednarz, 1988).

d) **Nest raiding:** When it comes to protecting their nest, three-spined sticklebacks work together in "gangs" of females to subdue males. This fish's males are the only ones who provide for the family. Enormous groups of females raid, destroy, and frequently eat the eggs found in the nests of males that guard the eggs. The nutritional content of the eggs consumed, mating with the male whose nest has been destroyed, and removing possible rivals for one's own progeny are the three advantages of such nest raiding. The males can, however, protect the nest against small parties. In the event of an assault by bigger numbers, they are defenseless. In this instance, the only female who gains the most is the one who starts the raid and mates with the male whose nest was destroyed.

e) **Allee effect:** Allee (1951) highlighted the detrimental effects of undercrowding, often known as the Allee effect. Through experiments, Allee demonstrated that goldfish (*Cyprinus sp.*) raised in a poisonous colloidal suspension of silver outlived those kept in isolation. This is because the fish in such conditioned waters make slime, which precipitates silver on the tank bottom to protect the fish. Planarian flatworms (*Planaria sp.*) are another example of an organism that groups to shield itself from ultraviolet light.

16.3 PATHS TO THE EVOLUTION AND MAINTENANCE OF COOPERATION

Over the past 40 years, cooperative behavior studies have grown in popularity. Four components have been identified as being in charge of the development and upkeep of animal cooperation since the groundbreaking research of W. D. Hamilton in 1963. Reciprocity, also known as reciprocal altruism, and kin selection, byproduct mutualism, and group selection are four significant avenues (Alcock 2005) (Figure 16.1).

A. **Reciprocal altruism/reciprocity:** In actuality, reciprocity is a unique form of mutualism in which the helping party suffers a temporary setback until it is returned, at which point it gains a net improvement in fitness. On the other hand, there are instances where assisting another individual actually results in the donor losing all chance of ever becoming a parent. Reciprocal altruism is the term used to describe this type of helpful conduct in evolutionary biology. Since altruistic behaviors break the rule that traits cannot spread throughout evolutionary time if they reduce an individual's reproductive success relative to that of other individuals, they represent a very intriguing Darwinian challenge for adaptationists, if they exist at all (Jamieson 1991). Robert Trivers proposed the evolution of cooperation between unrelated individuals in his study "The Evolution of Reciprocal Altruism," which was published in 1971. If there's a possibility that they may be the beneficiaries of these activities in the future, individuals will collaborate and act kindly toward one another. The possibility of cheating is a problem with reciprocal altruism. Cheating happens when someone doesn't give back. If individuals can recognize one another, have decent memory, and have plenty of opportunities for reciprocal charity, then there is little incentive to deceive. Cheaters risk having their activities discovered, banned, or shunned in such circumstances. Because of this, we do not observe the evolution of reciprocal altruism in scenarios where there is a "one off" advantage and two animals are unlikely to cross paths again. Instead,

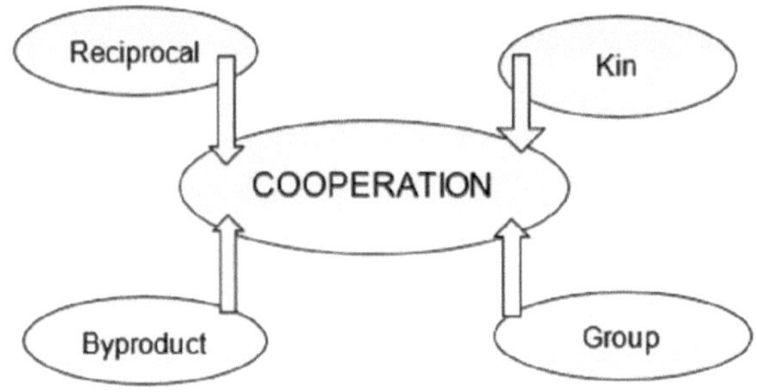

FIGURE 16.1 Probable pathways to cooperative behavior.

it develops under circumstances where the same animals connect over an extended length of time. According to Trivers, if animals assign prosocial behaviors like these to those who have already shown donor cooperation and altruism (doing good for others), then genes for these traits may be selected. Trivers showed that when selfless deeds are distributed at random among individuals in a sizable community, the genes that encourage such behavior will vanish because there is little chance that the one doing the deed will receive it back. But if benevolent deeds are distributed in a non-random manner among strangers, genes that encourage them may become more prevalent if a reciprocal relationship develops. The following are the factors that contribute to this likelihood: 1. Lifespan, with longer-living individuals having a higher likelihood of reuniting to reciprocate. 2. Dispersal rate: a low dispersal rate will make recurrent interactions more likely to occur. 3. Mutual reliance: grouping individuals together will raise the likelihood of reciprocation. But in any social structure, when dishonesty escalates to a certain point, good deeds become rare. The following field research can be used to emphasize the significance of reciprocal altruism.

1. Males in Africa's olive baboons (*Papio anubis*) form a coalition against a third male when they are unrelated. One of the assailants could be able to reach the female if the third male was in consort with her while she was in estrus. The two males would uphold the previously formed alliance, with the other member of the male pair assuming control of the next "stolen" female.
2. Wilkinson 1984 has described a remarkable example of reciprocal altruism in vampire bats, which feed on the blood of mammals, particularly domestic animals such as cattle and horses. According to Wilkinson (1984), the majority of female vampire bats (*Desmodus rotundus*) in a normal group are not closely linked to one another. Every sixty hours, female vampire bats in a nest regurgitate blood meals to other bats that have not been able to find food and will starve to death if they do not receive any. Bats feed their relatives, but they also feed other bats, especially those that have fed them in the past. Starving bats are more likely to be fed than well-nourished ones. Only bats that were frequently seen together (more than 60% of sightings) exchange regurgitated blood meals. They come back to the same roost daily and form lasting relationships with a specific individual over time, recalling the advantages bestowed upon them. Additionally, bats can tell if a fellow bat is starving or well-fed, and they can assist in a starvation bath without suffering too much damage to themselves. The majority of the interactions we typically observed were between mothers and daughters. Wilkinson went on to say that the pattern of transaction was greatly influenced by both relatedness and association. Wilkinson proposed three lines of evidence to support his theory that reciprocity must endure: i) the pairings must endure long enough to allow for reciprocation; ii) the recipient's benefit must outweigh the donor's expense; and iii)

FIGURE 16.2 Cooperation between vampire bats during feeding.

donors need to identify dishonest individuals and refuse to feed them. Additionally, Wilkinson proved that vampire bats fit these requirements (Figure 16.2).

3. Trivers (1971) also showed how reciprocal altruism could have led to the evolution of cooperation between members of different species. He gave the cleaning symbiosis in fish as one example. Numerous fish removed ectoparasites from other fish species. They entered the gill chamber or the host's mouth. The cleaners may now be recognized by their unique hue, thanks to evolution. The host indicates when it is ready to be cleaned and when it has had enough. Typically, cleaners and hosts meet at stations. Reciprocal altruism can occur when hosts use the same cleaner over and over again, and cleaners have lengthy lives. In the short term, hosts who consume cleaners after being cleaned by them stand to gain from an easy dinner, but in the long run, they risk being eliminated from selection as the supply of cleaners is diminished.
4. The tiny African mammal known as the meerkat forages in packs. Occasionally, meerkats will abandon their earthly hunting for insects and scale a tree or termite mound to scan their surroundings for potential predators. When the high sentinel spots a predator, it is the first to sound the warning, causing the other steel-foraging meerkats to quickly seek shelter. One reason for this behavior could be that sentinels put themselves in danger to protect others because they know that when they take turns being watchdogs, their colleagues will pay them back. Individuals finally repay the favors they get because they were benefited. Selection may favor making the first move if doing so has a little upfront cost but results in a large payoff from getting the favor returned. In a population, reciprocity can spread if the advantage outweighs the expense (Figure 16.3).

In actuality, reciprocity is a unique form of mutualism in which the helpful party suffers a temporary setback until receiving the same favors in return, at which point it gains a net rise in fitness. On the other hand, there are instances where assisting another individual actually results in the donor losing all chance of ever becoming a parent. Altruism is the term used to describe this type of helpful conduct in evolutionary biology.

W. D. Hamilton created a unique theory for the evolution of altruism that did not rely on arguments for the benefit of the collective. Rather, the underlying assumption of Hamilton's theory was that animals reproduce with the unintentional aim of more successfully propagating their

FIGURE 16.3 A meerkat sentinel (Pinterest, courtesy Graham Towler Locksmiths).

alleles than those of other animals. This final objective is directly aided by personal reproduction. However, assisting genetically related individuals—that is, one's relatives—to survive and procreate can offer a side path leading directly to the same goal (Abbot, Withgott, and Moran 2001). The idea of the coefficient of relatedness is crucial to comprehending why the likelihood that an allele received from a recent common ancestor by two individuals is present in the other individual. Any parent's donated egg or sperm has a one in two chance of carrying the allele; hence, any offspring of that parent will have a 50% chance of acquiring any gene. The coefficient of relatedness (r) between parent and offspring is therefore ½ or 0.5. The coefficient of relatedness varies for different category relatives. The coefficient of relatedness for an uncle and his nephew is $\frac{1}{2} \times \frac{1}{2} = \frac{1}{4}$, or 0.25. Because there is a one in two probability that an individual and his sister will share this gene, there is a one in two chance that the sister will pass that allele onto any given offspring. In contrast, there is zero relatedness between an unconnected individual and another. We can predict the outcome of a rare altruistic allele that is in competition with other selfish alleles by knowing the coefficient of relatedness between the altruists and the member they assist. Whether the altruistic allele becomes more common if its carriers choose not to reproduce in favor of helping their relatives procreate is the crucial question.

Comparing the genetic effects of those who help strangers with individuals who help their own family members offers another perspective on this issue. No form of a gene is likely to benefit more than any other if it is supplied randomly. However, if close relatives preferentially support one another, any unusual family alleles they may carry may survive better and become more common than other gene variants in the community. When one considers things this way, it becomes evident that our type of natural selection can take place when genetically distinct individuals have differing influences on the success of near relatives' reproduction. Jerry Brown refers to this selection process as "indirect selection," which, in comparison to direct selection, accelerates individuals' reproduction (Figure 16.4).

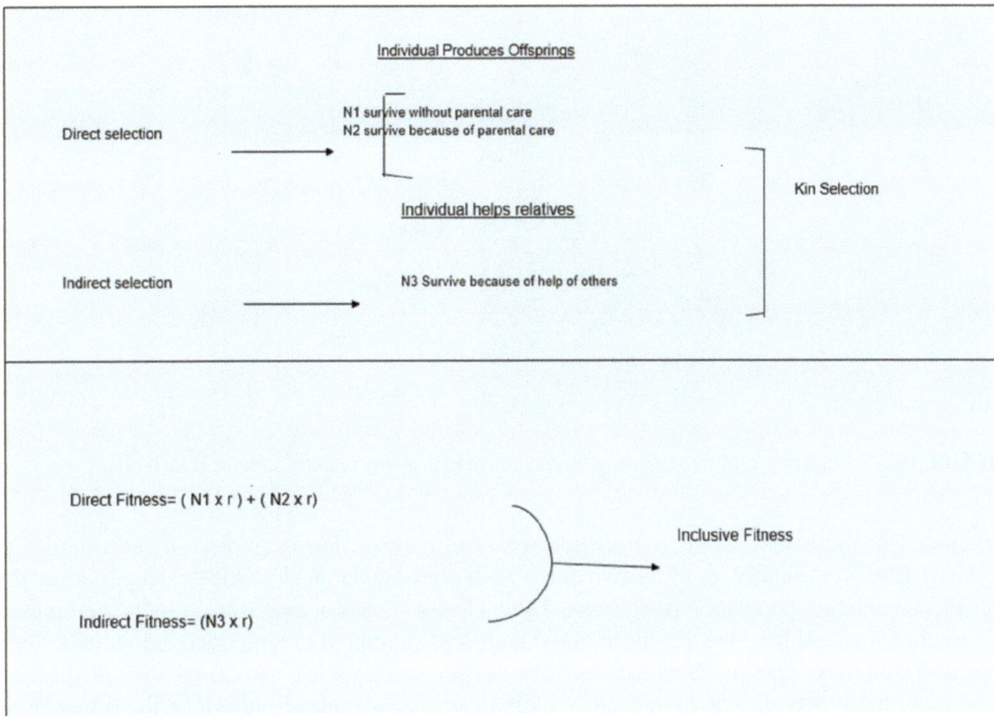

FIGURE 16.4 The components of selection and fitness.

B. **Kin selection:** W. D. Hamilton, an English population geneticist, introduced the idea of kin selection in 1963, and it is arguably one of the greatest contributions to sociobiology. He postulated the kin selection mechanism to explain the evolution of altruistic behavior even in situations where an individual's fitness was reduced. When a predator is around, many animals may make piercing alarm sounds to alert other creatures to their presence. The point is, why do animals put their own lives at danger to issue warnings? In the instance of Belding's ground squirrels, for instance, it has been noted that females rather than males typically sound the alert (Sherman 1981). The female ground squirrels remain in their birth location to mature, while the males in Belding's abandon their homes (Holcamp 1984). Because of this male dispersal, there is an imbalance, and the females are encircled by family. Males, on the other hand, are usually in places where total strangers are present. Because of this imbalance, females become those who sound the alarm to alert close relatives. In addition to encouraging open-minded social behavior, kinship also serves as a deterrent to antisocial behavior. Kinship has the unique ability to influence animal behavior more than other factors. Merely knowing which members of a group are blood relations gives ethologists and behavioral ecologists enough information to make some basic behavioral predictions (Figure 16.5).

Because the fitness obtained from aiding non-descent kin in their survival (indirect fitness) and direct fitness obtained from personal reproduction (direct fitness) can both be expressed in identical genetic units, we can calculate an individual's total gene contribution to the following generation, resulting in a quantitative measure known as inclusive fitness. Rather than providing an exact estimate of an individual's lifetime genetic contribution, inclusive fitness is a notion that allows us to compare the evolutionary outcomes of two different hereditary features. Put differently, the significance of inclusive fitness arises when assessing the comparative genetic effectiveness of multiple

FIGURE 16.5 Belding's ground squirrel gives an alarm call (www.mun.ca, creator Ron Wolf).

behavioral methods that are in competition with one another. For example, we can evaluate the inclusive fitness consequences of the two qualities to see whether an altruistic strategy is better than one that encourages personal reproduction. The inclusive fitness of an altruistic individual must be higher than it would have been if those individuals had attempted to reproduce individually for an altruistic trait to be adaptive. Stated differently, the uncommon allele associated with altruism will only grow more prevalent if the altruist's indirect fitness advantage outweighs the direct fitness it forfeits due to its selfless actions. When the relatedness is calculated, it becomes clear why kin selection influences animal behavior in such a potent way. However, the next stage of the process involves choosing genes based on how many copies of themselves they pass down to the next generation. This is because relatives are more likely than strangers to share any given gene. A particular gene may also have a higher chance of spreading to future generations if it has an effect on not only the individual in which it resides but also that individual's blood relatives. This kind of representation of fitness is called inclusive fitness.

Hamilton (1963) proposed a hypothesis that included the individual as a unit of selection in addition to the gene. According to the theory of kin selection, the success of a gene that results in altruistic conduct depends not on whether the gene benefits the individual who carries it, but rather on the gene's value to itself. The frequency of that gene in the gene pool will rise if the individual who benefits from the act is a relative of the altruist (non-relatives are unlikely to carry the same genes). The likelihood of a relative harboring these genes decreases with distance. Consequently, the ratio of receiver gain to altruist expense must be larger. A common term for this kind of interaction is Hamilton's rule. The expression for it is $b/c > 1/r$, where b is the advantage that other individuals obtain from the transaction that the gene codes for, and c is the expense incurred by the individual exhibiting the characteristic. The percentage of genes that two individuals have in common due to their shared ancestry is known as the coefficient of relationship, or r.

An uncommon analysis of the pied kingfisher can be used to demonstrate the usefulness of Hamilton's rule (Reyer 1984). These lovely African birds build their colonial nests in tunnels on the shores of big lakes. Males over a certain age have trouble finding a partner. They take on the role of main caregivers, bringing fish to their mother and their siblings while fending off dangerous snakes and mongooses that pose a threat to the nest. By providing for their siblings, are these males doing the best possible job of passing on their genes? They do possess alternative choices. They may act as secondary helpers by supporting unrelated nesting couples, or they could act as delayers by waiting for the next mating season to begin. Here, we must ascertain the advantages and disadvantages of their course of action. Compared to delayers and more secondary helpers, primary helpers put in greater effort. The higher price paid by first responders translates into a decreased likelihood that they will survive to visit their breeding areas the following year. Only 54% return as opposed to 74% of secondary assistants and 70% of delayers returning. Moreover, only two out of every three main caregivers that survive find a partner in their second year and procreate directly. In contrast, 91% of secondary helpers who return are successful breeders. A lot of one-time secondary helpers mate with the female they assisted the year before, implying that better access to a possible partner is the ultimate reward for their original kindness. With the use of this data, we are able to determine the direct fitness cost in terms of decreased personal reproduction in the second year of life for the selfless primary helpers. For the sake of simplicity, we will limit our comparison to lone primary helpers who, if they survive and find a mate, assist their parents in raising siblings in the first year before breeding independently in the second year. Secondary helpers that assist non-relatives in the first year when there are no other helpers around, and if they survive and find a partner, reproduce independently in the second year. They invest all of their energy into assisting their parents in raising children, even if it means they will have fewer opportunities for personal reproduction the next year. However, in the second year, primary assistants perform better than delayers. Due to their increased chance of acquiring a mate and higher survival rate, secondary helpers perform even better. Reyer proved this using Hamilton's rule. In year 1, the primary helpers assist more non-descendant kin individuals at the expense of future personal reproduction in year 2. These additional family members give indirect fitness advantages that more than make up for the loss of direct fitness that primary helpers endure in their second year since they carry some of the helpers' alleles in relation to auxiliary personnel (Figure 16.6).

C. **Byproduct mutualism:** Another step in the evolution of cooperation is byproduct mutualism. Here, cooperation happens when there is an immediate cost or penalty for acting in a non-cooperative manner, and the immediate net benefit of acting collaboratively outweighs the cost of cheating. One excellent example is when two people are stuck in a cave and the only way out is for them to both push the big boulder that's in the way of the entrance. Since both parties would bear a greater expense if one of them did not collaborate, there is no incentive for either to behave otherwise in this situation. For there to be byproduct mutualism, there must be a chance that a defector may fall prey to its own deception. There are two main ways that byproduct mutualism is different from reciprocal altruism: 1. Because the environment under byproduct mutualism favors either everyone cooperating or no one cooperating, there is no incentive to cheat. Therefore, byproduct mutualism

FIGURE 16.6 Pied kingfisher (Pinterest.com, courtesy Vista Create, *Animal Behaviour* 10e, Figure 3.14, 2013 Sinauer Associates, Inc.).

appears to have no costs for either party. Conversely, in reciprocal altruism, the actor's expense appears to exceed the gains. 2. The most prevalent kind of reciprocal altruism scorekeeping is necessary. On the other hand, collaboration through byproduct mutualism does not involve scorekeeping. Therefore, reciprocal altruism is comparable to mutualism or byproduct mutualism, with the exception that there is a temporal lag.

To support the aforementioned claims, nesting flocks of quail that bundle up in chilly weather could serve as an excellent illustration of byproduct mutualism. Since everyone benefits instantly from being protected from freezing, there is no apparent cost to individuals and no waiting period in this straightforward aggregate.

Another instance of byproduct mutualism is in the Costa Rican jungles, where two male long-tailed manakins (*Chiroxiphia linearis*) collaborate in a sophisticated show to draw in a receptive female (Snow 1956). The performance consists of two male "leapfrog hops" and "butterfly" flights performed in tandem with a call that sounds like the word "Toledo." While subordinate males do the majority of the work display, dominant males receive nearly all copulations (McDonald and Potts, 1994). The most obvious issue in this situation is: Why do submissive males cooperate and forego procreation? The reason for this is that although the beta males' direct fitness increases, the reward is not realized for five to thirteen years. Alpha males would eventually become beta males. Since the alpha male never pays back the debt offered by the beta male, this cooperative conduct cannot be seen as reciprocal altruism. As a result, by working together to draw in females, both alpha and beta males gain from the rise in their direct fitness. Even while beta males might not seem to be getting anything at first, they might develop into alpha males and take over the territory in the end (Figure 16.7)

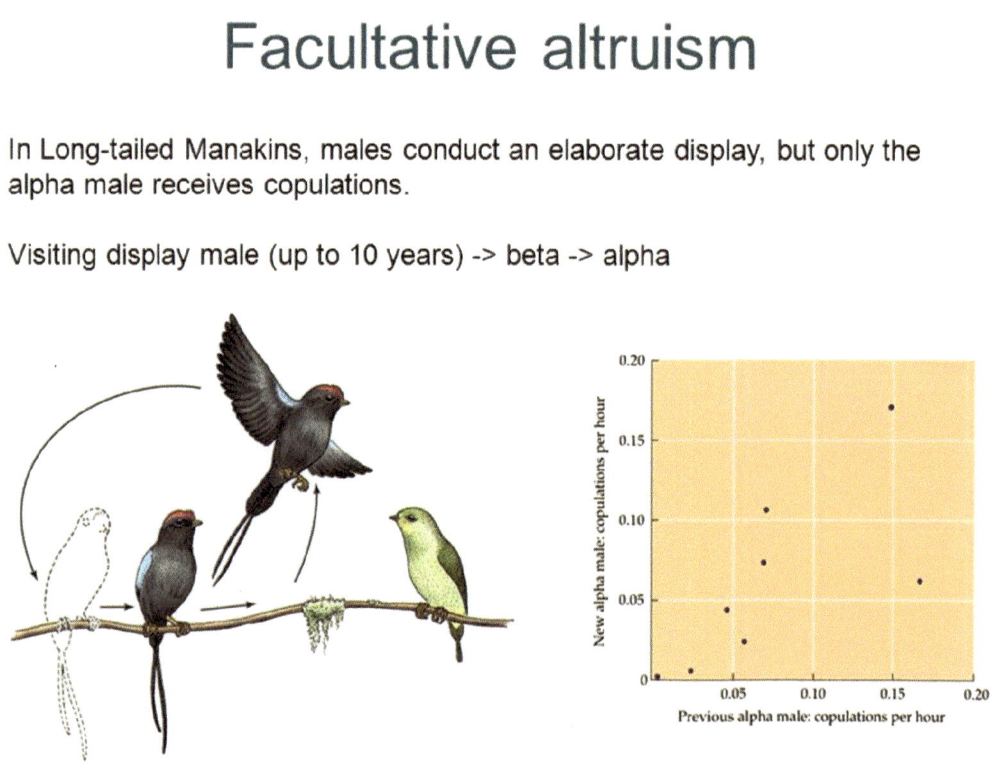

FIGURE 16.7 Long-tailed manakin males conduct an elaborate display, but only the alpha male receives copulations (Fig. 13.10, chapter 13, Alcock J. 8th ed, Sinauer, USA; McDonald and Potts,1994).

D. **Group selection:** In general, there are two stages of natural selection at work: intragroup selection and intergroup selection. Since cooperative and altruistic individuals by definition bear a cost that others do not, selection tends to work against them in groups. Within-group selection always favors selfish types because they absorb the benefits of the actions of altruists and cooperators while not having to shoulder any of the costs. Conversely, as groups with more cooperators typically outproduce those with fewer, between-group selection actually encourages collaboration. Generally speaking, this is achievable by multiplying the overall number of offspring or by quickly colonizing new areas. When an individual within a group of unrelated animals sounds the alarm, for instance, the group pays a price since that individual may become the primary target of a predator that is homing in on the alarm call. Since the alarm cry allows others to escape, their sacrifice benefits the group as a whole. Groups with a high number of alarm callers might outproduce those with a lower number of cooperators. The frequency of cooperators within groups must vary for such group level benefits to materialize. Herds of wild buffalo are another example, where groups with more male defenders have an advantage over groups with fewer defenders. As in the previous case, this supplies the intergroup element required for cooperation to develop.

BIBLIOGRAPHY

Abbot, P., J. H. Withgott, and N. A. Moran. 2001. Genetic conflict and conditional altruism in social aphid colonies. *Proceedings of National Academy of Science* 98: 12068–12071.

Alcock, J. 2001. *The triumph of sociobiology.* New York: Oxford University Press.

Alcock, J. 2005. *Animal behaviour: An evolutionary approach.* Sinauer Associates, Inc.

Alexander, R. D. 1974. The evolution of social behavior. *Annual Review of Ecology and Systematics* 5: 325–383.

Allee, W. C. 1951. *Cooperation among Animals, with Human Implications.* New York: Henry Schuman.

Allen L., B. Beckwith, J. Beckwith, and S. D. Culver. 1976. Sociobiology- Another biological determinism. *Bioscience* 26: 182–186.

Anderson, M. 1984. The evolution of Eusociality. *Annual Review of Ecology and Systematics* 15: 165–189.

Axelrod, R. and W. D. Hamilton. 1980. The evolution of cooperation. *Science* 211: 1390–1396

Bednarz, J. C. 1988. Cooperative Hunting Harris' Hawks (Parabuteo unicinctus). *Science* 239: 1525–1527.

Barland, T. M., N. C. Bennett, J. U. M. Jarvis, and C. G. Faulkes. 2002. Eusociality in African mole rats: New insights from patterns of genetic relatedness in the Damaraland mole rat (*Cryptomys damarensis*). *Proceedings of the Royal Society of London B* 269: 1025–1030.

Bernasconi, G. and J. E. Strassman. 1999. Cooperation among unrelated individuals: The ant foundress case. *Trends in Ecology and Evolution* 14: 477–482.

Brooks, D. R. and D. A. McLennan. 1991. *Phylogeny, ecology and behavior.* Chicago, IL: The University of Chicago Press.

Brown, J. L. 1975. *The evolution of behavior.* New York: W. W. Norton.

Brown, J. L. 1987. *Helping and communal breeding in birds: Ecology and evolution.* Princeton, NJ: Princeton University Press.

Burda, H., R. L. Honeycutt, S. Begall, O. Locker-Grutgjn, and A. Scharff. 2000. Are naked and common mole rats eusocial and if so, why? *Behavioral Ecology and Sociobiology* 47: 293–303.

Clutton-Brock, T. H., and P. H. Harvey. 1984. Comparative approaches to investigating adaptation. In *Behavioral ecology: An evolutionary approach*, eds. Krebs, J. R., and N. B. Davies. Oxford: Blackwell, pp. 18–30.

Charles Darwin, M. A. (1859). The origin of species. By means of natural selection, or the reservation of favoured races in the struggle for life. John Murray, Albemarle Street. London, p. 171.

Foster, M. S. 1977. Odd couples in manakins: A study of social organization and cooperative breeding in *Chiroxiphia linearis. American Naturalist* 111: 845–853.

Hamilton, W. D. 1963. The evolution of altruistic behavior. *The American Naturalist* 97: 354–356.

Hatchwell, B. J. and J. Komdeur. 2000. Ecological constraints, life history traits and the revolution of cooperative breeding. *Animal Behavior* 59: 1079–1086.

Holcamp, K. E. 1984. Natal dispersal in Belding's ground squirrels (*Spermophilus beldingi*). *Behavioral Ecology and Sociobiology* 16: 21–30.

Jamieson, I. G. 1991. The unselected hypothesis for the evolution of helping behavior: Too much or too little emphasis on natural selection? *American Naturalist* 138: 71–282.

Kunz, T. H., Allgaier, A. L., Seyjagat, J., and Caliguiri R. 1994. Allomaternal care: helperassisted birth in the Rodrigues fruit bat, Pteropus rodricensis (Chiroptera: Pteropodidae). *Journal of the Zoological Society of London* 232:691–700.

Maestripieri, D. 1993. Vigilance costs of allogrooming in macaque mothers. *The American naturalist* 141:744–753.

McDonald, D. B., and W. K. Potts 1994. Cooperative display and relatedness among males in a lek-mating bird. *Science* 266:1030–1032.

Reyer, H. U. 1984. Investment and relatedness: A cost/benefit analysis of breeding and helping in the Pied Kingfisher. *Animal Behavior* 32: 1163–1178.

Sherman, P. W. 1981. Kinship, demography and building's ground squirrel nepotism. *Behavioral Ecology and Sociobiology* 8: 250–259.

Snow, D. W. 1956. Courtship ritual: The dance of manakins. *Animal Kingdom* 50: 80–91.

Trivers, R. 1971. The Evolution of Reciprocal Altruism. *Quarterly Review of Biology* 46: 35–57.

Wilkinson, G. S. 1984. Reciprocal food sharing in the vampire bat. *Nature* 308: 181–184.

17 Social Dominance and Territoriality

Cherryl Dimphna Miranda and Vallabhaneni Srikanth

17.1 INTRODUCTION

Dominance is typically defined as asymmetry in aggression by one animal toward another animal. In other cases, dominance refers to individuals that win short-term dyadica contests (Drews, 1993).

Animal groups are frequently arranged hierarchically, with dominant members having greater access to resources and opportunities for reproduction than members of lesser ranks. Early dominance hierarchy formation can be impacted by a variety of interrelated elements, such as the unique characteristics of an animal, social norms, and self-organizing social dynamics. After they are created, hierarchies are usually upheld for a long time because individuals recognize and follow established dominance connections, which save time, energy, and lower the danger of harm.

Signals of dominance:

There are various signals of dominance exhibited by a variety of living creatures. These dominance expressions give reliable signaling to recipients. Signals are acts or structures that signalers perform in order to communicate information to recipients in such a way that the information induces a response from the recipient and the response has, typically, beneficial fitness consequences for both the signaler and the recipient. These signals need not be prominently noticeable every time. Fundamentally, signals can either inform the recipient by giving it relevant information that aids in making the right response, or they can occasionally be deceptive by eliciting a response that is advantageous to the signaler but not necessarily to the recipient. This manipulation problem does not exist if the sender and the receiver have compatible interests; nevertheless, if the sender and the recipient have divergent interests, manipulation may be an issue.

Badges of status: Numerous animals have distinguishing characteristics that serve as status symbols, such as color patches on their fur or lizard tails. According to Dawkins and Krebs (1978), badges of status are random structures (such as color patches) that communicate status signals and are inexpensive to produce, and may be subject to deception. In contrast, weaponry like antlers directly influences the outcome of a battle and provides the bearer a genuine advantage. Badges of status in black bibs of male house sparrows or the black head feathers of Harris' sparrows distinguish dominant from subordinate animals based on the size of the badge they possess.

Dominance hierarchies:

Many social animals identify patterns of authority; the two most fundamental are the connection of the stronger to the weaker or aggressive to submissive and the parent-child or provider-dependent relationship. The group of these connections is frequently referred to as a social or dominance hierarchy. There may not be any rank relationships between subordinates, and the single dominating individual who is dominating all others is part of what is known as a basic dominant hierarchy. Numerous social groupings feature hierarchies of dominance that are linear or nearly linear. The hierarchical structure of dominance differs among communities. Most frequently, there are

DOI: 10.1201/9781003710271-17

a number of rank relationships, such as an "alpha" who is superior to every other member of the group, a "beta" who is superior to every member except the "alpha," and down the line to the least ranked who is subservient to all members. Thus, a system is referred to as a simple and linear dominance hierarchy if dominant males can be ranked in a single sequence (Tibbetts et al., 2022).

There are two concepts in overall dominance hierarchies.

i. **Establishing dominance**: The initial establishment of hierarchies is important under dominance relationships, which depend on resource holding potential and motivation. Individual characteristics are evaluated through direct antagonistic encounters, signs of fighting skills, and social knowledge obtained from seeing how others interact. Conventions are distinctive qualities that distinguish a specific individual as the next dominant one without expressing innate qualities that enable them to prevail in competitions.
ii. **Maintaining dominance**: Information regarding dominance rank is provided via signals of dominance. Individual identification signals are distinct characteristics that recipients pick up on and link to specific details about the sender, such as dominance rank. Dominant individuals use self-reinforcing behavioral patterns, threats, and punishment to preserve their rank.

When individuals engage in low-stakes competitive interactions in big or unstable groups, they frequently build dominance connections without necessarily forming permanent hierarchies. However, hierarchies are usually preserved over time in taxa that inhabit smaller groups with stable membership. Prolonged rank-based conflict is expensive to any individual as it involves energy, time, and risk of injury; hence, stable hierarchies are always preferred.

17.2 TYPES OF DOMINANCE HIERARCHIES

The following three fundamental forms of dominance hierarchies have been identified by Barnard and Burk (1979).

1. **Statistical hierarchy**. Those animals who have no recollection of previous interactions and fight again whenever they happen to meet just due to differences in physical attributes; some usually win while other individuals lose. This kind of hierarchy is probably uncommon in nature due to the costs of fighting. However, they might exist in certain lower invertebrates. Sea anemone aggression, for instance, seems to follow a statistical hierarchy in which success is based on relative body size.
2. **Confidence hierarchy**. Winning a fight gives the winner more energy for the next fight, increasing the likelihood that it will win again, and vice versa. In the end, the confidence from the previous fight results in a hierarchy of linear dominance. This type of hierarchy has been documented in mice, where fight results are primarily predicted by an individual's performance in their previous bouts rather than the attributes of their opponents.
3. **Assessment hierarchy**. Animals base their decisions about whom to fight and whom to avoid on their recollections of previous interactions. Based on characteristics of challengers that are correlated with winning or losing, fights turn into calculated risks. In nature, assessment hierarchies are the most prevalent type. Given that fighting is likely to be costly in terms of time, energy, and potential injury, we would anticipate that natural selection would favor individuals who are sensitive to any warning that indicates how a fight will turn out. In fact, it appears that this type of hierarchy accounts for almost all of the documented hierarchies in organisms.

17.3 MAINTENANCE OF ESTABLISHED DOMINANCE

Stable dominance is especially important for individuals living in small groups. Longevity, stress reactions, weaning percentage, young animal's weight gain, animal survivability, and reproductive success are all negatively impacted by social instability. In unstable communities, individual animals in the subordinate ranks occasionally become aggressive in order to test their dominant peers, as it would benefit them to rise in their position. Nevertheless, due to the costs of ongoing conflict, members of stable groups usually acknowledge and respect established dominance relationships.

Two main mechanisms that keep hierarchies stable are a variety of behaviors, such as threats, punishment, and self-reinforcing behavioral differences, which uphold the supremacy of the dominant. The other is a society set up that possesses signals that convey information about rank and keep hierarchies stable, particularly signals of dominance and individual identity. Among the many behaviors used by dominants to punish their subordinates, some are eviction, aggression, social stress, and infanticide. In a stable community that is established, dominant individuals have the right to retaliate violently against subordinates who do not support the group by providing alloparental care or by foraging.

The position of supremacy in a hierarchical system is, in fact, frequently self-sustaining due to rank-related differences in physical condition, work, and nutrition. Usually, superiors have the edge on food resources, and nutrition actually decides the reproductive health of individuals. Thus, reproductive and behavioral dominance hierarchies are maintained by the differences between dominants and subordinates, making it more challenging for people to rise in rank after serving as subordinates for an extended length of time.

Individual recognition is the most popular method for determining dominance position in stable groupings. In the case of dominance relationships among vertebrates, where a small number of individuals interact repeatedly, individual recognition is a relatively common mechanism. For instance, it's believed that in many stable bird flocks, individual recognition through plumage and calls maintains hierarchies. Primates use information from their senses of smell, sight, and sound to individually identify group members. Baboons identify each member of their group uniquely and categorize individuals based on their kinship and rank. According to the social intelligence hypothesis, animals with complex social structures benefit from higher cognitive capacities because these individuals are better able to keep track of multiple relationships between individuals and respond appropriately in social situations.

In addition to influencing the behavior of the recipient, signals of dominance convey information about the signaler's level of dominance. Due to the relational nature of dominance, trustworthy signals of dominance exhibit phenotypic changes in tandem with changes in an individual's rank. A simple method that requires little cognitive work to quickly and accurately determine dominance rank is the use of signals of dominance, as it takes little memory to react. Even a small interaction between two animals early in life may create lasting differences in their dominance status.

Individual dominance relationships in any group are formed very quickly, and while there are occasional changes, they are rare. In pigs, such relationships are established before the animals are eight weeks old (Beilharz & Cox, 1967). In calves, the ranking order is established between 3 and 6 months of age. Social dominance does not only rank animals in a group but also indicates the degree of differences between the animals. Dominance is established in a number of ways: with no threat of aggression; with aggression; with a fight; or after a fight. Some animals may not exhibit aggression even in a competitive situation (Beilharz & Zeeb, 1982, Ramseyer et al., 2009).

Dominance development is basically a learning process which is influenced by several environmental determinants and genetic causes. Through the dispersal and migration of the animals, this dominance relationship frequently serves as a control mechanism for the social organization in the ecology of species.

17.4 TERRITORIALITY

A territory is an area held and defended by an organism or group of organisms of the same or different species. Territoriality or territorial behavior is a set of actions related to defending one's territory or a self-boundary regulation mechanism (Altman, 1975) and is observed in specific mammal species, as well as in birds, fish, and various other animal taxa (Noble, 1939). It is mostly predicated on a favorable response to a specific location and a negative response to another individual, through aggressive interactions and the use of visual, acoustic, or olfactory cues (Brown & Orians, 1970, Mayer et al., 2020).

Definitions of territoriality given by various authors:

Definition of Territoriality	Reference
Actions by which an organism creates physical barriers around itself, claims the area or space inside those barriers, and defends it against intruders.	Brower, 1965
Groups keep members of the same species from invading a space region.	Ardrey, 1966
Perception, utilization, and safeguarding of locations, people, things, and concepts, as well as transient or long-term defense and response.	Altman, 1970
It is an action taken either temporarily or permanently to stop others from using the areas and items.	Taylor, 1988
A person's outward manifestation of their sense of possession over a material or social object.	Brown, 2005
To assert, declare, maintain, consolidate, safeguard, and expand their control over territory, individuals, or groups that take part in a variety of territorially linked activities.	Peng, 2013

Vertebrate individuals, pairs, and family groupings, as well as higher invertebrates, typically confine their activities to a certain region known as their home range. This area is referred to as a territory if it is actively defended, which is dynamic; they frequently alter in size and shape in response to age, season, population density, and other variables. Broadly speaking, territories can be categorized into two main types:

A. **Exclusive territories**: These territories are defended against all intruders, typically by a single individual or a group. Exclusive territories are common among species where resources are limited, and competition for those resources is high. Examples include nesting sites in birds, feeding areas in carnivores, and breeding grounds in amphibians.
B. **Overlap territories**: In contrast to exclusive territories, overlap territories are areas where the territories of multiple individuals or groups intersect. Within overlap territories, individuals may still defend specific resources or areas within the territory, but they tolerate the presence of others to some extent. This phenomenon is often observed in social species where cooperation or tolerance among conspecifics provides benefits.

The dimensions of territory and the extent of the home range typically diminish as population density rises, as indicated by Lindeman et al. (2015). Conversely, territorial behavior can influence the dynamics and density of populations, such as by maintaining a stable population size through the inhibition of territory acquisition by other individuals in established territories, as highlighted by Both and Visser (2003). In vertebrates and some arthropods, territoriality is particularly noticeable due to their complex reproductive behavioral patterns, which include nest construction, egg laying, and the upbringing and defense of young. Most of the time, territorial behavior is limited to the mating season.

The purpose of territory formation varies across species, but it universally serves to ensure that mating pairs and their offspring are appropriately spaced to access necessary resources such as food and breeding areas, thus maximizing habitat utilization. Territory size fluctuates seasonally based on resource availability. Birds of prey and large carnivores claim territories spanning several square miles to fulfill their food needs, while species like herring gulls and penguins maintain territories of only a few square meters, primarily for breeding purposes.

Territories are typically established prior to breeding, predominantly by males, and defense intensifies during breeding season, particularly among males of the same species. Various behavioral activities, including threat displays, are associated with territory formation, often triggered by specific stimuli acting as releasers. Aggression levels peak toward the center of the territory, influenced in part by testosterone levels, which can impact territory size—evidenced by experiments increasing red grouse territory size via testosterone injections (Mougeot et al., 2004). Territory acquisition primarily involves threats, gestures, and postures rather than physical combat. Once secured, many species, particularly carnivores, mark territory boundaries with scent trails, accomplished through urination or rubbing against scent posts along the perimeter.

17.5 CLASSIFICATION OF TERRITORIES

Territories in relation to birds have been classified by Nice (1941) into several basic types:

A. Entire mating, feeding, and breeding area defended, e.g. song sparrow
B. Mating and nesting but not feeding area defended, e.g. scarlet finch
C. Mating area only defended, e.g. Gould's manakin
D. Nest only defended, e.g. night heron
E. Winter territories, e.g. black-bellied plover
F. Roosting territories, e.g. starling, tree creeper

Functions of territorial behavior:
Territorial behavior serves multiple functions crucial for an animal's survival and reproductive success:

a. **Resource acquisition**: One of the primary functions of territoriality is to secure access to essential resources such as food, shelter, and mates. By establishing and defending territories containing these resources, individuals ensure their own survival and reproductive success.
b. **Mate attraction and defense**: Territories play a vital role in mate attraction and breeding success. In many species, individuals use territories to display their quality or fitness to potential mates through elaborate courtship displays or territorial defense. Additionally, defending a territory can prevent rivals from accessing mates or competing for breeding opportunities.
c. **Protection of offspring**: Territorial behavior extends to the protection of offspring and nesting sites. By defending a territory, parents can create a safe environment for raising their young, shielding them from predators and other threats.
d. **Status and social structure**: Territories often reflect an individual's status within a social group or hierarchy. Dominant individuals may control larger or more resource-rich territories, while subordinate individuals occupy smaller or less desirable areas. Territorial disputes and interactions contribute to the establishment and maintenance of social structures within animal populations.

Mechanisms of territorial defense: Animals employ various mechanisms to establish and defend their territories:

1. **Physical defense**: Many species use direct physical aggression to defend their territories against intruders. This can involve chasing, fighting, or displaying threatening behaviors to deter competitors. In some cases, physical combat may escalate, leading to injuries or even death.
2. **Vocalizations and displays**: Vocalizations, such as territorial calls or songs, serve as auditory signals to deter intruders and assert ownership over a territory. Visual displays, including posturing, plumage displays, or territorial markings, are also common strategies to advertise ownership and intimidate rivals.
3. **Chemical signaling**: Chemical signals, such as pheromones or scent markings, play a crucial role in territorial communication among many species. By depositing scent marks or pheromones around their territory, individuals can convey information about their presence, identity, and reproductive status to potential intruders.
4. **Territorial boundaries**: Territorial boundaries may be marked and maintained through physical landmarks, such as scent posts, visual cues, or physical barriers like fences or burrows. Clear delineation of boundaries reduces the likelihood of territorial disputes and helps individuals recognize and respect neighboring territories.

Evolutionary significance of territoriality:
The evolution of territorial behavior is shaped by ecological, social, and evolutionary pressures:

i. **Resource partitioning**: Territoriality allows individuals to partition limited resources effectively, reducing competition and maximizing resource utilization within a population. This can promote species coexistence and niche differentiation, ultimately enhancing population fitness and survival.
ii. **Reproductive success**: Territorial behavior is strongly linked to reproductive success, as individuals with access to high-quality territories often have greater mating opportunities and produce more offspring. Natural selection favors traits and behaviors that increase an individual's ability to acquire and defend territories, leading to the widespread prevalence of territoriality across diverse taxa.
iii. **Population regulation**: Territorial behavior plays a role in population regulation by influencing population density, dispersal patterns, and interpopulation dynamics. Territories act as spatial units within which individuals compete for resources and mates, contributing to the maintenance of stable population sizes and genetic diversity.

Defending a territory:
Several animals employ different strategies to defend their territory, with fighting being one option. However, engaging in combat often proves to be suboptimal due to the significant energy expenditure involved and the potential for injury or mortality. Instead, many animals resort to various intimidation tactics, utilizing vocalizations, odors, or visual displays. For instance, bird songs, woodpeckers' drumming, and the loud calls of monkeys serve as long-range warnings, signaling to potential intruders the proximity of another's territory. Additionally, many animals use scent marking by spraying urine, leaving droppings, or rubbing scent glands along their territory boundaries. This method effectively deters approaching animals without the need for direct confrontation with the territory's defender.

Territoriality in domestic animals:
The primal predecessors of dogs and cats utilized territorial markings to assert their presence within an area, a behavior that has persisted in domesticated dogs and cats. Canine territorial markings encompass both olfactory and visual elements. Male dogs, particularly as they mature, exhibit

this behavior prominently by lifting one hind leg and spraying urine on objects, with males tending to urinate more frequently in less familiar areas. Female dogs engage in similar behavior, albeit less frequently, directing their urine at objects and potentially lifting their legs. Certain objects like tree bases, lampposts, or fire hydrants become prominent olfactory markers, attracting extensive sniffing and marking. In regions with winter snowfall, these urination sites are recognizable as "yellow snow." Additionally, ground scratching can leave a lasting visual indication on the ground.

Cats also employ a territorial marking system, with both males and occasionally females spraying urine on vertical surfaces. However, cats differ crucially from dogs in that they concentrate their urine spraying at the core of their territory. For domestic cats, this entails spraying surfaces such as walls, doorways, and furniture within their household confines.

Cattle display territorial aggression through acts such as butting or threatening to butt, particularly when another individual's head is nearby, often using sharp, oblique head swings. In compact environments like small paddocks, older bulls demonstrate strong territorial behavior, characterized by digging the soil with their forefeet and scooping loose soil over their backs as a territorial display. In densely crowded conditions within pens, beef cattle exhibit clear preferences for specific locations within their enclosures. They tend to congregate around the outer edges and corners of pens, while central areas are typically less occupied.

Horses exhibit territorial aggression by utilizing their typical offensive and defensive tactics, including biting, kicking with hind feet, and striking with forefeet when their space is encroached upon. They may employ a double kick with both hind feet aimed directly backward, especially after backing up toward the intruder. Territorial rituals are observed in equine behavior, with stallions often depositing feces in specific sites, leading to the accumulation of dung heaps. All horses designate certain spots for defecation within their territory, which are left ungrazed. Over time, the grazing territory of horses becomes divided into closely cropped areas known as "lawns" and dung areas called "roughs," which are typically left untouched except during scarcity of food resources.

Sheep use head movements to intimidate one another. In the absence of a submissive reaction, they might tug, push, or butt at wool.

Pigs use their heads to threaten others or to defend their territory. If they don't get their way, they may bark, push with their noses, or barge their shoulders. More than other farm animals, pigs exhibit intensive contact behavior and very little territorialism—unless they live in a feral or semi-feral environment.

Seasonal variations in territorial behavior are observed in wild domestic poultry. During the breeding season, dominant males hold clearly defined territories where they mate with females. These tend a brood close to their home range. Every male has a harem for the remainder of the year, and the harems use overlapping home ranges. Under intense conditions, hens also exhibit home range behavior. More cock-crowing happens during territory establishment.

In conclusion, territoriality is a multifaceted phenomenon deeply ingrained in the behavioral repertoire of animals. From its role in resource acquisition and mate attraction to its significance in social structure and evolutionary dynamics, territorial behavior shapes the ecology and evolution of animal populations across the globe. Understanding the mechanisms and functions of territoriality provides valuable insights into the complex interactions driving animal behavior and population dynamics in natural environments.

REFERENCES

Altman, I. (1970). Territorial behavior in humans: An analysis of the concept. In Pastalan, L. A., and Carson, D. A. (Eds.), *Spatial Behavior of Older People*, 169–182. University of Michigan Press, Ann Arbor, MI.

Altman I. (1975). *The Environment and Social Behaviour: Privacy, Personal Space, Territory, and Crowding*, p. 237. Brooks/Cole Publishing Company, Monterey, CA

Ardrey, R. (1966). The territorial imperative: A personal inquiry into the animal origins of property and nations. Robert Ardrey's Nature of Man series, Atheneum, New York, NY

Barnard, C. J., & Burk, T. (1979). Dominance hierarchies and the evolution of "individual recognition." *Journal of Theoretical Biology, 81*(1), 65–73.

Beilharz, R. G., & Cox, D. F. (1967). Social dominance in swine. *Animal Behaviour, 15*(1), 117–122.

Beilharz, R. G., & Zeeb, K. (1982). Social dominance in dairy cattle. *Applied Animal Ethology, 8*(1–2), 79–97.

Both, C., & Visser, M. E. (2003). Density dependence, territoriality, and divisibility of resources: From optimality models to population processes. *The American Naturalist, 161*, 326–336. https://doi.org/10.1086/346098

Brower, S. (1965). Territoriality - The exterior spaces: The signs welearn to read. *Landscape, 15*(1), 9–12.

Brown, G. (2005). *Territoriality in organizations: theory and measurement* (Doctoral dissertation, University of British Columbia).

Brown, J. L., & Orians, G. H. (1970). Spacing patterns in mobile animals. *Annual Review of Ecology, Evolution, and Systematics, 1*, 239–262.

Dawkins, R., & Krebs, J. R. (1978). Animal signals: information or manipulation? In Krebs, J. R. and Davies, N. B. (Eds.), *Behavioural Ecology: An Evolutionary Approach*, 1st edition, 282–309. Blackwell Scientific Publications, Oxford.

Drews, C. (1993). The concept and definition of dominance in animal behaviour. *Behaviour, 125*(3–4), 283–313.

Lindeman, A. A., Grant, J. W., & Desjardins, C. M. (2015). Density-dependent territory size and individual growth rate in juvenile Atlantic salmon (Salmo salar). *Ecology of Freshwater Fish, 24*, 15–22.

Mayer, M., Frank, S. C., Zedrosser, A., & Rosell, F. (2020). Causes and consequences of inverse density-dependent territorial behaviour and aggression in a monogamous mammal. *Journal of Animal Ecology, 89*(2), 577–588.

Mougeot, F., Irvine, J. R., Seivwright, L., Redpath, S. M., & Piertney, S. (2004). Testosterone, immunocompetence, and honest sexual signaling in male red grouse. *Behavioral Ecology, 15*(6), 930–937.

Nice, M. M. (1941). The role of territory in bird life. *American Midland Naturalist, 26*, 441–487.

Noble, G. K. (1939). The role of dominance in the social life of birds. *The Auk, 56*, 263–273.

Peng, H. (2013). Why and when do people hide knowledge? *Journal of Knowledge Management, 17*(3), 398–415.

Ramseyer, A., Boissy, A., Thierry, B., & Dumont, B. (2009). Individual and social determinants of spontaneous group movements in cattle and sheep. *Animal, 3*(9), 1319–1326.

Taylor, R. J. (1988). Territory size and location in animals with refuges: influence of predation risk. *Evolutionary Ecology*, 2, 95–101.

Tibbetts, E. A., Pardo-Sanchez, J., & Weise, C. (2022). The establishment and maintenance of dominance hierarchies. *Philosophical Transactions of the Royal Society B, 377*(1845), 20200450.

18 Food and Habitat Selection

Naveen Swaroop Murikipudi, Vasantha S. K. I., Varij Nayan and Aswani Kumar K.

18.1 INTRODUCTION

Habitat and food selection are crucial aspects of an animal's survival and overall ecological niche. Animals have evolved over time to adapt to specific habitats and find suitable food sources that meet their energy and nutritional needs. Let's explore both of these concepts in more detail.

Habitat selection refers to the process by which animals choose specific areas to live, reproduce, and carry out various activities. Animals select habitats that provide the necessary resources for their survival, reproduction, and protection from predators. These resources may include food, water, shelter, and suitable nesting or breeding sites.

18.2 FACTORS INFLUENCING HABITAT SELECTION

1. Food availability: Animals often choose habitats where their primary food sources are abundant and accessible.
2. Shelter and cover: Adequate shelter from harsh weather conditions and predators is essential for survival.
3. Breeding and nesting sites: Animals select areas that provide suitable conditions for reproduction and raising their young.
4. Temperature and climate: Animals may prefer habitats with temperature ranges that match their physiological tolerances.
5. Predator avoidance: Animals may select habitats that minimize the risk of encountering predators.
6. Competitors: Avoiding areas with intense competition for resources may play a role in habitat selection.
7. Human disturbance: Some animals may avoid areas with high human activity due to disturbance.

For example, desert animals like camels are adapted to arid environments with limited water availability, while aquatic animals like fish are adapted to live in water bodies where oxygen is available. Birds often choose habitats with suitable nesting sites and food sources for raising their young (Figure 18.1).

18.3 FOOD SELECTION

Food selection is the process through which animals choose their diet from available food sources. Animals have evolved various adaptations to acquire the necessary nutrients for growth, energy, and reproduction. The type of food an animal consumes is often closely linked to its physical traits, such as the shape of its beak, teeth, or digestive system.

DOI: 10.1201/9781003710271-18

FIGURE 18.1 Effect of climate on habitat selection.

18.3.1 Factors Influencing Food Selection

1. Nutritional needs: Animals choose foods that provide the necessary nutrients, including proteins, carbohydrates, fats, vitamins, and minerals.
2. Digestive adaptations: Animals have evolved specific digestive systems to process different types of food. For example, herbivores have specialized digestive systems for breaking down plant matter.
3. Availability: Animals select foods that are readily available in their habitats.
4. Palatability: Taste and sensory preferences influence food selection. Some animals have a preference for certain flavors.
5. Predator-prey relationships: Predators often select prey based on factors like size, behavior, and vulnerability.

Dietary selection factors have been extensively researched and are the subject of recent reviews (Milne et al., 1991). Simple observation shows that free-roaming herbivores choose a diet that is higher in nutrients and lower in secondary plant compounds than the average available plant material (Godde et al., 2021). Diet selection is the process by which herbivores balance the twin goals of nutrient maximization and toxin minimization. Research with pen-fed agricultural animals has advanced our understanding of the physiological basis of diet selection decisions, particularly in the last decade. Some of this research will be discussed here, as well as the extent to which the findings are applicable to free-roaming herbivores.

18.3.2 Grazing Behavior of Herbivores

Grazing animals influence sward structure and composition, for example, through selective feeding and excreta return patterns. They can also have an impact on the nutritive value and productivity of the herbage. The effect is likely to differ between animal species. Sheep are more selective than cattle and require higher nutritive value herbage. Sheep, as a result, prefer plants in the vegetative stage of development; when presented with plants in the reproductive stage, they prefer herbage with low acid detergent fiber (ADF) and high N concentration. Cattle, on the other hand, have larger mouths and are less selective. They typically consume a higher proportion of mature or even dead material. Diet digestibility is usually higher in sheep than in cattle. As a result of these differences in selective behavior, a more rapid decrease in the nutritive value of the herbage in sheep-grazed paddocks than in cattle-grazed paddocks appears likely over the course of the vegetation season. Large local N concentrations in cattle dung and urine patches have been found to cause large N losses and to reduce the rate of N fixation (Seither et al., 2012).

18.3.3 Forage Quality

Chemical and physical properties of forage play an important role in determining the rate at which it can be consumed and its palatability. Goats prefer plants with a narrow leaf surface and a small culm diameter, while plants with a high latex content and the presence of down and thorns are less consumed. Furthermore, taste, odor, and feel all play a role in determining preferences. It has also been suggested that goats prefer plants with high non-structural carbohydrate contents (e.g. grasses), whereas plants with high levels of secondary cytoplasmic metabolites such as tannins, alkaloids, terpenes, cyanogenic glycosides, and so on (e.g. legumes) are disliked. Furthermore, juvenile animals may be more sensitive to secondary metabolites, which could explain the higher grass/legume ratio of these animals' intake rates during single-food presentations. It has been reported that goats prefer more digestible material.

18.4 SPATIAL AND SOCIAL RELATIONSHIP OF ANIMALS

The spatial–social interface can be defined as the intersection of social and spatial aspects of individuals' phenotypes and environments. Animals move around their preferred habitat and are said to be constantly exhibiting spatial and social behaviors to associate with preferred social partners. Their spatial and social behavior apprises of their movement, which generates the route taken by the animal across the landscape and the conspecific/congenerics with whom the animal was associated (Webber et al., 2023). The evolution of social systems, population dynamics, and pathogen transmission are certain repercussions of behavioral variation at the spatial–social interface.

18.5 DEFINITION OF SPATIAL–SOCIAL INTERFACE

The relationship between social and spatial phenotypes and environments operates as four interconnected components. Individual social phenotypes and the social environment are linked to individual spatial phenotypes and the underlying spatial environment. The phenotype of an individual can be defined as a set of measurable characteristics that contemplate underlying interactions between an individual's genotype and its environment. Social and spatial phenotypes are quantifiable traits that include mate preferences, social network centrality, home range size, and dispersal distance. An environment is a set of spatial/social components that an animal comes across and is characterized by risks, resources, and conditions (Matthiopoulos, Fieberg & Aarts, 2020).

18.6 SPATIAL ENVIRONMENT

The spatial environment of an individual has two components: geographic and environmental space (Van Moorter et al., 2016). Geographic space is about the position, size, shape, and location of animals' home ranges and the physical characteristics of their ranges in two- or three-dimensional space (Moorcroft, 2012). Environmental space comprises abiotic and biotic factors like environmental conditions (temperature or humidity), risks (predators), resources (mates or food), and their spatial distributions (Matthiopoulos et al., 2020). An individual's fitness partly depends on its spatial environment, which is dependent on the animal's ability to move. The fitness of migrants (long-distance) depends on climatic conditions, while for small rodents, it depends on responses to local weather conditions. The physiological and energetic requirements of an animal are met by its surrounding biotic and abiotic environment. Thus, environmental space limits geographic ranges to locations within the species and an individual's ecological niche. The geographic and environmental spaces dominate availability and the suite of resources, risks, and conditions that the animal must face to survive, grow, and reproduce (Webber et al., 2023).

18.7 SOCIAL ENVIRONMENT

The social environment is defined as the size and composition of a group and the type of interactions among individuals within a group/population. The social environment can differ among individuals within the same group, for example when different individuals have different interactions with one another. It can also be comprised of different types of relationships, such as both conspecifics and heterospecifics (Goodale et al., 2020).

18.8 SPATIAL PHENOTYPE

The spatial phenotype of an individual includes aspects of space use and is defined as the movement within the spatial environment. Spatial phenotypes are contextual and evolve from the geographic space an individual occupies, which correlates to environmental space. Spatial phenotypes could be either pattern/process-based. Pattern-based ones also emerge from geographic space and include those that directly relate to movement, e.g. daily movement distances (Michelangeli et al., 2022), dispersal (Cote et al., 2017), or migration (Shaw, 2020). Process-based spatial phenotypes include habitat selection (Leclerc et al., 2016) and environmental niche specialization (Carlson et al., 2021).

18.9 SOCIAL PHENOTYPE

The social phenotype of an individual is the social position of an individual in a society that emerges from its social interactions with other individuals. It is a reflection of its position in society (Farine, Montiglio, & Spiegel, 2015). A few of quantifiable social phenotypes are mating strategies (Fisher, Rodríguez-Muñoz & Tregenza, 2016), aggression level (Kilgour et al., 2018), and cooperative behavior (Dakin & Ryder, 2018), all of which are influenced by the social environment. Social phenotypes can also be pattern/process-based (Cantor et al., 2021).

Pattern-based social phenotypes refer to who individuals associate with, like mate choice (Wang et al., 2022) or differences in the social rules that individuals use when moving in a group (del Mar Delgado et al., 2018) and the social properties that arise from variation in social behavior, like the social position of individuals in their group (Farine, 2017). Process-based social phenotypes include decisions that emerge in retaliation to changes in the social environment, like group dispersion in response to competition (Fretwell & Lucas, 1969) and niche specialization (Montiglio, Ferrari, & Reale, 2013).

18.10 HABITAT

The term habitat was conceptually and technically defined according to Morris (2003) as discrete spatial units where the species of interest is known to/could occur. In contrast, Matthiopoulos et al. (2020) defined habitat as a point in environmental space, defined by a set of conditions, resources, and risks for the species of interest. Conceptually, habitat for a species represents all the environmental factors that could influence the fitness of the animal at a given point (all conditions, resources, and risks). Analyst-defined habitat is the way in which the species experiences its environment.

Technically, in an HSA, habitat is defined by user-chosen environmental covariates measured at a point, or commonly within analyst-defined pixels (making the HSA an analysis in discrete space) in a geographic information system (GIS) (Northrup et al., 2022).

18.11 HABITAT SELECTION

Habitat selection is a fundamental animal behavior. It shapes a wide range of ecological processes, which include animal movement, nutrient transfer, trophic dynamics, and population distribution. The study of habitat selection is important to understand how individual animals interact with the environment to produce population-level patterns of distribution and abundance (Matthiopoulos et al., 2020). Habitat selection also drives an array of important ecological and evolutionary processes, including trophic structuring (Stephen et al., 2024), spatial patterns of relatedness and dispersal (Shafer et al., 2013), and the formation of ecological traps (Thorsen et al., 2022). The relationship between species and their habitats is fundamental for a range of problems in applied ecology, which includes assessing and predicting the effects of climate and land-use change (Bradbury et al., 2014), modeling evolution, and conservation and management of species.

18.12 PHYSIOLOGICAL CONSEQUENCES OF HABITAT SELECTION

Early ecology was about why organisms are specific to a particular habitat, as well as referring to limiting physical factors like temperature, pH, and salinity, which influence the same. Early ecology was mostly focused on the physiology of the organism and the physical environment. The micro-climatic conditions an animal experiences and its habitat influence its physiological capacities and obviously its demographic and ecological performance. As a result, the ecology of organisms can be significantly affected by the physiological consequences of habitat selection. When the animal is out of its comfort zone (TNZ), there are profound effects on its behavior and ecology. Extreme temperatures can cause stress and sometimes lead to death. The food intake of animals is inversely related to temperature, and the energy available for growth, behavior, and reproduction is maximum at its TNZ (Figure 18.2).

18.13 FUNCTIONAL RESPONSES IN HABITAT SELECTION

Habitat selection, the process by which animals choose resources, provides an overview of the mechanisms through linking patterns of animal behavior and the underlying resource availability (Johnson, 1980). The selection process is assumed to result from innate or learned preferences and is commonly studied by monitoring the spatial behavior of animals (Aarts et al., 2012). Habitat selection is often quantified by statistically comparing environmental attributes at a sample of animal locations (habitat use) to another sample of environmental attributes at surrounding locations (Lele et al., 2013) via a resource selection function (Boyce and McDonald, 1999). However, there are certain challenges in this method, as the animals exhibit context-dependent habitat selection.

Habitat varies across a sample of individuals, as each individual experiences different environments and hence behaves differently. Characterizing a functional response in habitat use was an important contribution and was derived from predation studies indicating that predator consumption rates change according to variation in prey abundance (Lele et al., 2013). It evaluates the assumption of constant habitat selection within use, availability designs, and additional behavioral mechanisms that contribute to the distributions of animals.

FIGURE 18.2 Effect of food intake on reproduction.

18.14 HABITAT-SELECTION MODELING: CONCEPTUAL AND METHODOLOGICAL ADVANCES

Ecological studies, although focused on habitat selection from the beginning, have developed with conceptual and methodological advances over the last 20 years. Recent technological advancements in animal tracking (Kays et al., 2015), coupled with diverse and flexible statistical techniques (Hooten et al., 2017), and ever-increasing computational power, offer the opportunity for unprecedented insight and inference into the habitat-selection patterns of animals (Northrup et al., 2022). These advances have led to a considerable increase in the number of studies developing and refining habitat-selection theory and statistical models for quantifying habitat selection. Chief among the methods used to make inferences about habitat-selection patterns is a suite of models referred to as selection functions (e.g., resource-selection functions, step-selection functions) (Fortin et al., 2005). But with these methods, it was difficult to draw general conclusions about ecological (Rather et al., 2020) and evolutionary processes (Leclerc et al., 2016) related to habitat selection. The most applied modeling framework for understanding the habitat-selection process is habitat-selection analyses (HSAs) (Northrup et al., 2022).

18.15 HABITAT SELECTION ANALYSIS

Conceptually, we define a habitat selection analysis as any analysis that attempts to capture the following, highly simplified process: as an animal encounters an available habitat, it selects habitat and uses it in some way. Any HSA consists of four components that we discuss in the sections that follow:

(1) A precise, technical definition of habitat,
(2) A probability density function quantifying the distribution of available habitat,

(3) A probability density function quantifying the distribution of used habitat, or, more commonly, a sample from that distribution, and
(4) A function, referred to as a selection function, quantifying selection of habitat relative to availability.

18.15.1 Selection Functions

The term "selection function" is used in habitat-selection studies and is defined as a function that describes how one probability distribution is changed into another probability distribution (McDonald et al., 1995). Manly (1985) and later McDonald et al. (1995) describe the use of selection functions in modeling natural selection, whereby a population with some distribution of phenotypes changes to a population with a different distribution of phenotypes based on the selection function. The selection function can be defined specifically to HSAs as a weighting function describing the relative probability of selecting a location or unit, based on environmental covariates. However, the general idea developed in the foundational literature (McDonald et al., 1995) holds: the selection function translates a probability distribution describing the habitat that is available to an animal into a probability distribution describing the habitat used by the animal.

18.16 IMPACT OF CLIMATE CHANGE ON ANIMAL HABITATS

Climate change represents change in climate over a few decades/millions of years which may be due to natural causes and/or as a result of human activities. Entire ecosystems are altered along with all the living organisms. Climate change threatens the global environment, biodiversity, and sustainable human development, primarily by changing global thermal regimes and the water cycle, causing acidification (Huang et al., 2021). Climate change is a result of an increase in greenhouse gas emissions in the atmosphere, which is a consequence of various anthropogenic activities. The sustainability of animals, including humans, is affected by global warming, as climate change has an adverse impact on water resources and the agricultural economy. Environmental factors like rising temperatures and sea levels have impacted terrestrial and marine ecosystems massively, leading to extinction or endangered species. Factors like deforestation, large-scale developments, construction of dams and other big developmental projects, and the clearing of forest land for agriculture and other non-agricultural uses have profoundly impacted biodiversity. Climate change has direct impacts on biodiversity by shaping species' habitat range (Colwell et al., 2008), survival (Chamaille-Jammes et al., 2006), resource availability (Visser and Both, 2005), and interactions among species (Moritz and Agudo, 2013). Globally, anthropogenic climate change has already led to warmer temperatures, changing precipitation, and more frequent extreme weather events (Bindoff et al., 2007; IPCC, 2014).

The protection of biodiversity is of paramount importance today. It is therefore crucial to understand the impacts of climate change on biodiversity, and accordingly mitigation strategies should be undertaken to decrease the impact of climate change on biodiversity. Planning and management of biological resources to secure their wide use and continuous supply, and maintaining their quality, value, and diversity is the need of the hour for the conservation of biodiversity (Prakash, 2021). Preparing for impacts of climate change must be based on insight from real world actions, taking into consideration the complex interactions between species and the physical environment (Ockendon et al., 2014).

The impact of climate change on animal habitats cannot be ignored. As temperatures rise and weather patterns shift, animal species may face significant challenges in finding suitable habitats to meet their needs. Climate change can alter the availability and distribution of resources, such as food and water, as well as disrupt critical ecological processes. For example, changes in temperature and precipitation patterns can affect vegetation growth and phenology, which in turn can impact the availability of food sources for herbivores and the animals that depend on them. Furthermore, climate change can also lead to alterations in habitat quality. These changes can include the loss or degradation of important habitat features, such as nesting sites or breeding grounds, making it more difficult for animals to successfully reproduce or find suitable mates.

The consequences of these habitat disruptions can be far-reaching, impacting not only individual animals but also entire populations and ecosystems. Therefore, it is crucial for ecologists and researchers to gain a better understanding of the underlying mechanisms by which climate change affects animal populations. This will allow for the development of effective strategies to mitigate negative impacts and promote the resilience and adaptation of animal species.

REFERENCES

Aarts, G., Fieberg, J., & Matthiopoulos, J. 2012. Comparative interpretation of count, presence–absence and point methods for species distribution models. *Methods in Ecology and Evolution, 3*, 177–187.

Boyce, M. S., & McDonald, L. L. 1999. Relating populations to habitats using resource selection functions. *Trends in Ecology & Evolution, 14*, 268–272.

Bradbury, R. B., Bradley, C., Butchart, S. H. M., Doswald, N., Foden, W., Gill, D. J. C., Green, R. E., Sutherland, W. J., Tanner, E. V. J., & Pearce-Higgins, J. W. (2014). Mechanisms underpinning climatic impacts on natural populations: altered species interactions are more important than direct effects. *Global Change Biology, 20*, 2221–2229.

Cantor, M., Maldonado-Chaparro, A. A., Beck, K. B., Brandl, H. B., Carter, G. G., He, P., Hillemann, F., Klarevas-Irby, J. A., Ogino, M., Papageorgiou, D., Prox, L., & Farine, D. R. (2021). The importance of individual-to-society feedbacks in animal ecology and evolution. *Journal of Animal Ecology, 90*, 27–44..

Carlson, B. S., Rotics, S., Nathan, R., Wikelski, M., & Getz, W. (2021). Individual environmental niches in mobile organisms. *Nature Communications, 12*, 4572.

Chamaille-Jammes, S., Massot, M., Aragon, P., & Clobert, J. (2006). Global warming and positive fitness response in mountain populations of common lizards Lacerta vivipara. *Global Change Biology,* 12, 392–402.

Colwell, R. K., Brehm, G., Cardelús, C. L., Gilman, A. C., & Longino, J. T. (2008). Global warming, elevational range shifts, and lowland biotic attrition in the wet tropics. *Science, 322*, 258–261.

Cote, J., Bestion, E., Jacob, S., Travis, J., Legrand, D., & Baguette, M. (2017). Evolution of dispersal strategies and dispersal syndromes in fragmented landscapes. *Ecography, 40*, 56–73.

Dakin, R., & Ryder, T. B. (2018). Dynamic network partnerships and social contagion drive cooperation. *Proceedings of the Royal Society B: Biological Sciences, 285*, 20181973.

del Mar Delgado, M., Miranda, M., Alvarez, S. J., Gurarie, E., Fagan, W. F., Penteriani, V., di Virgilio, A., & Morales, J. M. (2018). The importance of individual variation in the dynamics of animal collective movements. *Philosophical Transactions of the Royal Society B: Biological Sciences, 373*, 20170008.

Farine, D. R. (2017). A guide to null models for animal social network analysis. *Methods in Ecology and Evolution, 8*, 1309–1320.

Farine, D. R., Montiglio, P.-O., & Spiegel, O. (2015). From individuals to groups and back: the evolutionary implications of group phenotypic composition. *Trends in Ecology & Evolution, 30*, 609–621.

Fisher, D. N., Rodrı´guez-Mun˜oz, R., & Tregenza, T. (2016). Comparing preand post-copulatory mate competition using social network analysis in wild crickets. *Behavioral Ecology, 27*, 912–919

Fortin, D., Beyer, H. L., Boyce, M. S., Smith, D. W., Duchesne, T., & Mao, J. S. (2005). Wolves influence elk movements: behavior shapes a trophic cascade in Yellowstone National Park. *Ecology, 86*, 1320–1330.

Fretwell, S. D., & Lucas, H. L. J. (1969). On territorial behaviour and other factors influencing habitat distribution in birds. *Acta Biotheoretica, 19*, 16–36.

Godde, C. M., Mason-D'Croz, D., Mayberry, D. E., Thornton, P. K., & Herrero, M. (2021). Impacts of climate change on the livestock food supply chain; a review of the evidence. *Global Food Security, 28*, 100488.

Goodale, E., Sridhar, H., Sieving, K. E., Bangal, P., Colorado, Z. G. J., Farine, D. R., Heymann, E. W., Jones, H. H., Krams, I., Martı´nez, A. E., Montan˜o-Centellas, F., Mun˜oz, J., Srinivasan, U., Theo, A., & Shanker, K. (2020). Mixed company: a framework for understanding the composition and organization of mixed-species animal groups. *Biological Reviews, 95*, 889–910.

Hooten, M. B., Johnson, D. S., McClintock, B. T., & Morales, J. M. (2017). *Animal movement: statistical models for telemetry data.* CRC Press, Boca Raton, FL.

Huang M., Ding, L., Wang, J., Ding, C., & Tao, J. (2021). The impacts of climate change on growth: A summary of conducted studies and current knowledge. *Ecological Indicators, 121.* doi:10.1016/j.ecolind.2020.106976.

IPCC. (2014). Climate change 2014: synthesis report. In: Pachauri, R., & Meyer, L. (Eds.), *Contribution of Working Groups I, II and III to the Fifth Assessment Report of the Intergovernmental Panel on Climate Change.* Geneva.

Johnson, D. H. (1980). The comparison of usage and availability measurements for evaluating resource preference. *Ecology, 61*, 65–71.

Kays, R., Crofoot, M. C., Jetz, W., and Wikelski, M. (2015). Terrestrial animal tracking as an eye on life and planet. *Science, 348*, aaa2478.

Kilgour, R. J., Mcadam, A. G., Betini, G. S., & Norris, D. R. (2018). Experimental evidence that density mediates negative dependent selection on aggression. *Journal of Animal Ecology, 87,* 1091–1101.

Leclerc, M., Vander Wal, E., Zedrosser, A., Swenson, J. E., Kindberg, J., & Pelletier, F. (2016). Quantifying consistent individual differences in habitat selection. *Oecologia, 180*, 697–705.

Lele, S. R., Merrill, E. H., Keim, J., & Boyce, M. S. (2013). Selection, use, choice and occupancy: clarifying concepts in resource selection studies. *Journal of Animal Ecology, 82*, 1183–1191.

Manly, B. F. (1985). *The statistics of natural selection on animal populations.* Springer Science & Business Media, Berlin.

Matthiopoulos, J., Fieberg, J., & Aarts, G. (2020). *Species-habitat associations: spatial data, predictive models, and ecological insights.* University of Minnesota Libraries Publishing, Saint Paul, MN. https://bookdown.org/jfieberg/ SHABook/ Accessed 1.10.2021.

McDonald, L. L., Gonzalez, L., & Manly, B. F. 1995. Using selection functions to describe changes in environmental variables. *Environmental and Ecological Statistics, 2*, 225–237.

Michelangeli, M., Payne, E., Spiegel, O., Sinn, D. L., Leu, S. T., Gardner, M. G., & Sih, A. (2022). Personality, spatiotemporal ecological variation and resident/explorer movement syndromes in the sleepy lizard. *Journal of Animal Ecology, 91*, 210–223

Milne, A. C., McNeill, G. & Zakary, A. 1991. Weight change as an indicator of energy imbalance during 7 day weighed food intake studies. *Ecology of Food and Nutrition*, 281–289.

Montiglio, P. O., Ferrari, C., & Reale, D. (2013). Social niche specialization under constraints: personality, social interactions and environmental heterogeneity. *Philosophical Transactions of the Royal Society B: Biological Sciences, 368*(1618), 20120343.

Moorcroft, P. R. (2012). Mechanistic approaches to understanding and predicting mammalian space use: recent advances, future directions. *Journal of Mammalogy, 93*, 903–916.

Moritz, C., & Agudo, R. (2013). The future of species under climate change: resilience or decline? *Science, 341*, 504–508.

Morris, D. W. (2003). Toward an ecological synthesis: a case for habitat selection. *Oecologia, 136*, 1–13.

Northrup, J. M., Vander Wal, E., Bonar, M., Fieberg, J., Laforge, M. P., Leclerc, M., ..., & Gerber, B. D. (2022). Conceptual and methodological advances in habitat-selection modeling: guidelines for ecology and evolution. *Ecological Applications, 32*(1), e02470.

Ockendon, N., Baker, D. J., Carr, J. A., White, E. C., Almond, R. E. A., Amano, T., Bertram, E., Bradbury, R. B., Bradley, C., Butchart, S. H. M., Doswald, N., Foden, W., Gill, D. J. C., Green, R. E., Sutherland, W. J., Tanner, E. V. J., & Pearce-Higgins, J. W. (2014). Mechanisms underpinning climatic impacts on natural populations: Altered species interactions are more important than direct effects. *Global Change Biology, 20*(7), 2221–2229.

Rather, T. A., Kumar, S. & Khan, J. A. (2020). Multi-scale habitat selection and impacts of climate change on the distribution of four sympatric meso-carnivores using random forest algorithm. *Ecol Process, 9*, 60.

Shafer, A. B. A., Nielsen, S. E., Northrup, J. M., & Stenhouse, G. B. 2014. Linking genotype, ecotype, and phenotype in an intensively managed large carnivore. *Evolutionary Applications, 7*(2), 301–312.

Shaw, A. K. (2020). Causes and consequences of individual variation in animal movement. *Movement Ecology, 8*, 1–12.

Seither, M., Wrage, N., & Isselstein, J. (2012). Sward composition and grazer species effects on nutritive value and herbage accumulation. *Agronomy Journal, 104*(2), 497. doi:10.2134/agronj2011.0322

Stephens, R. B., Millspaugh, J. J., McRoberts, J. T. et al. (2024). Scale-dependent habitat selection is shaped by landscape context in dispersing white-tailed deer. *Landsc Ecol, 39*, 84.

Thorsen, N. H., Hansen, J. E., Støen, OG. et al. (2022). Movement and habitat selection of a large carnivore in response to human infrastructure differs by life stage. *Mov Ecol, 10*, 52.

Van Moorter, B., Rolandsen, C. M., Basille, M., & Gaillard, J.-M. (2016). Movement is the glue connecting home ranges and habitat selection. *Journal of Animal Ecology, 85*, 21–31.

Visser, M.E., Both, C., 2005. Shifts in phenology due to global climate change: the need for a yardstick. *Proceedings of the Royal Society B: Biological Sciences, 272*, 2561–2569.

Wang, D., Forstmeier, W., Farine, D. R., Maldonado-Chaparro, A. A., Martin, K., Pei, Y., Alarc_on-Nieto, G., Klarevas-Irby, J. A., Ma, S., Aplin, L. M., & Kempenaers, B. (2022). Machine learning reveals cryptic dialects that explain mate choice in a songbird. *Nature Communications, 13*, 1630.

Webber, Q. M., Albery, G. F., Farine, D. R., Pinter-Wollman, N., Sharma, N., Spiegel, O., ..., & Manlove, K. (2023). Behavioural ecology at the spatial–social interface. *Biological Reviews, 98*(3), 868–886.

19 Altered Ecosystems and Animal Behaviour

Hammad Ahmad Khan and Zain ul Abdin

19.1 INTRODUCTION

The global environment has been experiencing drastic changes mainly due to human activities. Of these, ecological pollution, habitat annihilation, invasive species, and changed weather conditions have jeopardised the environment (Ceballos *et al.* 2015; Blowes *et al.* 2019). Present environmental changes impact faunal diversity directly and indirectly and are influenced by abiotic conditions (Gilman *et al.* 2010). Such effects are likely to cause irreversible interactions to ecosystem function and structure, population dynamics, environmental resilience, and possible reduced ecosystem efficiency (Ives and Carpenter, 2007).

The strong impact of ecological changes is well recognised by a variety of animal communities, the reflex action of the environmentally disturbed habitats. Concurrently, their physiological profiles also indicate altered phylogenetic distribution patterns (Wong and Candolin, 2015). Stressed behavioural actions induced in varied communities represent inhibited growth rates and decreased maturity of mechanisms, with cascading effects (Bartley *et al.*, 2019). The cascading effects trigger large impacts on various animal species and accelerated interactions of environmental changes (Ockendon *et al.*, 2014).

It is customary that the mainstream of behavioural actions is dependent on the type of species and also on population abundance for the specific habitat. As assessed, the predicted impacts of ecological factors may prove sufficiently strong and, therefore, refer to keystone, dominant, and foundation species, and ecosystem engineers (Power *et al.*, 1996; Avolio *et al.*, 2019; Ellison, 2019; Wright and Jones, 2006). Communally, they are also called 'biotic multipliers'. Their varying responses depict the environmental fluctuations concurrent with ecosystem health (Urban *et al.*, 2017). This chapter reports on the interactions of animal communities for sustainable ecosystems, where their actions are altered for survival.

19.2 BIOTIC MULTIPLIERS

Biotic multipliers are regarded as the species that regulate the richness and periodicities of different animal species for single or multiple habitats. Predictable alterations of animal behaviour can trigger cascade impacts on the keystone, dominant, and ecosystem engineers. There is, therefore, the probability of overlapping by vibrant species of various groups of animals (Visser *et al.*, 2012).

19.3 KEYSTONE SPECIES

Keystone species are largely considered the most functional species that significantly impact ecosystem biomass (Power *et al.*, 1996). From an ecological aspect, the removal of a keystone species can have dramatic effects. The importance of keystone species is related to their connection to ecosystem predators, where they help to streamline large prey populations with top-down effects, as seen with woodland wolves (*Canis lupus*). The ecological ranking of such species is not dependent

DOI: 10.1201/9781003710271-19

on larger sizes; for example the small starfish (*Pisaster* ochraceus) is considered the first inter-tidal keystone species (Power *et al.*, 1996).

19.4 ECOSYSTEM ENGINEERS

These are largely responsible for making adjustments to specific habitats while reducing resource availability to others, viz. beavers (*Castor* sp.) and earthworms (*Lumbricoides* sp.). They are also regarded as bringing physical changes in both abiotic and biotic components due to certain atypical structural organs and are known as autogenic engineers. They cause the conversion of living and non-living components of ecosystems mostly mechanically and, therefore, are referred to as allogenic engineers.

19.5 DOMINANT SPECIES

The occurrence of a prevalent species in any ecosystem on the animal community structure is fairly significant. Dominant species are well connected to others in the ecosystem and are considered 'contributors' in ecosystem functioning. They are important to the bioturbation and carbon sink, and thereby resist major alterations in their population in any ecosystem, such as aquatic zooplankton in sufficiently large water bodies without predatory fish Avolio *et al.*, 2019; Taylor *et al.*, 2006).

19.6 ENVIRONMENT CHANGE AND ANIMAL BEHAVIOUR

Different living organisms display varied responses to environmental changes. These can be swift or delayed depending on their physiological profiles and learning skills. Current environmental changes significantly affect their plasticity and individual dispositions in social interactions (McCallum *et al.*, 2021). Different responses are required over different time periods in a variety of communities. Concomitantly, the induced behavioural actions also rely heavily on ecological conditions, causing steady alterations in many ecosystems (Jackson *et al.*, 2021).

19.6.1 Harvesting

Partial removal of animal communities from any habitat affects animal behaviour. In ecosystems with specific characteristics, the removal of some animals often leads to the dominance of bolder individuals (Palkovacs *et al.*, 2018; Sbragaglia *et al.*, 2021a). For example, freshwater fish that are removed from the aquatic environment can trigger behavioural and ecological responses in the remaining fish populations (Sbragaglia *et al.*, 2021b). This can cause social interactions to change, indicated by aggression, altruism, and mate selection, which occurs due to population density, community structure, and animal species distribution. Concurrently, the demise of other species and changes in their population structure affect intra-specific and inter-specific interactions (Hoover and Tylianakis, 2012).

19.6.2 Habitat Change

Alterations in habitat characteristics viz. environmental modifications, and fragmentation and obliteration of the specific habitat, bring about behavioural changes. It is thought that the removal of important croplands in any ecological habitat influences predators and consumers to change their predation methods for their prospective feeding (Michael and Adams, 2009). Thus, the behavioural response of animals can impact dispersal patterns as well as natality and mortality rates (Michael and Adams, 2009).

19.6.3 Ecological Pollution

Environmental degradation over the last few decades has caused serious impairments of flora and fauna worldwide. Environmental pollution affects habitat selection, physiological and hormonal changes that alter consumer behaviour, and reduced investment in breeding behaviour (Wingfield, 2013; Candolin and Wong, 2019). Moreover, there are indirect effects of environmental contamination that have been reported in species interactions (Willems *et al.*, 2022), specifically in the trophic levels of various pollutants and of heavy metals, pharmaceutical waste, and micro-plastics, which can accumulate in the body tissues of animals (Arnot and Gobas, 2006; Previsic *et al.*, 2021; Gall *et al.*, 2015).

19.6.4 Non-Native Species

Such species are pivotal in the successful promotion of other species in new habitats (Pysek, 2020). Such species impact the behaviour of other species in various ways, viz. predators, prey, competitors, parasites, and mutualists (Simberloff *et al.* 2013). Introduced species can outcompete native species for resources, prey, and habitat choice, often with aggression. Altered behaviour has been recorded for native species. Concomitantly, the indirect impacts of invaders are likely to influence foraging and constitute large food webs and species interactions (Ricciardi *et al.* 2013; Candolin *et al.* 2018).

19.6.5 Climate Change

Environmental change regarding increased temperature, precipitation, relative humidity, and variable wind patterns causes significant modification of global ecology. Direct and ancillary effects are experienced in various habitats (Buchholz *et al.*, 2019; Rosenblatt and Schmitz, 2016) to deter the physiological mechanisms of the ecosystem. Altered environmental temperature and water access significantly affect the nutrient recycling of vegetation to directly modify foraging patterns of herbivores (Pelletier *et al.*, 2012). Metabolic fluctuations among ectotherms altering their behaviour have proven to be critical. Higher temperatures also expand the feeding and foraging periods of predators (Rosenblatt and Schmitz, 2016).

Weather changes also affect the life history patterns of several animal communities. They are likely to affect breeding behaviour and non-consecutive and consecutive patterns to effect population dynamics and species interactions (Buchanan and Partecke, 2012; Kharouba *et al.*, 2018).

19.6.6 Evolutionary Behaviour

Several animal communities manifest responses regarding changes in the environment. Sometimes such evolutionary changes can be considered lethal, and the animals suffer due to a lack of adaptations (Candolin and Wong, 2012). Certain genetic-based adaptations appear to rely on different factors, such as the rate of genetic change of species, type of genetic variance as depicted in animal behaviour, consecutive and non-consecutive dispersal patterns, and the frequency of mutations to signify gene flow (Bell, 2017). Adaptive behaviour influenced by certain animal communities triggers genetic adaptations seen through shielding effects to inhibit population decline and to provide additional time for genetic changes.

The focal species considered to organise and achieve better adaptations should have large population densities. Therefore, their ability to show better adaptations in human-based environments and changing conditions is unpredictable. Nonetheless, such species with short generation times possess substantial genetic diversity and, thus, are more likely to become adaptive to ecological changes (Hoffmann and Sgro, 2011). Animal species such as large predators have longer generation times and alter their behavioural response prior to the inception of the environmental change in their respective ecosystem.

19.6.7 Species Interaction Web

There seems to be increasing denial of the fact that a single species can impact others through their interconnected network. In many animals, behavioural responses change their own distribution patterns in many ecological habitats. Consider the example of enhanced foraging activity of consumers in anthropogenic and eutrophied habitats. The nutritional biomass of vibrant species would perhaps reduce the effects of eutrophication for the others (Candolin, 2019). The effects of animal behaviour feedback of different species are either trait- or population-density mediated (Werner and Peacor, 2003; Schmitz and Trussell, 2016). Logically, trait-mediated effects occur due to the direct impact of a specific species transferred to the others in a given time scale. However, density-mediated effects are well appreciated, as the behaviour is changed with respect to the natality, mortality, and dispersal patterns, impacting other species (Carpenter and Kitchell, 1988; Lima and Dill, 1990). Both trait- and density-mediated effects result in top-down and bottom-up cascade effects that pass through the web of species, with lateral impacts and nutrient recycling in abiotic conditions (Leroux and Loreau, 2010; Heath *et al.*, 2014).

19.6.8 Lateral Effects

Interactions between different trophic structures can be regarded as lateral interactions, viz. commensal, competitive, and mutualistic to assess the trophic structures. In any ecosystem comprising limited resources and more competitors, stress is likely to be induced in the habitat and, therefore, change the trophic interactions (Kéfi *et al.*, 2012; Davenport and Chalcraft, 2013). An example is that of freshwater insect consumers, which regulate their top-down effect for their prey choice of zooplankton (Sentis *et al.*, 2017). Considering the sensitivity of trophic and non-trophic interactions to ecological changes, they impose ecologically unacceptable effects on animal communities. Non-trophic interactions rely on species traits and have the ability to streamline the ecological animal communities with their waning effects and positive population interactions (Veselý *et al.*, 2019; Rall *et al.*, 2008).

19.6.9 Phenology

Man-made environmental changes that directly affect animal behaviour, such as the migration of species and breeding performance, cause disparity in optimised abiotic and biotic factors (Renner and Zohner, 2018). Some birds that breed ahead of their breeding season owing to climate change are likely to suffer from food scarcity, with drastic affects such as reduced breeding with variable time scales (Visser *et al.*, 2012). In any habitat, the main cause of mismatches occurs due to differences either in responding directly or the time intervals for different species owing to the environmental constraints (Both *et al.*, 2009; Chen *et al.*, 2019). Phonological mismatches have also been recorded when environmental fluctuations occur during different times of the year to enable dominance of one trophic level over the other (Straile *et al.*, 2015)

Finally, mismatches in any ecosystem occur due to the abiotic factors that limit the fitness of some species while others may not respond to environmental change. Consider the example of grazing individuals, which optimise their skimming impacts in response to increased temperature; primary producers would fail to respond due to a lack of desirable nutrients and induce overgrazing (Thackeray *et al.*, 2016).

19.6.10 Individual Behaviour Variations

Considering the large animal populations, individual members depict some peculiarities in behaviours. These also depend on environmental change, genetic traits, physiological profiles, sociability, and sexual behaviour. Such variabilities influence the response to resource availability and competitiveness with others (Belgrad and Griffen, 2016; Sommer and Schmitz, 2020; Violle *et al.*, 2012). Customarily, species with more variable responses can increase their ecological niches due to added competitive advantage. Similarly, several changes among individuals at their different growth stages also impact species interactions over large areas (Pelletier *et al.*, 2012).

19.6.11 Restructuring the Species Interaction

The restructuring of the different animal species networks or webs is important in animal behaviour. Recently, increased global warming worldwide has induced more predatory attacks from consumers at different trophic levels (Shurin *et al.*, 2012; Dell *et al.*, 2014). Reformation of species amplifies under swift environmental changes, such that the negative impact of temperature warming for the predominant tadpole (Hyla versicolor) can reduce its survival with its predatory competitors, the dragonfly, to become comparable in feeding proportions on the existing food resources, therefore reducing its survival value (Redolf and Roman, 2018). Similarly, the restructuring networks of animal species perhaps also trigger the incursion of new or non-native species to jeopardise sustainability of existing species (Pysek *et al.*, 2020; Chapple *et al.*, 2012).

19.7 TOP-DOWN EFFECTS

Considering that high-ranked predators of ecosystems are effective regulators of food webs, their exclusion from ecosystems creates cascading effects on the lower tropic levels (Ripple *et al.*, 2014; Rasher *et al.*, 2020). Therefore, the extermination and increased numerical numbers of predominant consumers seem to initiate further extinctions from animal communities. Important examples are those of wolves (*Canis lupus*) preying on moose (*Alces alces*) in various ecosystems of Michigan in the United States. This has enabled the growth of various populations as a consequence of the wolves' decline (Donohue *et al.*, 2017; Wilmers *et al.*, 2006). The main cause of the decline of important consumers for multiple ecosystems occurs due to their lower population diversity and perhaps owing to their functional redundancy (Duffy, 2003). Sensitivity patterns of important predators for environmental fluctuations at lower trophic levels occur based on their large body size, as well as their greater metabolic and resource requirements (Urban *et al.*, 2017). Certainly, the joint interaction of the cascading effects on large consumers with minimum ability to endure ecological fluctuations is apparent (Shurin *et al.*, 2012). An example is that of the sea otter (*Enhydra lutris*), a keystone predator of the sub-arctic marine ecosystem, with a decline due to erosion of calcareous coral reefs and reduced population abundance (Rasher *et al.*, 2020). Nonetheless, some major consumers show resistance to environmental changes and therefore persist, such as zooplanktons, which dominate the abundance of phytoplankton to reduce their populations (Velthuis *et al.*, 2017). Thus, such consumers with their resistance to ecological changes transform the environment to be suitable for their requirements. An important example in this regard is of freshwater sculpin (*Cottus nozawae*), predominantly considered a predator in temperate aquatic water bodies, as it protects the impact of heat stress on the benthic algae through the inhibition of macroinvertebrates (Ross *et al.*, 2022).

19.8 BOTTOM-UP EFFECTS

Such effects are triggered by the presence of nutrients in the environment. They provide substance to community regulations and balance of ecosystems. Moreover, there is lower producer redundancy, thereby providing higher consumers with specific dietary components (Hunter and Price,

1992). Concurrently, any ecological fluctuations that cause changes in the nutritional requirements of plants also alter growth patterns of consumers, affecting their population dynamics (Cross *et al.*, 2015).

19.8.1 Biogeochemical Cycles

Many behavioural alterations, which are responsible for ecosystem mechanisms like energy flow among the different food webs and varied habitats, are mediated through all consumers and also detritivores (Ren *et al.*, 2022). It is important to note that the changed behavioural actions of consumers lead to transformed biomass and respiratory processes (Ren *et al.*, 2022). Further modifications also cause reduced metabolism and impact the availability of oxygen in the environment (Ren *et al.*, 2022: Kristensen, 2000).

Herbivores can experience changes in their consumption rate of the existing vegetation resource within and below the levels of vegetation. Feeding proportions of phytoplankton by zooplanktons in aquatic resources are affected by reduced light penetration and water turbidity. In deeper waters, such activities are likely to cause a loss of oxygen (hypoxia) and change benthic organisms, inhibiting the supply of nutrients and impacting the growth performance of algal blooms.

19.8.2 Biodiversity: Ecosystem Stability

Variability of living organisms in the biosphere ensures sustainable ecosystem biodiversity; nonetheless, their continuous behavioural alterations weaken ecosystem biodiversity indices (Hautier *et al.*, 2015). In recent times, specifically the decline of larger predators from various habitats, followed by their reduced top-down control, this can deteriorate ecological functioning and, therefore, lead to predation moderation and cause ecosystem functioning to be vulnerable (Loreau and de Mazancourt, 2013).

Altered ecosystems due to dominant and co-dominant species and cascades challenge ecosystem sustainability (Gaston, 2010; Dunne *et al.*, 2002). Concurrently, the specialised interactions of important species seem more sensitive to ecological alterations with their declined resilience (Rooney and McCann, 2012; Oliver *et al.*, 2015). Undeniably, such ecosystems depicting specific and weak connections appear to be largely disturbed. The loss of behavioural interactions for different animals, therefore, inhibits the tolerance limits of ecological systems in wake of massive human-based disturbances.

19.8.3 Ecosystem Management

Such management measures are concerned with the provision of food security, suitable regulation of plant and animal pests, occurrence of environmental parasites, pure air to breathe, and sufficient water resources for the population (Pecl *et al.*, 2017). Predominantly, the food security depends on the behaviour of several woodland species, and lack of feeding resources in ecosystems deems it to be weak and vulnerable, not only due to reduced plantations but also to reduced numbers of animals (Wang *et al.*, 2018).

Altered animal behaviours also impact parasites and predators. This can further deteriorate the ecological conditions of the environment (Budria and Candolin, 2014). Conclusively, the resilience of ecosystem processes to human-based ecological changes relies on the delicacy of species interactions, including their numbers and traits along with their connections (Ross *et al.*, 2021). Unquestionably, more vibrant species indicating several behavioural connections perhaps facilitate better-adapted and resilient ecosystems (Ross *et al.*, 2021).

REFERENCES

Arnot, J.A. and F.A. Gobas. 2006. A review of bioconcentration factor (BCF) and bioaccumulation factor (BAF) assessments for organic chemicals in aquatic organisms. *Environ. Rev.* 14:257–297.

Avolio, M.L. Forrestel, E.J. Chang, C.C. La Pierre, K.J. Burghardt, K.T. and M.D. Smith. 2019. Demystifying dominant species. *New Phytol.* 223:1106–1126.

Bartley, T.J. McCann, K.S. Bieg, C. Cazelles, K. Granados, M. Guzzo, M.M. MacDougall, A.S. Tunney, T.D. and B.C. McMeans. 2019. Food web rewiring in a changing world. *Nat. Ecol. Evol.* 3:345–354.

Belgrad, B.A. and B.D. Griffen. 2016. Predator–prey interactions mediated by prey personality and predator hunting mode. *Proc Royal Soc B, Biol. Sci.* 283:20160408.

Blowes, S.A. Supp, S.R. Antão, L.H. Bates, A. Bruelheide, H. Chase, J.M. Moyes, F. Magurran, A. McGill, B. Myers-Smith, I.H. and M. Winter, 2019. The geography of biodiversity change in marine and terrestrial assemblages. *Science* 366:339–345.

Both, C. Van-Asch, M. Bijlsma, R.G. Van Den Burg, A.B. and M.E. Visser. 2009. Climate change and unequal phenological changes across four trophic levels: constraints or adaptations. *J. Animal Ecol.* 78:73–83.

Buchanan, K. and J.J. Partecke. 2012. *The endocrine system: can homeostasis be maintained in a changing world.* Oxford University Press, 3:32–45.

Buchholz, R. Banusiewicz, J.D. Burgess, S. Crocker-Buta, S. Eveland, L. and L. Fuller. 2019. Behavioural research priorities for the study of animal response to climate change. *Anim. Behav.* 150:127–137.

Budria, A. and U. Candolin. 2014. How does human-induced environmental change influence host-parasite interactions. *Parasitology* 141:462–474.

Candolin, U. 2019. Mate choice in a changing world. *Biol. Rev.* 94:1246–1260.

Candolin, U. Bertell, E. and J. Kallio. 2018. Environmental disturbance alters the ecological impact of an invading shrimp. *Funct. Ecol.* 32:1370–1378.

Candolin, U. and B.B.M. Wong. 2012. *Behavioural responses to a changing world: mechanisms and consequences.* Oxford University Press.

Candolin, U. and B.B.M. Wong. 2019. Mate choice in a polluted world: consequences for individuals, population and communities. *Philos.Trans. R. Soc. B Biol.Sci.*

Carpenter, S.R. and J.F. Kitchell. 1988. Consumer control of lake productivity. *Bioscience* 38:764–769.

Ceballos, G. Ehrlich, P.R. Barnosky, A.D. Garcia, A. Pringle, R.M. and T.M. Palmer. 2015. Accelerated modern–human induced species loss: Entering the sixth mass extinction. *Sci. Adv.* 1:e1400253.

Chapple, D.G. Simmonds, S.M. and B.B. Wong. 2012. Can behavioral and personality traits influence the success of unintentional species introduction. *Trends Ecol. Evol.* 27:57–64.

Chen, C. Golz, R. Biere, A. and J.A. Harvey. 2019. Differential effects of climate warming on reproduction and functional responses on insects in fourth trophic level. *Funct. Ecol.* 33:693–702.

Cross, W.F. Hood, J.M. Benstead, J.P. Huryn, A.D. and D. Nelson. 2015. Interactions between temperature and nutrients across levels of ecological organization. *Glob. Chang. Biol.* 21:1025–1040.

Davenport, J.M. and D.R. Chalcraft. 2013. Nonconsumptive effects in a multiple predator system reduce the foraging efficiency of a keystone predator. *Ecol. Evol.* 3:3063–3072.

Dell, A.I. Pawar, S. and V.M. Savage. 2014. Temperature dependence of trophic interactions are driven by asymmetry of species responses and foraging strategy. *J. Anim. Ecol.* 83:70–84.

Donohue, I. Petchey, O.L. Kéfi, S. Génin, A. Jackson, A.L. Yang, Q. and N.E. O'Connor. 2017. Loss of predator species, not intermediate consumers, triggers rapid and dramatic extinction cascades. *Glob. Chang. Biol.* 23:2962–2972.

Duffy, J.E. 2003. Biodiversity loss, trophic skew and ecosystem functioning. *Ecol. Lett.* 6:680–687.

Dunne, J.A. Williams, R.J. and Martinez, N.D., 2002. Network structure and biodiversity loss in food webs: robustness increases with connectance. *Ecol. Lett.* 5:558–567.

Ellison, A.M. 2019. Foundation species, non-trophic interactions, and the value of being common. *Iscience* 13:254–268.

Gall, J.E. Boyd, R.S. and N. Rajakaruna. 2015. Transfer of heavy metals through terrestrial food webs: a review. *Environ. Monit. Assess.* 187:1–21.

Gaston, K.J. 2010. Valuing common species. *Science* 327:154–155.

Gilman, S.E. Urban, M.C. Tewksbury, J. Gilchrist, G.W. and R.D. Holt. 2010. A framework for community interactions under climate change. T*rends Ecol. Evol.* 25:325–331.

Hautier, Y. Tilman, D. Isbell, F. Seabloom, E.W. Borer, E.T. and P.B. Reich. 2015. Anthropogenic environmental changes affect ecosystem stability via biodiversity. *Science* 348:336–340.

Heath, M.R. Speirs, D.C. and J.H. Steele. 2014. Understanding patterns and processes in models of trophic cascades. *Ecol. Lett.* 17:101–114.

Hoffmann, A.A. and C.M. Sgro. 2011. Climate change and evolutionary adaptation. *Nature* 470:479–485.

Hoover, S.E.R. and J.M. Tylianakis. 2012. Species Interactions. In: Behavioural Responses to a Changing World.U. Candolin & R. Wong (eds). Oxford University Press. pp.129–142.

Hunter, M.D. and P.W. Price. 1992. Playing chutes and ladders: heterogeneity and the relative roles of bottom-up and top-down forces in natural communities. *Ecology* 73:724–732.

Ives, A.R. and S.R. Carpenter. 2007. Stability and diversity of ecosystems. *Science* 317:58–62.

Jackson, M.C. Pawar, S. and G. Woodward. 2021. The temporal dynamics of multiple stressor effects: from individuals to ecosystems. *Trends Ecol. Evol.* 36:402–410.

Kéfi, S. Berlow, E.L. Wieters, E.A. Navarrete, S.A. Petchey, O.L. Wood, S.A. Boit, A. Joppa, L.N. Lafferty, K.D. Williams, R.J. and N.D. Martinez. 2012. More than a meal integrating non-feeding interactions into food webs. *Ecol. Lett.* 15:291–300.

Kharouba, H.M. Ehrlén, J. Gelman, A. Bolmgren, K. Allen, J.M. Travers, S.E. and E.M. Wolkovich. 2018. Global shifts in the phenological synchrony of species interactions over recent decades. *Proc. Natl. Acad. Sci. USA*. 115:5211–5216.

Kristensen, E., 2000. Organic matter diagenesis at the oxic/anoxic interface in coastal marine sediments, with emphasis on the role of burrowing animals. In *Life at interfaces and under extreme conditions: proceedings of the 33rd European marine biology symposium*, held at Wilhelmshaven, Germany, 7–11 September 1998, Springer, pp.1–24.

Leroux, S.J. and M. Loreau. 2010. Consumer-mediated recycling and cascading trophic interactions. *Ecology* 91:2162–2171.

Lima, S.L. and L.M. Dill. 1990. Behavioral decisions made under the risk of predation: a review and prospectus. *Can. J. Zool.* 68:619–640.

Loreau, M. and C. De-Mazancourt. 2013. Biodiversity and ecosystem stability: a synthesis of underlying mechanisms. *Ecol. Lett.* 16:106–115.

McCallum, E.S. Dey, C.J. Cerveny, D. Bose, A.P. and T. Brodin. 2021. Social status modulates the behavioral and physiological consequences of a chemical pollutant in animal groups. *Ecol. Appl.* 31:e02454.

Michel, M.J. and M.M. Adams. 2009. Differential effects of structural complexity on predator foraging behavior. *Behav. Ecol.* 20:313–317.

Ockendon, N. Baker, D.J. Carr, J.A. White, E.C. Almond, R.E. Amano, T. Bertram, E. Bradbury, R.B. Bradley, C. Butchart, S.H. and N. Doswald. 2014. Mechanisms underpinning climatic impacts on natural populations: altered species interactions are more important than direct effects. *Glob. Chang. Biol.* 20:2221–2229.

Oliver, T.H. Heard, M.S. Isaac, N.J., Roy, D.B. Procter, D. Eigenbrod, F. Freckleton, R. Hector, A. Orme, C.D.L. Petchey, O.L. and V. Proenca. 2015. Biodiversity and resilience of ecosystem functions. *Trends Ecol. Evol.* 30:673–684.

Palkovacs, E.P. Moritsch, M.M. Contolini, G.M. and F. Pelletier. 2018. Ecology of harvest-driven trait changes and implications for ecosystem management. *Front. Ecol. Environ.* 16:20–28.

Pecl, G.T. Araújo, M.B. Bell, J.D. Blanchard, J. Bonebrake, T.C. Chen, I.C. Clark, T.D. Colwell, R.K. Danielsen, F. Evengård, B. and L. Falconi. 2017. Biodiversity redistribution under climate change: impacts on ecosystems and human well-being. *Science* 355:9214.

Pelletier, F. Garant, D. Candolin, U. and B.B.M. Wong. 2012. Population consequences of individual variation in behaviour. In *Behavioural responses to a changing world: mechanisms and consequences*. Oxford University Press, pp.159–174.

Power, M.E. Tilman, D. Estes, J.A. Menge, B.A. Bond, W.J. Mills, L.S. Daily, G. Castilla, J.C. Lubchenco, J. and R.T. Paine. 1996. Challenges in the quest for keystones: identifying keystone species is difficult but essential to understanding how loss of species will affect ecosystems. *BioScience* 46:609–620.

Previsic, A. Vilenica, M. Vuckovic, N. Petrovic, M. and M. Rozman. 2021. Aquatic insects transfer pharmaceuticals and endocrine disruptors from aquatic to terrestrial ecosystems. *Environ. Sci. Technol.* 55:3736–3746.

Pysek, P. Hulme, P.E. Simberloff, D.Bacher, S., Blackburn, T.M. and J.T. Carlton. 2020. Scientists' warning on invasive alien species. *Biol. Rev.* 95:1511–1534.

Rall, B.C. Guill, C. and U. Brose. 2008. Food-web connectance and predator interference dampen the paradox of enrichment. *Oikos* 117:202–213.

Rasher, D.B. Steneck, R.S. Halfar, J. Kroeker, K.J. Ries, J.B. Tinker, M.T. Chan, P.T. Fietzke, J. Kamenos, N.A. Konar, B.H. and J.S. Lefcheck. 2020. Keystone predators govern the pathway and pace of climate impacts in a subarctic marine ecosystem. *Science* 369:1351–1354.

Ren, L. Jensen, K. Porada, P. and P. Mueller. 2022. Biota-mediated carbon cycling-A synthesis of biotic-interaction controls on blue carbon. *Ecol. Lett.* 25:.521–540.

Ricciardi, A. Hoopes, M.F. Marchetti, M.P. and J.L. Lockwood. 2013. Progress toward understanding the ecological impacts of nonnative species. *Ecol. Monogr.* 83:263–282.

Ripple, W.J. Estes, J.A. Beschta, R.L. Wilmers, C.C. Ritchie, E.G. Hebblewhite, M. Berger, J. Elmhagen, B. Letnic, M. Nelson, M.P. and O.J. Schmitz, 2014. Status and ecological effects of the world's largest carnivores. *Science* 343:1241484.

Rooney, N. and K.S. McCann. 2012. Integrating food web diversity, structure and stability. *Trends Ecol. Evol.* 27:40–46.

Rosenblatt, A.E. and O.J. Schmitz. 2016. Climate change, nutrition, and bottom-up and top-down food web processes. *Trends Ecol. Evol.* 31:965–975.

Ross, S.R.J. Arnoldi, J.F. Loreau, M. White, C.D. Stout, J.C. Jackson, A.L. and I. Donohue. 2021. Universal scaling of robustness of ecosystem services to species loss. *Nat. Commun.* 12:5167.

Ross, S.R.J. Garcia Molinos, J. Okuda, A. Johnstone, J. Atsumi, K. Futamura, R. Williams, M.A. Matsuoka, Y. Uchida, J. Kumikawa, S. and H. Sugiyama. 2022. Predators mitigate the destabilising effects of heat-waves on multitrophic stream communities. *Glob. Chang. Biol.* 28:403–416.

Rudolf, V.H. and A. Roman. 2018. Trophic structure alters consequences of environmental warming. *Oikos* 127:1646–1656.

Sbragaglia, V. Jolles, J.W. Coll, M. and R. Arlinghaus. 2021a. Fisheries-induced changes of shoaling behaviour: mechanisms and potential consequences. *Trends Ecol. Evol.* 36:885–888.

Sbragaglia, V. Lopez-Olmeda, J.F. Frigato, E. Bertolucci, C. and R. Arlinghaus. 2021b. Size-selective mortality induces evolutionary changes in group risk-taking behaviour and the circadian system in a fish. *J. Anim. Ecol.* 90:387–403.

Schmitz, O.J. and G.C. Trussell. 2016. Multiple stressors, state-dependence and predation risk-foraging trade-offs: toward a modern concept of trait-mediated indirect effects in communities and ecosystems. *Curr. Opin. Behav. Sci.* 12:6–11.

Sentis, A. Binzer, A. and D.S. Boukal. 2017. Temperature-size responses alter food chain persistence across environmental gradients. *Ecol. Lett.* 20:852–862.

Shurin, J.B. Clasen, J.L. Greig, H.S. Kratina, P. and P.L. Thompson. 2012. Warming shifts top-down and bottom-up control of pond food web structure and function. *Philos. Trans. R. Soc. B Biol. Sci.* 367:3008–3017.

Simberloff, D. Martin, J.L. Genovesi, P. Maris, V. Wardle, D.A. Aronson, J. Courchamp, F. Galil, B. Garcia-Berthou, E. Pascal, M. and P. Pysek. 2013. Impacts of biological invasions: what's what and the way forward. *Trends Ecol. Evol.* 28:58–66.

Sommer, N.R. and O.J. Schmitz. 2020. Differences in prey personality mediate trophic cascades. *Ecol. Evol.* 10:9538–9551.

Straile, D. Kerimoglu, O. and F. Peeters. 2015. Trophic mismatch requires seasonal heterogeneity of warming. *Ecology* 96:2794–2805.

Taylor, B.W. Flecker, A.S. and R.O. Hall-Jr. 2006. Loss of a harvested fish species disrupts carbon flow in a diverse tropical river. *Science* 313:833–836.

Thackeray, S.J. Henrys, P.A. Hemming, D. Bell, J.R. Botham, M.S. Burthe, S. Helaouet, P. Johns, D.G. Jones, I.D. Leech, D.I. and E.B. Mackay. 2016. Phenological sensitivity to climate across taxa and trophic levels. *Nature* 535:241–245.

Renner, S.S., and C.M. Zohner. 2018. Climate change and phenological mismatch in trophic interactions among plants, insects, and vertebrates. *Annual Review of Ecology, Evolution, and Systematics* 49:165–182.

Urban, M.C. Zarnetske, P.L. and D.K. Skelly. 2017. Searching for biotic multipliers of climate change. *Integr. Comp. Biol.* 57:134–147.

Velthuis, M. Domis, L.N. Frenken, T. Stephan, S. Kazanjian, G. Aben, R. Hilt, S. Kosten, S. Van-Donk, E. and D.B. Van de Waal. 2017. Warming advances top-down control and reduces producer biomass in a freshwater plankton community. *Ecosphere* 8:e01651.

Violle, C. Enquist, B.J. McGill, B.J. Jiang, L.I.N. Albert, C.H. Hulshof, C. Jung, V. and J. Messier. 2012. The return of the variance: intraspecific variability in community ecology. *Trends Ecol Evol.* 27:244–252.

Visser, M.E. Marvelde, L. and M.E. Lof. 2012. Adaptive phenological mismatches of birds and their food in a warming world. *J. Ornithol.* 153:75–84.

Wang, B. Shuman, J. Shugart, H.H. and M.T. Lerdau. 2018. Biodiversity matters in feedbacks between climate change and air quality: a study using an individual-based model. *Ecol. Appl.* 28:1223–1231.

Werner, E.E. and S.D. Peacor. 2003. A review of trait-mediated indirect interactions in ecological communities. *Ecology* 84:1083–1100.
Willems, J.S. Phillips, J.N. and C.D. Francis. 2022. Artificial light at night and anthropogenic noise alter the foraging activity and structure of vertebrate communities. *Sci. Total Environ.* 805:150223.
Wilmers, C.C. Post, E. Peterson, R.O. and J.A. Vucetich. 2006. Predator disease out-break modulates top-down, bottom-up and climatic effects on herbivore population dynamics. *Ecol. Lett.* 9:383–389.
Wingfield, J.C. 2013. The comparative biology of environmental stress: behavioural endocrinology and variation in ability to cope with novel, changing environments. *Anim. Behav.* 85:1127–1133.
Wong, B.B. and U. Candolin. 2015. Behavioral responses to changing environments. *Behav. Ecol.* 26:665–673.
Wright, J.P. and C.G. Jones. 2006. The concept of organisms as ecosystem engineers ten years on: progress, limitations, and challenges. *Bioscience* 56:203–209.

Index

Note: Page numbers in *italics* refer to figures, and page numbers in **bold** refer to tables

A

I

J

K

U

V

W

Y

Z